Scientific Tools for Conservation

Infrared Spectroscopy in Conservation Science

Michele R. Derrick

Dusan Stulik

James M. Landry

The Getty Conservation Institute Los Angeles

Tevvy Ball, *Managing Editor*
Sylvia Tidwell, *Copy Editor*
Anita Keys, *Production Coordinator*
Hespenheide Design, *Designer*

Printed in the United States of America
10 9 8 7 6 5 4 3 2 1

Illustration credits: Figures 3.8, 4.27, 4.28, 4.29, 5.16, 5.24, and 6.9: Courtesy of the *Journal of the American Institute for Conservation of Historic and Artistic Works*. Figures 4.20, 6.1, 6.2, 6.3, 6.10, 6.14, 6.15, 6.20, 6.21, 6.23, and 6.24: From "Infrared Microspectroscopy in the Analysis of Cultural Artifacts," by Michele R. Derrick, in *Practical Guide to Infrared Microspectroscopy*, ed. Howard Humecki, pp. 287–322 (New York: Marcel Dekker, Inc., 1995). Reproduced by permission.

The Getty Conservation Institute

The Getty Conservation Institute works internationally to further appreciation and preservation of the world's cultural heritage for the enrichment and use of present and future generations. The Institute is an operating program of the J. Paul Getty Trust.

The Institute's Scientific Tools for Conservation series provides practical scientific procedures and methodologies for the practice of conservation. The series is specifically directed to conservation scientists, conservators, and technical experts in related fields. Previously published in the series is *Thin-Layer Chromatography for Binding Media Analysis* (1996), by Mary F. Striegel and Jo Hill.

Library of Congress Cataloging-in-Publication Data
Derrick, Michele R., 1955–
Infrared spectroscopy in conservation science / Michele R. Derrick, Dusan Stulik, James M. Landry.
p. cm.—(Scientific tools for conservation)
Includes bibliographical references and index.
ISBN 0-89236-469-6
1. Infrared spectroscopy. 2. Art—Expertising—Methodology Case studies. I. Stulik, Dusan, 1956– . II. Landry. James M. III. Title. IV. Series.
N8558.2.I53D47 1999
702′.8′8—dc21 99-37860
CIP

Contents

Foreword

Since the founding of the conservation field's flagship journal, *Studies in Conservation,* in the early 1950s, the number of museum laboratories has increased remarkably. Where once a single optical microscope may have sat in dusty isolation, today chemists, engineers, physicists, and other scientists make use of a wide range of analytical instruments, often working side by side with conservators and curators. It is clear that those who benefit from this development are people responsible for managing and interpreting vast arrays of collection types, but the public is also a prime recipient of this benefaction.

This relatively recent infusion of science into the museum environment has significantly enhanced preservation practices, affording a deepened understanding of materials' properties and degradation processes. The simple optical microscope now coexists with many other instruments, which allow advanced techniques for separation and chemical identification, high magnification and analysis of microstructure, and imaging using nuclear magnetic resonance, thermal neutrons, and a range of electromagnetic radiation.

Today the most widely used method in most museum laboratories is infrared (IR) spectroscopy. It is extremely cost-effective, and it has directly contributed to the current enhanced interest in organic materials in art and archaeology. Recent years have witnessed the development of a robust network of IR users, who share spectra and insights on a regular basis. The activities of such a group, informal as it may be, help to refine the method, broaden its utility, and provide a specialized training environment for its novitiates. Many benefit from this—curators, museum directors, collectors, and, of course, the museum-going public.

For all these reasons we are proud to include this volume as the second to be published in the Getty Conservation Institute's Scientific Tools for Conservation series, which is designed to present methods and procedures of practical use to conservators, conservation scientists, and others engaged in the preservation of the cultural heritage.

We are grateful to the authors, Michele R. Derrick, Dusan Stulik, and James M. Landry, for their dedication in writing this book. With its focus on the practical applications of IR spectroscopy—including a detailed presentation of ten case studies—the present volume hopes to make this important technology more readily available to those who can benefit from it the most.

Tim Whalen
Director
The Getty Conservation Institute

Preface

The purchase and use of infrared (IR) spectrometers in art conservation labs have grown more than tenfold in the last decade. This expansion can be traced to decreased instrument costs, enhanced interest in organic materials in art and archaeological objects, and increased requests for scientific analyses by conservators and curators. Additionally, the current computer-driven, user-friendly IR instruments make it extremely easy to perform an IR spectral analysis on a sample. But herein lies a problem, since simplification of a method can result in lost intricacies and misinformation. Thus, critical evaluation and availability of information on a technique are important in order for pitfalls to be recognized and for any of the technique's significant limitations to be understood.

This book fills a gap, since currently there is no available compilation of information that deals directly with the IR analysis of historical and artistic materials. These materials can pose significant problems for analysts, as they tend to be natural products that are often mixtures and are sometimes affected by age. Scientific books and reference sources rarely provide key information necessary for dealing with such materials.

This text provides practical information on the use of IR spectroscopy for the analysis of museum objects, disseminating many sample handling and spectral acquisition techniques specifically applicable to their analysis, along with discussions of these techniques' potential problems. This book is meant to be a learning tool as well as an information resource. While the present volume provides a comprehensive overview of the technique, individual chapters may be read independently for specific information.

Chapter 1 defines IR spectroscopy and provides a historical perspective on its development as a modern analytical technique.

Chapter 2 provides an overview of the position and relationship of the IR spectral region to the entire electromagnetic spectrum. It also supplies a simplified version of the theory of molecular interactions that produce IR spectral patterns. Included is a basic description of the phenomena that occur in a material to produce a unique IR spectrum.

Sample collection and preparation are key steps in the analysis of all materials. Chapter 3 provides basic and widely applicable information on sampling methodology and implementation. This chapter is important not just for analysts but also for anyone who may be requesting, collecting, or submitting samples for analysis. Emphasis is placed on the effect of sampling on analytical results. Comparative tables are included that illustrate the capabilities and sample requirements of some common analytical techniques.

Once the sample is collected, several options exist for its IR analysis. Chapter 4 reviews the theory and operation of both transmission and reflectance methods, as well as provides references as to how each of the techniques has been used in the analysis of art materials. Many excellent books that provide more detailed information about IR methods and instrumentation are also included in an additional reading list.

Chapter 5 focuses on the spectral interpretation of conservation and artist materials and provides schemes or flowcharts to facilitate the characterization of unknowns. Examples of basic peak identification for the major classifications of materials found in works of art are given. Problems relating to material mixtures are discussed, as are several options for the mathematical manipulation of spectra.

Chapter 6 incorporates reviews of existing conservation and related literature. Case studies illustrate several types of problems and materials and show approach, utility, and limitations of the techniques. Professionals who are not spectroscopists may find individual sections useful to their understanding of the applications and restrictions of the different procedures.

One of the most important factors in IR spectral identification is access to reference spectra corresponding to appropriate materials. Appendix I is a list of commercial sources that supply digitized and hard-copy spectral collections. Appendix II provides IR spectra of common art and conservation materials. There is a list of references, and additional readings on related subjects are provided at the end of each chapter.

IR spectroscopy is a useful and fascinating challenge that can provide the answers to many of the problems encountered in the analysis of works of art. It is hoped that the practical information provided in this book will stimulate interest in, and perhaps lay the groundwork for, many future IR applications.

Michele R. Derrick

Acknowledgments

Funds for the preparation of this manuscript were provided in part by a Kress Foundation conservation publication fellowship awarded to Michele Derrick in 1995. Portions of this book were presented at a workshop held jointly by the Getty Conservation Institute (GCI) and Loyola Marymount University, Los Angeles, in 1991.

The authors would like to thank the following organizations and people for providing samples for analysis: the J. Paul Getty Museum, Los Angeles (Andrea Rothe and Brian Considine); the Museum of Modern Art, New York (Eugena Ordonez and Patricia Houlihan); the Fine Arts Museum, Houston (Christopher Shelton); the Bureau of Land Management, Needles, California (Tom Holcomb and Claire Dean); and the Museum of Natural History, New York (David Thomas). The authors would also like to thank Miguel Angel Corzo, the former director of the GCI, for his support of the book; Neville Agnew for initiating this series of books; Kevin Thompson for providing several historic books on optics; Eric Hansen, Richard Newman, and Charles Selwitz for their helpful suggestions; Janice Carlson for her extensive and meticulous technical editing; Tanya Kieslich and Margaret Bolton for preparing some of the drawings in Chapter 4; Herant Khanjian for photographs in Chapter 4; and the GCI Information Center for obtaining numerous and often obscure offprints.

Additionally, Michele Derrick gives her deepest appreciation and gratitude to her family, John, Erin, and Kristin Stacy, for their patience during the lengthy period of manuscript writing and rewriting.

Chapter 1

History of Infrared Spectroscopy

In simplified terms, one dictionary defines *infrared* as "those invisible rays just beyond the red end of the visible spectrum . . . [that] have a penetrating heating effect: used in cooking, photography, etc." (Guralnik 1984). While this may seem obvious, since infrared (IR) is such an integral part of our lives, it was not until 1800 that IR was recognized as a distinct region of the energy spectrum. This discovery was made by Sir William Herschel, an astronomer, when he measured the heating effect of sunlight (Herschel 1800). However, the field of IR spectroscopy—i.e., the study of wavelengths of light in the IR region of the spectrum and their interaction with various materials—did not develop at that time due to difficulties in building suitable detectors to measure IR radiation. Eventually, beginning in 1903, William W. Coblentz conducted comprehensive experiments leading to the accurate measurement of IR spectra for hundreds of inorganic and organic compounds (Coblentz 1905). Interest increased in the potential of IR spectroscopy for analytical chemistry when the first prototypes of IR spectrometers were built in the 1930s. Commercial development of IR spectrometers quickly followed, stimulated by the need for rapid analytical methods in the synthetic rubber industry. This development expanded the acquisition of IR spectra and motivated research into deeper theoretical studies of all features of IR spectra. The next jump came as advances in electronics during and after World War II furnished the thermocouple detector technology that led to development of stable double beam spectrophotometers (Wright and Hersher 1947). By the 1950s, IR spectroscopy was established as a key analytical method in academic and industrial labs.

Most IR instrumentation used through the 1970s was based on prism or grating monochromators. A major breakthrough in IR technology was the introduction of Fourier transform infrared (FT-IR) spectrometers. In the late 1800s, Albert Michelson developed the interferometer for studying the speed of light (Michelson 1891). Two years later, Lord Rayleigh recognized that the output from an interferometer could be converted to a spectrum by a mathematical procedure developed about seventy years earlier by the French mathematician Fourier (Rayleigh 1891). However, because of the numerical complexities of the Fourier procedure, it was not until 1949 that Peter Fellgett actually calculated a spectrum from an interferogram (Fellgett 1970). Fellgett, an astronomer, used Fourier transform spectroscopy to isolate the weak signals of distant stars from environmental background noise and in doing so discovered the multiplex advantage, later named in his honor. A few years later, Jacquinot, a French scientist, pointed out the throughput advantage of interferometry (Jacquinot 1954). These advantages for FT-IR over dispersive instruments translated into practical improvements such as high-speed data collection, increased resolution, lower detection limits, and greater energy throughput.

Acceptance of FT-IR spectroscopy, however, was slowed by the complexity of the calculations required to transform an interferogram into a spectrum. Then, in 1964, the discovery of the fast Fourier transform (FFT) algorithm by James Cooley and John Tukey reduced the time for the computer calculation of the transform from hours to just a few seconds (Cooley and Tukey 1965). Even so, it was still a time-consuming process for spectroscopists to record an interferogram on a paper tape or punched cards, walk it over to the computing center and wait for calculation of the spectrum (Griffiths and de Haseth 1986). The next significant change came in 1969, when the first commercial FT-IR with a dedicated minicomputer was developed and sold by Digilab (see Suppliers). For the first time, spectroscopists were able to see a spectrum plotted shortly after the interferogram was collected; correspondingly, it took less total time to obtain a spectrum using an FT-IR spectrometer than using a dispersive instrument. Since then, further developments in computer technology, together with substantial price decreases, have been responsible for the large number of commercial FT-IR spectrometers on the market and for the wide application of FT-IR spectroscopy in all branches of science and technology.

Particularly useful for the field of art conservation was the coupling of the optical microscope with the IR spectrometer. Initial studies were done at Oxford by Barer, Cole, and Thompson, who used a microscope designed with all-reflecting optics attached to a dispersive IR spectrometer (Barer, Cole, and Thompson 1949). Although the system performed well, the sensitivity achievable by a microscope in combination with a wavelength dispersive IR spectrometer was low, and interest in the technique waned. Revival of IR microspectroscopy was stimulated by the development of FT-IR instrumentation, with its increased energy throughput. In 1981, at McCrone Associates, Robert Z. Muggli successfully adapted a microscope to an FT-IR spectrometer (Palenik 1992). In 1983 Digilab introduced the first commercial microscope designed specifically for FT-IR spectrometers by Spectra-Tech (see Suppliers). Following that time, a number of other instrument companies have developed IR microspectrophotometers and microaccessories for IR spectrometers. Availability of the IR microscopes precipitated a new range of analytical applications, and IR microspectroscopy has become a powerful tool in many branches of basic and applied research.

The historical background for IR spectroscopy is shown in Figure 1.1.

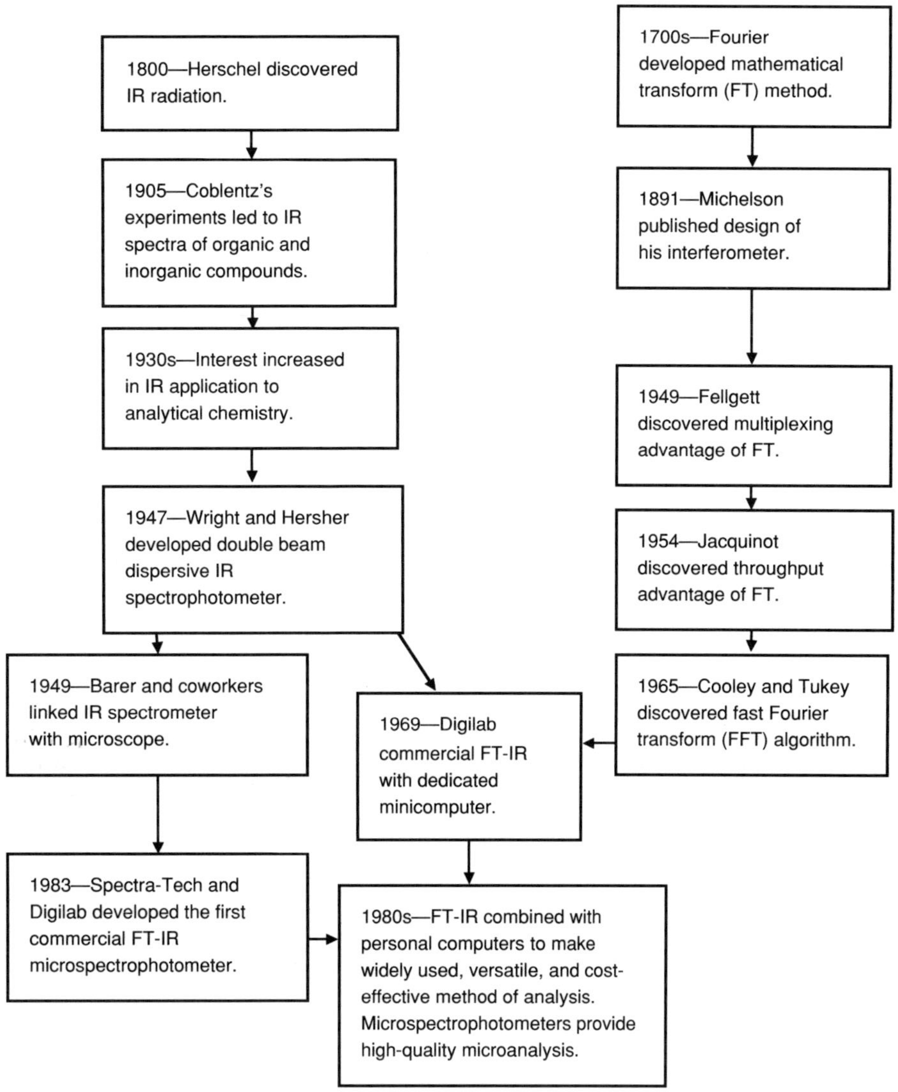

Figure 1.1
Significant events leading to the development of current FT-IR spectrometers and microspectrometers.

Additional Reading

Colthup, N. B., L. H. Daly, and S. E. Wiberly
1990. *Introduction to Infrared and Raman Spectroscopy*. 3d ed. Boston: Academic Press.

Griffiths, P. A., and J. A. de Haseth
1986. *Fourier Transform Infrared Spectrometry*. New York: Wiley-Interscience.

Silverstein, R. M., F. C. Bassler, and T. C. Morrill
1981. *Spectrometric Identification of Organic Compounds*. 4th ed. New York: John Wiley and Sons.

Smith, A. L.
1979. *Applied Infrared Spectroscopy: Fundamentals, Techniques, and Analytical Problem Solving*. Chemical Analysis series, vol. 54. New York: John Wiley and Sons.

Chapter 2
Infrared Absorption Theory

Electromagnetic Radiation

In simplest terms, spectroscopy is defined as the interaction of light with matter. Light, in this context, is the broad spectrum of continuous energy called the electromagnetic spectrum. The major regions of the electromagnetic spectrum are shown in Figure 2.1. There is no uniform naming system for the spectral regions, as the specific names for each type of radiation—such as visible, X ray, and radio—were assigned when it was discovered. Even so, the designations are commonly known and useful for quick orientation. The same physical properties govern all radiation, regardless of its spectral region.

The types of radiation are generally grouped by the kinds of chemical and physical effects they can produce on matter. For example, in a magnetic field, exposure to the low-energy radio frequency radiation only reorients nuclei, while exposure to the slightly higher-energy microwave region changes electron spin states of molecules with unpaired electrons. Microwave radiation can also change the rotational energy of molecules; this effect is used to heat food quickly in a microwave oven. In the middle regions of the electromagnetic spectrum, absorption of IR radiation causes changes in the vibrational energy of molecules. Visible (Vis) and ultraviolet (UV) radiations alter the electron energies of loosely held outer electrons of atoms and molecules. Higher-energy X rays can cause electron transitions between inner electron levels, and gamma radiation produces changes within atomic nuclei. As all compounds absorb radiation in multiple regions of the spectrum, the information on molecular activity in each region provides complementary data for material characterization.

The following overview of the relationship of radiant energy to its effects on matter has a focus on IR absorption theory. For more information, several excellent references are listed at the end of the chapter. In particular, the book by Colthup, Daly, and Wiberley (1990) provides good, in-depth coverage of IR theory and molecular structure correlations.

Wave theory

All energies of the electromagnetic spectrum can be considered to be waves that move at the speed of light, with the types of radiation differ-

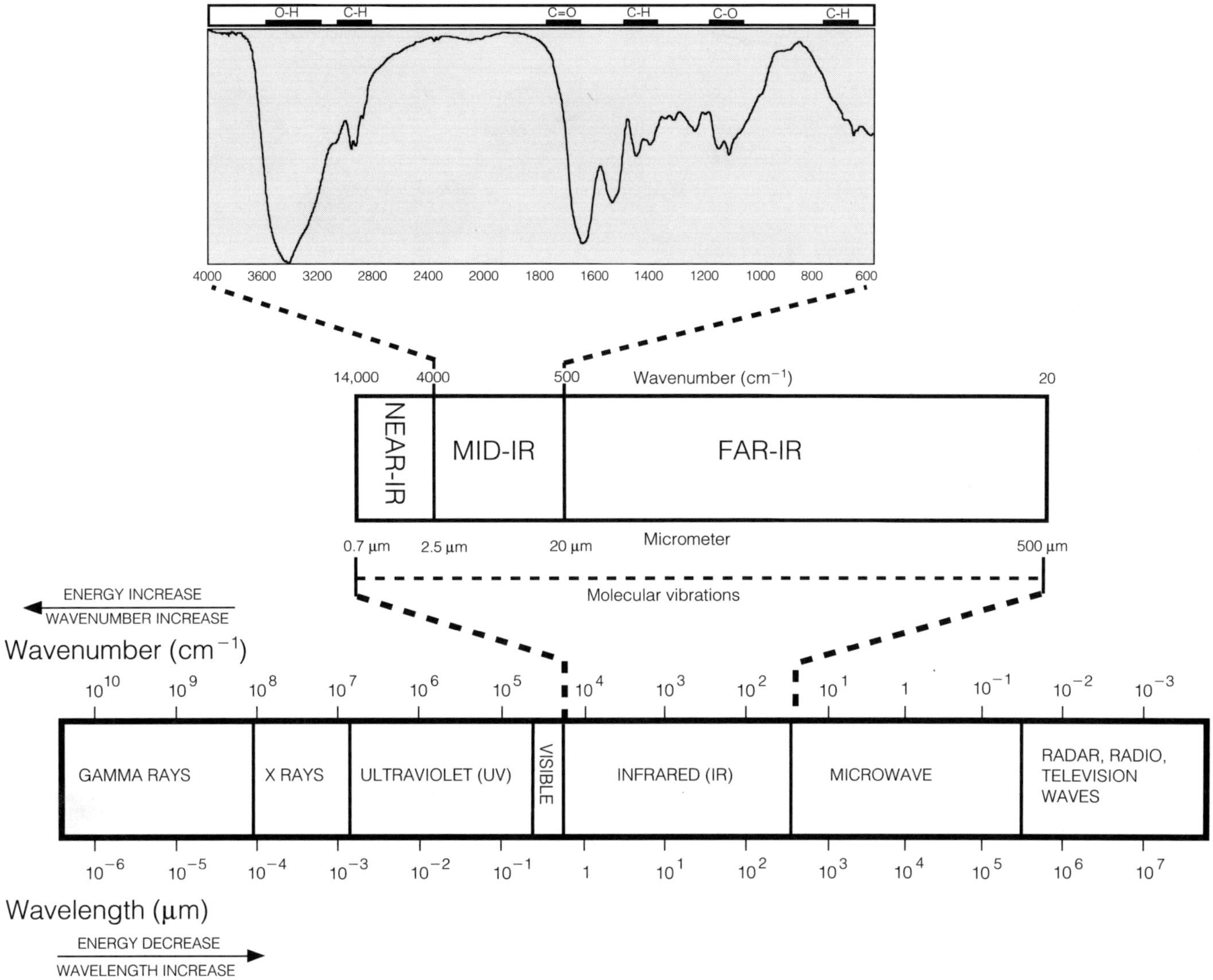

Figure 2.1
Spectral regions of electromagnetic radiation, with expansion of IR region.

ing only in amplitude, frequency, and wavelength (Fig. 2.2). The amplitude is the height, or maximum size, of the wave, which corresponds to the intensity, or volume, of the signal. The frequency, ν, is the number of oscillations, or waves, per unit time—that is, cycles per second. The wavelength, λ, is the distance between two successive maxima or minima of a wave—that is, the length of one wave. The wavelength of the radiation is inversely proportional to frequency. Thus, high-frequency radiation has short wavelengths.

Energy, according to Planck's law, is directly proportional to frequency. Since frequency is inversely proportional to wavelength, it follows that energy of electromagnetic radiation and wavelength are also inversely related. Thus, longer wavelengths have lower energy and frequency, while shorter wavelengths have higher energy.

Electromagnetic radiation can also be characterized by the number of waves per unit length. This is termed wavenumber, $\bar{\nu}$:

$$\bar{\nu} = \frac{1}{\lambda}$$

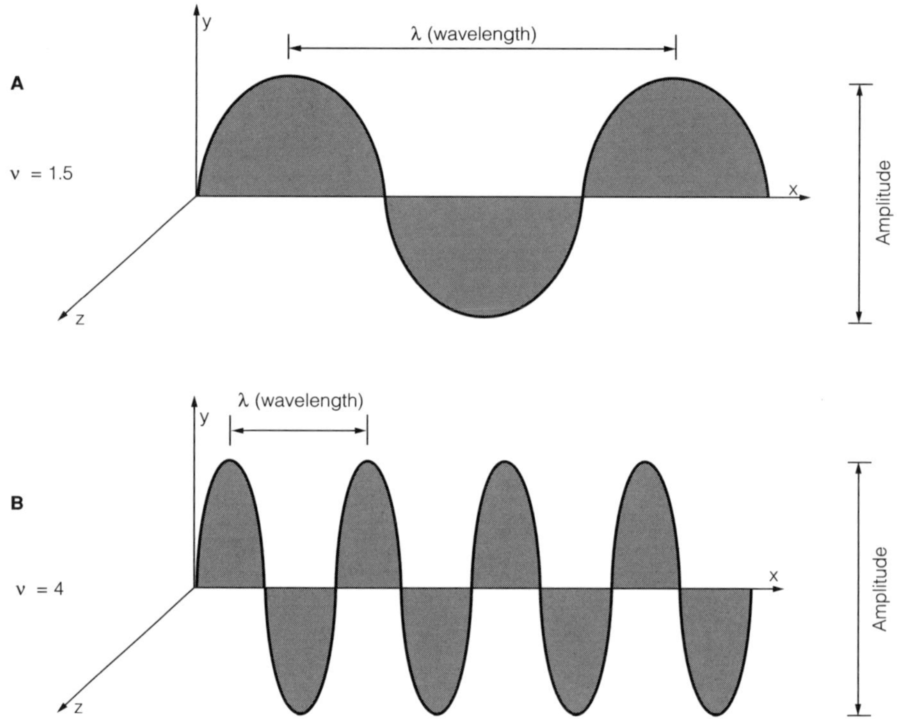

Figure 2.2
Two frequencies of electromagnetic waves at a 1 second interval. Radiation waves can be characterized by their amplitude (height), wavelength (distance between two maxima), and frequency (number of oscillations per unit time).

where: $\bar{\nu} = \nu / c$ (cm^{-1}) = frequency (Hz, sec^{-1}) / velocity of light in vacuum (3×10^{10} cm/sec); and λ = wavelength (cm). While wavenumber is usually expressed in cm units [($1/\lambda$ cm^{-1}], it could be expressed in any reciprocal distance units. Because of the simple inverse relationship, wavenumbers can readily be converted to wavelength units when needed. For example, 1000 wavenumbers (cm^{-1}) is the same as the wavelengths of 0.001 cm or 0.01 μm or 10 nm.

Frequency and wavenumber units have two advantages over wavelength units. The first is that they remain constant, regardless of the media traversed by the radiation, whereas the wavelength is reduced when radiation passes through a medium with a refractive index greater than that of a vacuum. This change in wavelength due to refractive index is ignored except for high-accuracy experiments, since the refractive index of air is near unity under normal conditions. The second advantage of the use of frequency and wavenumber units over wavelength is that they are directly proportional to energy. Thus, a transition that requires greater energy will occur at a higher wavenumber. For this reason, wavenumber units are commonly used in IR spectroscopy, as opposed to wavelength units (i.e., nanometers or micrometers), which are commonly used in visible and UV spectroscopy.

Absorption Theory

Toward the end of the nineteenth century, it became increasingly evident that the classical laws of physics describing natural phenomena—such as time, gravity, and momentum—did not successfully account for effects observed when light interacted with matter. At the 1904 St. Louis World's Fair, top scientists debated for and against the existence of atoms. Two years later, modern atomic theory was established when experiments by Ernest Rutherford showed that atoms existed and that each consisted of a positively charged nucleus surrounded by a cloud of negatively charged electrons.

Experimentally, the emission spectrum of hydrogen consists of a number of very sharp lines at distinct energies. In order to explain the fact that the hydrogen atom emits only these characteristic frequencies, Niels Bohr postulated in 1913 that the electrons of an atom occupy specific energy states or levels that are defined by the radius of the orbit of the electron around the nucleus. He further suggested that to move between different energy states, atoms must absorb or emit energy and that the amount of energy absorbed or emitted must be equal to the difference in energy between the two levels (Fig. 2.3). One photon of energy, $h\nu$, is emitted when an electron falls from a higher (E_h) to a lower (E_l) energy state. This is also called the energy of transition, ΔE. Alternatively, the electron can absorb a photon of light (energy) and move from a lower to a higher energy state.

Each element (hydrogen, helium, etc.) has electrons at unique energy levels corresponding to its atomic structure. Even though an element is exposed to radiation of all wavelengths, only the wavelengths (energy photons) that match the levels (energy states) within that atom can interact. The resultant pattern of energy lines, called a spectrum, coincides with the absorption or emission of the photons specific to that particular element. This phenomenon is responsible for the emission lines produced by excited atoms and molecules, as well as for the absorption bands in all regions of the electromagnetic spectrum. It forms the basis for all atomic and molecular spectroscopy.

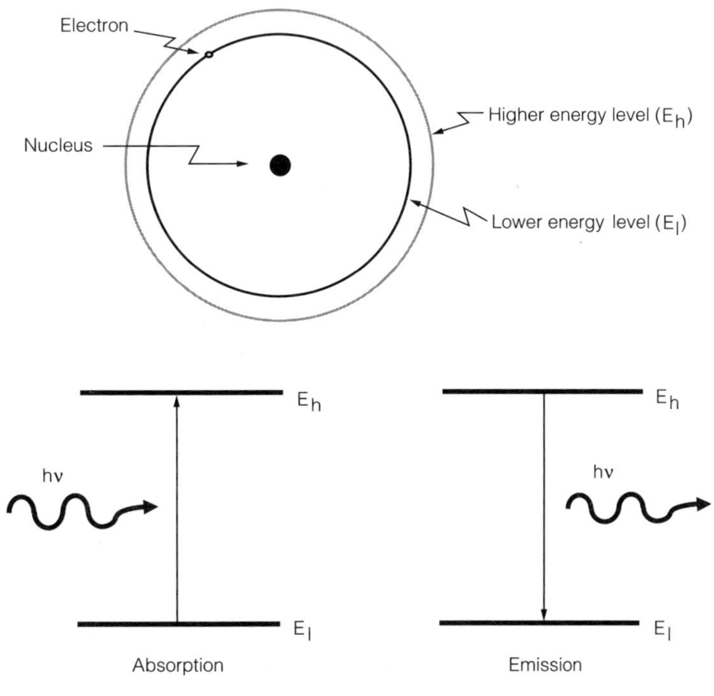

Figure 2.3
Bohr atomic model for hydrogen. Bohr theorized that energy whose frequency matches the energy difference between electronic levels can be absorbed or emitted as the electron transitions between the two levels ($h\nu$ = one photon).

Molecular absorptions

Scientists expanded and advanced Bohr's theories to include multielectron atoms and molecules. By the end of the 1930s, detailed models were in place that accounted for the placement of electrons in orbitals. These provided key links to understanding elemental bonding, molecular structures, and chemical reactions; this new knowledge in turn led to a greater understanding of molecular spectroscopy.

Specific wavelengths of energy correspond to all molecular transitions or motions: electronic, translational, rotational, and vibrational. In electronic motion, the electrons change energy levels or directions of spin. For translational motion, the entire molecule shifts to a new position, while for rotational motion, the molecule rotates around its center of mass. Vibrational energy is required for individual atoms within a molecule to change position relative to one another without moving or rotating the molecule.

The energy of IR radiation is too low to affect the electrons within an atom. IR radiation does, however, correspond to the energy required for translational, rotational, and vibrational energy transitions. Since a molecule's movements are unique to its structure, the measurement of these transitions makes IR a powerful tool for compound characterization.

The primary transitions in the IR region are vibrational. Rotational and translational transitions are weak and often difficult to measure in the IR region without the aid of high-resolution instruments. One common exception is for water vapor (regions 4000–3600 and 1800–1400 cm^{-1}), where the unrestricted motion of some gas-phase molecules produces sharp rotational absorption bands that can easily be seen (Fig. 2.4). The occurrence of many unresolvable rotational transitions in the IR region is one reason that a single, associated vibrational transition appears as an absorption envelope or band rather than a sharp line, as predicted by theory. Figure 2.5 illustrates the relationship of the vibrational energy levels to rotational energy levels.

Degrees of freedom

In a molecule, the atoms are constrained by molecular bonds to move together in certain specified ways, called degrees of freedom. When a particular molecular structure is known, its constraints allow prediction of expected molecular transitions.

To determine the degrees of freedom for any molecule, first consider the molecule positioned in a three-dimensional Cartesian coordinate system (x, y, and z), with its center at the point of origin (0, 0, 0). Then designate each atom by its coordinates in space (i.e., atom 1 = x_1, y_1, z_1). For a molecule with N (any number) atoms, the total number of coordinates (x_n, y_n, z_n) specified will be 3N. These 3N coordinates are the maximum number of potential transitions possessed by that molecule and are called its degrees of freedom. For example, a molecule with 5 atoms has 15 degrees of freedom.

The 3N degrees of freedom can be assigned to the translational, rotational, and vibrational motions of the molecule (see Table 2.1

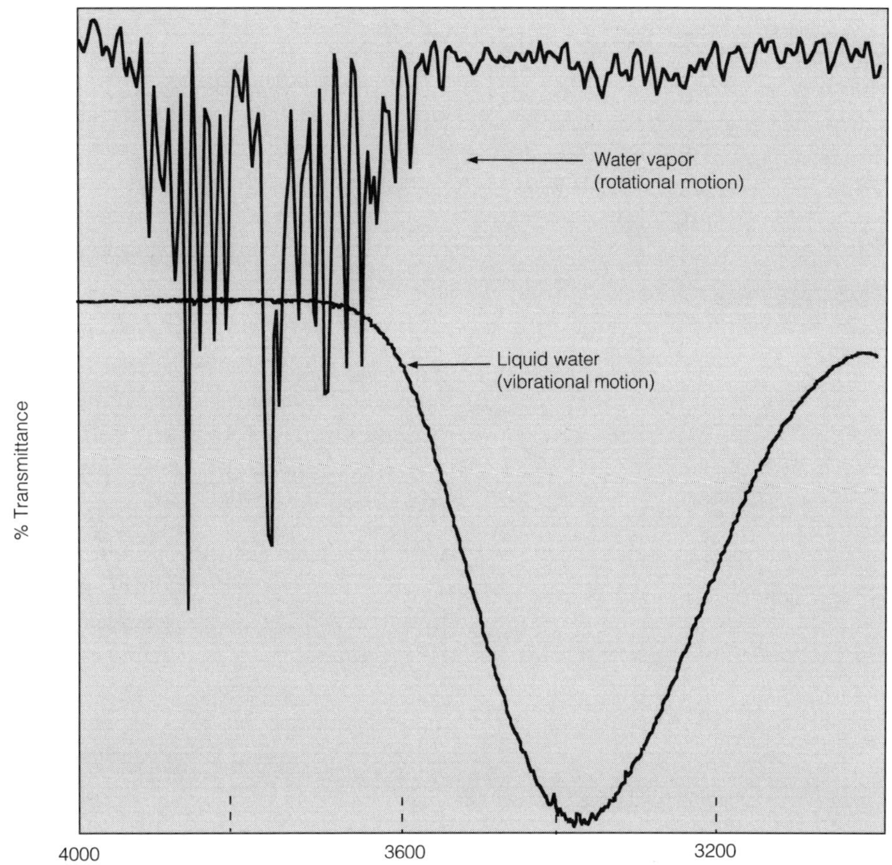

Figure 2.4
IR spectrum of water vapor rotational absorption bands versus a spectrum of the major vibrational absorption band for liquid water.

and Fig. 2.6). All molecules have three translational degrees of freedom—that is, the center of mass of the molecule can move in three directions (x, y, and z). If the molecule is nonlinear, then it also has three rotational degrees of freedom as it spins around each of the three axes (x, y, and z). Linear molecules have only two rotational degrees of freedom, since two rotation directions are equivalent. Thus, the total of translational and rotational degrees of freedom is 6 (5 for linear molecules). All remaining motions, 3N – 6, are vibrational degrees of freedom (3N – 5 in the case

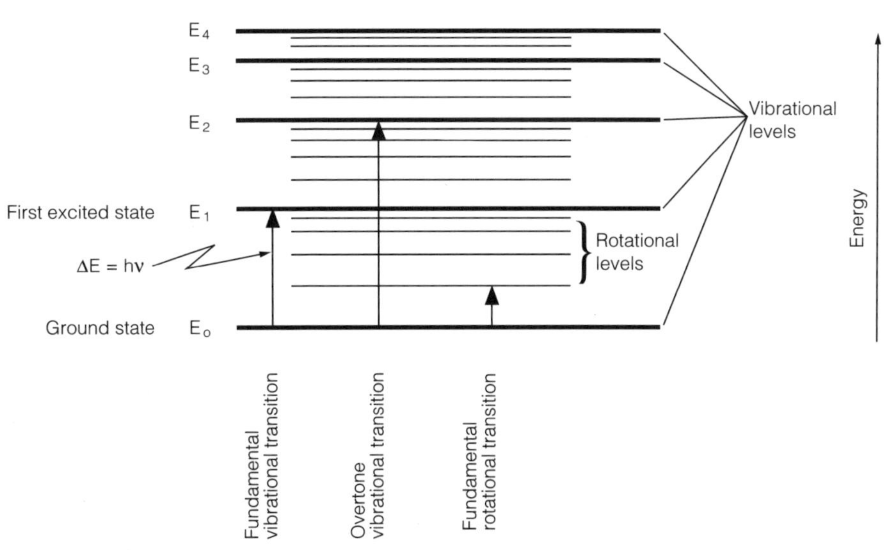

Figure 2.5
Energy transition levels for vibrational and rotational transitions. Vibrational transitions produce the primary IR absorption bands. A fundamental transition occurs when the absorbed photon increases the energy level from the ground state (E_0) to the first excited state (E_1). An overtone occurs when the transition covers two transition levels. The very small energy difference between rotational levels results in very sharp, closely spaced rotational bands in a spectrum.

Table 2.1
Degrees of freedom corresponding to type of motion for any molecule with N atoms. The total number of degrees of freedom for any molecule is 3N.

Motion	Degrees of freedom
Translation (position of the molecule)	3
Rotation (about the center of the molecule)	3 (nonlinear) 2 (linear)
Vibration	3N – 6 (nonlinear) 3N – 5 (linear)

of linear molecules). This is an important number, since vibrational transitions are the strongest and most important in IR spectroscopy. For example, a nonlinear molecule with 3 atoms, such as SO_2, will have 3 fundamental vibrations $[(3 \times 3) - 6 = 3]$, as shown in Figure 2.6. All vibrational motions of the atoms can be described completely in terms of these 3N – 6 or 3N – 5 fundamental vibrations, which are called the normal modes of vibration for the molecule.

The most common vibrations are stretching, torsional, and bending modes. A stretching vibration increases or decreases the length of the bonds between the atoms. A torsional, or skeletal, vibration

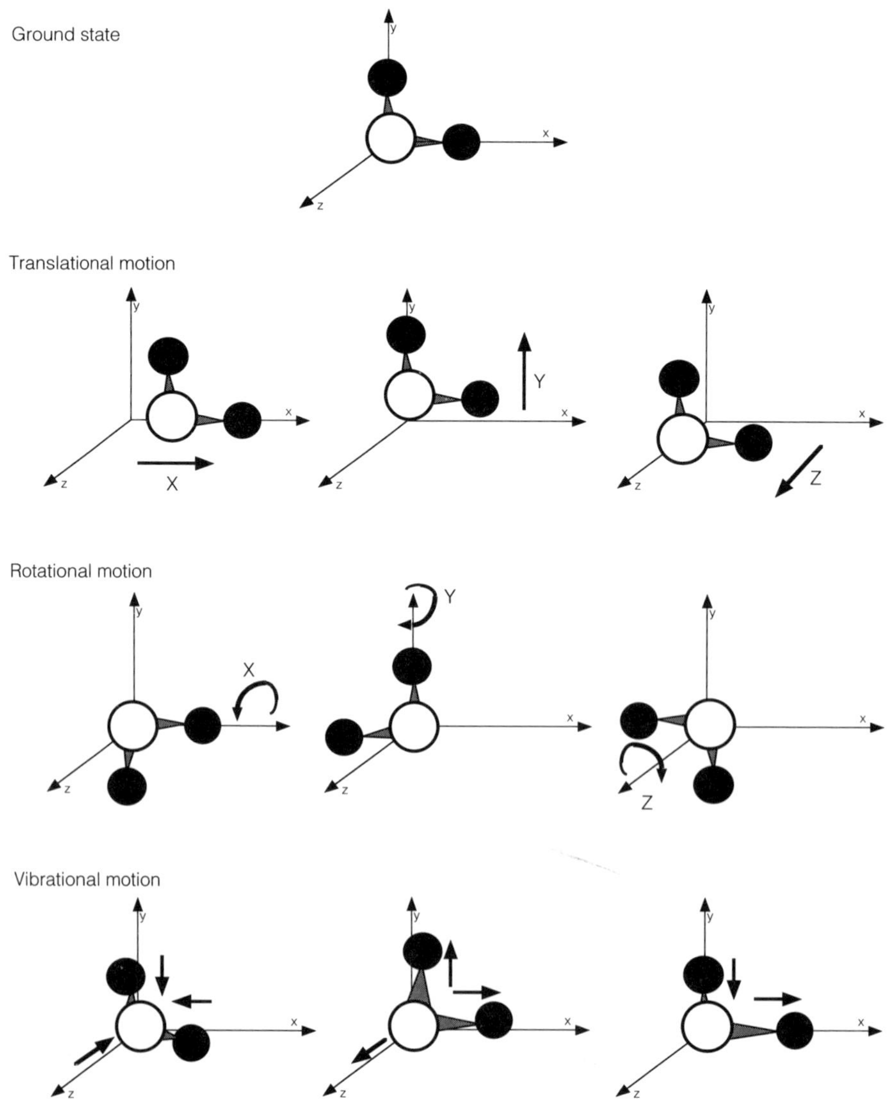

Figure 2.6
Degrees of freedom for molecular motion of a triatomic molecule such as SO_2. The molecule at rest is shown at the top. The molecule can move in three translational motions (x ,y, and z), as well as rotate around the three axes. Of the nine total motions allowed by the degrees of freedom for a triatomic molecule, only three are left for vibrational motion. These are shown as stretching and bending vibrations.

involves the twisting of the backbone of the molecule. A bending vibration changes the bond angles of the atoms relative to one another or to the remainder of the molecule. Bending vibrations can be further classified as scissoring, rocking, wagging, and twisting. Vibrations are also characterized by their symmetry—that is, they can be symmetric or asymmetric.

Selection rules

The degrees of freedom specify the maximum number of fundamental vibrations for a molecule. A fundamental vibration—also called a first-order vibration—corresponds to a change from the molecular ground state to the first energy level. A first-order vibration produces the strongest energy absorptions, but only those that are active in the IR region will be seen as an absorption band in the IR spectrum. Selection rules, based on the symmetry of the molecule, determine whether a given vibration will be seen in the spectrum. The primary selection rule, or requirement, for active IR absorptions is that the vibration must change the dipole moment of the molecule. In a heteronuclear molecule, nonsymmetrical vibrations change the distance between the two nuclei and thus its dipole moment. This vibrating dipole moment creates a dipolar electric field that in turn absorbs a discrete unit of energy unique to that transition. Vibrations in a homonuclear diatomic molecule or symmetrical vibrations in a heteronuclear molecule do not change the dipole moment and thus are not seen in the IR spectrum.

In practice, the number of absorption bands observed in an IR spectrum is usually smaller than the number of fundamental frequencies. In addition to the primary selection rule, other factors affecting a reduction in the number of absorption bands are (1) degeneracy, where two vibrational modes may occur at identical frequencies, (2) overlapping or weak absorptions, or (3) vibrational modes outside the instrument analysis range.

Alternatively, the number of observed bands may be increased by the detection of weak, or so-called forbidden, absorptions, in which the change in the vibrational level is greater than one. These are known as overtones. Two or more fundamental vibrations may also interact to produce discrete absorption bands that occur at a frequency corresponding to the sum or the difference of the individual band frequencies. Such absorptions are known as combination bands.

Group frequencies

Within any molecule, a given functional group (a combination of atoms such as a carbonyl group or an amide group) is responsible for IR absorptions at or near the same frequency, regardless of the rest of the molecule. The position (i.e., frequency or wavenumber) of an absorption band depends on the mass of the atoms in the absorbing group, along with the strength and angles of the connecting bonds. A mathematical equation for the vibrational frequency of a two-body system (Hooke's law) can be used to predict the absorption band position for simple molecules (Smith 1979). However, only a limited number of small molecules have vibrational spectra simple enough for complete theoretical analysis and interpretation. The majority of IR spectra-structure correlations are empirical, having been determined by the analysis of a large number of compounds.

The vibrational frequencies for any particular functional group are characteristic of that group—e.g., most carbonyl stretches occur between 1650 and 1750 cm^{-1}, and most carbon-hydrogen stretches occur near 3000 cm^{-1}. These characteristic vibrations are termed group frequencies and are used for the identification of materials and for the determination of structure in an unknown pure compound (see Chap. 5 on spectral interpretation).

Molecules rarely consist of just a two-atom pair but, rather, consist of multiple groups of atoms, each involved in its own vibrational transitions. The energy of a vibration and, thus, the position of the band in the IR spectrum are sometimes influenced by the atoms surrounding the vibrational group. A highly electronegative atom (e.g., chlorine) near a functional group can cause shifts in the electron distribution (inductive effects) that raise the frequency (wavenumber) of vibration. The presence of heavier atoms near a functional group (e.g., nitrogen next to carbonyl in an amide group) will dampen the oscillations (resonance) and lower the frequency (wavenumber) of vibration. Strong coupling between stretching vibrations occurs only when there is an atom common to the two vibrations. Interaction between bending vibrations requires a common bond between the vibrating groups. Little or no interaction is observed between groups separated by two or more bonds.

Infrared Spectra

An IR spectrum displays detector response as percent transmittance (%T) on the y-axis, and IR frequency in terms of wavenumber (cm^{-1}) on the x-axis, as shown in Figure 2.7. The detector response indicates the extent of interaction of the IR electromagnetic radiation with the sample as it is

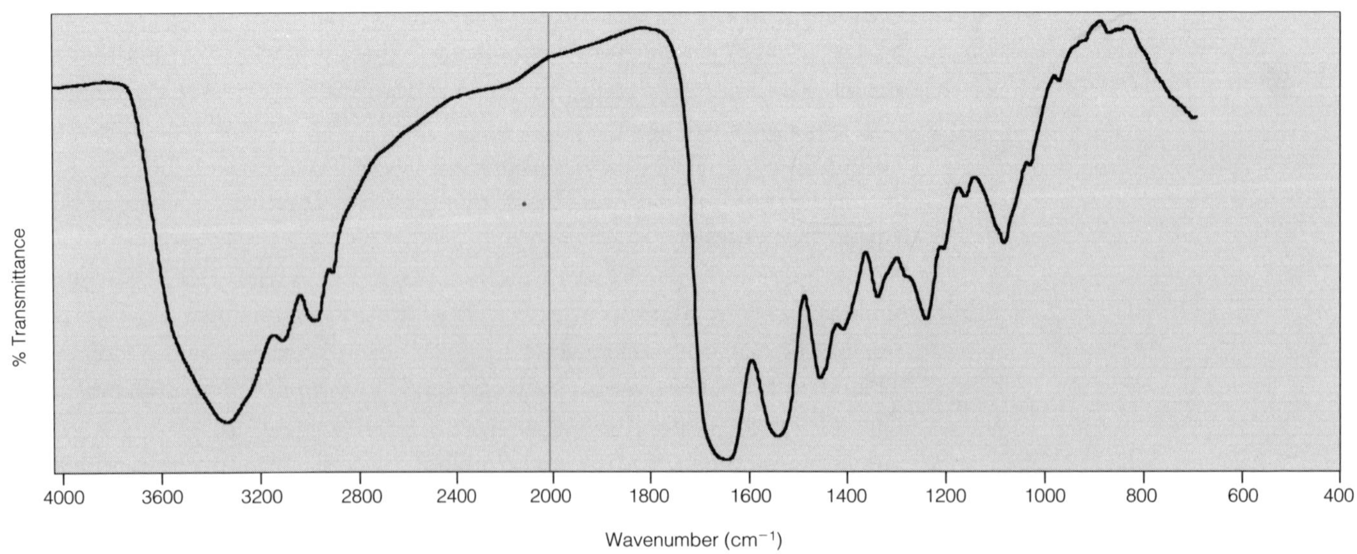

Figure 2.7
IR spectrum of gelatin plotted as percent transmittance (%T) on the y-axis, and IR frequency in terms of wavenumber (cm^{-1}) on the x-axis.

proportional to the resultant intensity of IR radiation that reaches the detector after passing through the sample.

Two types of interactions—absorption and transmission—are important in the typical IR experiment. When the molecule in the sample compartment of the spectrometer is exposed to a source of continuous IR radiation, the photons of discrete energy units that are absorbed by the molecule do not reach the detector. The IR spectrum reveals these missing photons, or absorptions, as a series of well-defined, characteristic, and reproducible absorption bands. Photons that are not absorbed by the sample are transmitted to the detector essentially unaltered.

For a given wavelength or frequency of IR radiation striking a sample, these two interactions are inversely related through the following equation:

$$A = \log 1/T$$

where: A = absorbance and T = transmittance (%T/100).

Infrared Regions

As was seen in Figure 2.1, the IR spectral region of the electromagnetic spectrum extends from the red end of the visible spectrum to the microwave region; it includes radiation with wavenumbers ranging from about 14,000 to 20 cm^{-1}, or wavelengths from 0.7 to 500 μm. Because of application and instrumentation reasons, it is convenient to divide the IR region into the near (NIR), middle (IR or mid-IR), and far (FIR) subregions. The majority of analytical applications are found in the middle region, extending from 4000 to 500 cm^{-1} (2.5 to 20 μm).

Near-infrared region

The near-IR (NIR, NIRS) region extends from the visible region at 14,000 cm^{-1} (0.7 μm) to the mid-IR region at 4000 cm^{-1} (2.5 μm). Because it is accessible with quartz optics, near-IR instrumentation is often combined with UV-Vis spectrometers (UV-Vis-NIR). Spectra generated in the near-IR region consist of many overtones and combinations of the mid-IR region fundamental vibration modes. Since all organic species absorb in the NIR and produce many overlapping bands, single-band spectroscopy and qualitative band assignments are nearly impossible. NIR is useful for quantitative work, including in situ monitoring of reactions.

Mid-infrared region

The spectral range of greatest use for chemical analysis is the mid-IR (MIR) region. It covers the frequency range from 4000 to 500 cm^{-1} (2.5–20 μm). This region can be subdivided into the group frequency region, 4000–1300 cm^{-1} (2.5–8.0 μm) and the fingerprint region, 1300–500 cm^{-1} (8.0–20 μm).

In the group frequency region, the main absorption bands may be assigned to vibrational modes corresponding to individual functional groups:

NH-OH (4000–3000 cm^{-1})
C-H stretch region (3000–2800 cm^{-1})
window region (2800–1800 cm^{-1})
carbonyl region (1800–1500 cm^{-1})

Both the presence and absence of these characteristic group frequency bands are useful for characterizing molecular structure.

The absorption bands in the fingerprint region of the spectrum are the results of single-bond as well as skeletal vibrations of polyatomic systems. Multiple absorptions in this region make it difficult to assign individual bands, but the overall combined pattern is very characteristic, reproducible, and useful for material identification when it is matched to reference spectra.

Far-infrared region

The far-IR (FIR) region is generally designated as 500–20 cm^{-1} (20–500 μm). In this region, the entire molecule is involved in low-frequency bending and torsional motions, such as lattice vibrations in crystals. These molecular vibrations are particularly sensitive to changes in the overall structure of the molecule that are difficult to detect in the mid-IR region. For example, the far-IR bands of isomers and long-chain fatty acids can often be differentiated in solid-state materials. FIR is also useful in the identification and differentiation of many minerals and colorants.

Summary

Light and matter can interact. The examination of this interaction is termed spectroscopy. The interactions are characterized by the energy of the radiation and its effects on materials. IR radiation supplies sufficient energy to produce translational, rotational, and vibrational motion in molecules. The measurement of the characteristic IR energies (photons) that correspond to these transitions results in a spectrum. Based on its atomic structure, each molecule produces a unique and characteristic IR spectrum. The specific number and position of absorption bands for any molecule are governed by its degrees of freedom, its functional groups, and the IR selection rules. A spectral pattern, sometimes called a fingerprint, is used to identify an unknown material when the absorptions in its spectrum are matched with the absorptions in the spectrum of a known material. Additionally, since functional groups (combinations of atoms) produce absorptions at or near the same frequency, regardless of the rest of the molecule, the presence or absence of certain functional groups can be determined by interpretation of the IR spectrum.

Additional Reading

Brill, T. B.
1980. *Light: Its Interaction with Art and Antiquities*. New York: Plenum Press.

Colthup, N. B., L. H. Daly, and S. E. Wiberley
1990. *Introduction to Infrared and Raman Spectroscopy*. 3d ed. Boston: Academic Press.

Schutte, C. J. H.
1976. *The Theory of Molecular Spectroscopy*. Vol. 1. New York: Elsevier.

Silverstein, R. M., F. C. Bassler, and T. C. Morrill
1991. *Spectrometric Identification of Organic Compounds*. 5th ed. New York: John Wiley and Sons.

Smith, A. L.
1979. *Applied Infrared Spectroscopy: Fundamentals, Techniques and Analytical Problem Solving*. Chemical Analysis series, vol. 54. New York: John Wiley and Sons.

Chapter 3
Sample Collection and Preparation

Much effort has been put forth to optimize instrumentation for material analysis, as well as to maximize the amount of information that can be gleaned from data. The most important step of any analysis, however, is the collection and preparation of a sample. Although numerous software manipulations can sometimes recover useful data from an ill-prepared sample, it is best to concentrate on using good sample handling and experimental technique. The extra time required will be rewarded by the generation of high-quality data. This chapter details collection and preparation procedures for many types of samples.

Sampling Methodology

Sampling is defined as the process of selecting and collecting the sample for analysis. Analytical chemists receive training in the importance of correct sampling and valid data treatment. For example, Majors states, "Collect the wrong sample, or collect the right sample incorrectly, and you trivialize all that follows, rendering your data worthless" (Majors 1992).

In the ideal situation, the analyst will be involved in and present at each sampling step. When this is not practical, the sampler should provide as much information as possible to the analyst. In the analysis of art objects, conservators are often the most qualified to remove samples because of their skill and knowledge of the history, problems, and objectives related to the sampling. When sampling and analysis functions are shared between more than one person, it is important to keep communication open. A sampling strategy should be established, based on thorough discussion of analytical questions and experimental choices.

When dealing with works of art, the possibility for sampling may be limited; thus, it is important to draft the sampling strategy accordingly (Reedy and Reedy 1988). Sampling consists of two steps: first, the sampling design and, second, the implementation. The purpose of the sampling design step is to determine how to obtain a representative sample or set of samples related directly to the analysis question. The second sampling step, implementation, involves the actual removal and preparation of the sample or samples, with the goals of avoiding sample loss and contamination.

Sampling design

The design, or planning, step requires intimate knowledge of the object as well as of the reason for analysis. Since cultural objects are irreplaceable and their preservation the ultimate goal in any conservation examination, samples are removed only when necessary. Nondestructive techniques (X-ray fluorescence [XRF], X radiography, IR and ultraviolet [UV] photography, etc.) should receive priority and are often used to survey objects. Physical testing methods—such as measurement of color, hardness, and porosity—may at times be in situ analyses. However, most chemical analysis techniques require sample removal. Table 3.1 compares the capabilities of IR spectroscopy to other commonly used chemical analysis techniques. Each technique has its own advantages, and when techniques are used together, they can supply complementary information on a sample.

The range of materials used in art objects is nearly universal. Inorganic substances are found as base materials (stone, metal, glass, ceramic), as colorants (pigments), as thickeners and fillers, as polishers (talc, alum, carbonates), as stabilizers and neutralizers, and as unwanted reaction products (corrosion, weathering crusts, salt deposits). Organic materials in objects may have either natural or synthetic sources. Both plants and animals generate natural products (cellulose, hair, skin, resin, gum, dye, oil, protein, wax) that are themselves complex mixtures of chemical compounds, even before being prepared for use as art materials. In contrast, synthetic organic materials (found as fibers, colorants, binders, adhesives, plasticizers, coatings, backings, supports) are, in

Table 3.1

A comparison of commonly used chemical analysis techniques.

Technique	Acronym	Description	Minimum sample size	Sample preparation	Sensitivity	Specificity
Polarized light microscopy	PLM	Identification of material based on physical properties	5 μm	easy	n/a	none
IR spectroscopy	IR	Compositional analysis of organic and inorganic compounds	10 μg	easy	10%	none
X-ray fluorescence	XRF	Elemental analysis	non-destructive 1 mm spot	none	0.1%	elements only (heavier than potassium)
Energy dispersive spectroscopy	EDS	Elemental analysis (similar to XRF but attached to scanning electron microscope)	1 μm spot	easy	0.1%	elements only (heavier than carbon)
X-ray diffraction	XRD	Compositional analysis of crystalline materials	10 μg	easy	5%	crystalline materials
Inductively coupled plasma	ICP	Quantitative/qualitative analysis of elements down to trace levels	1 mg	slow	0.01 ppm	elements only (heavier than nitrogen)
Chromatography		Quantitative/qualitative analysis of organic components in a mixture				General class of materials in mixture must be known before starting analysis
Thin-layer	TLC	(TLC usually only qualitative)	1 mg	slow	5%	
Liquid	HPLC	High performance liquid chromatography	1 pg	slow	0.01 ppm	
Gas	GC		1 μg	slow	0.01 ppm	

general, originally produced as well-defined, purified materials; they may later be combined for use in commercial products. In most cases, IR spectroscopy, since it is a nondiscriminatory technique, is one of the first analysis methods chosen. Table 3.2 indicates the general order in which common analytical methods are usually applied. Information on the microscopic characterization of particles, another common first-choice technique, can be found in McCrone and Delly (1973) and Aldrich and Smith (1995).

In the analysis of complex materials, such as paints, several analytical methods are usually needed to produce a complete characterization of a sample's components. Elemental analysis of the inorganic

Table 3.2

Common analytical methods and the general order in which they are applied to various organic and inorganic materials (IR = infrared spectroscopy [compositional analysis of organic/inorganic compounds]; XRF = X-ray fluorescence [nondestructive elemental analysis]; EDS = energy dispersive spectroscopy [same information as XRF but requires sample]; PLM = polarized light microscopy [size, color, shape, crystallinity, refractive index, etc.]; XRD = X-ray diffraction [compositional analysis of crystalline materials]; ICP = inductively coupled plasma [quantitative elemental analysis]; UV/Vis = ultraviolet/visible spectroscopy [precise color measurement]; fluorescence = UV-induced fluorescence [both autofluorescent and with reactive dyes]; color = visual color assessment; solubility = solvent solubility testing; microchemical tests = chemical reactivity tests conducted on a microscopic scale).

Material	First-choice analysis methods	Other useful analysis methods
Organic		
Coatings	IR	solubility, chromatography, fluorescence
Adhesives	IR	solubility, chromatography, fluorescence
Consolidants	IR	solubility, chromatography, fluorescence
Binders	IR	solubility, chromatography, fluorescence
Plastics	IR	solubility, chromatography, fluorescence
Polymer additives	IR	solubility, chromatography
Sizes/finishes	IR	solubility, chromatography, fluorescence
Solvents	IR	chromatography
Air pollutants	IR, chromatography	specific compound detectors
Synthetic fibers	PLM, IR	
Natural fibers	PLM, IR	(with mordants, also use XRF or EDS)
Dyes	IR, chromatography, UV/Vis	fluorescence
Wood	PLM	
Organic/inorganic		
Paints	XRF(EDS), IR, PLM	XRD, solubility, chromatography, fluorescence
Lacquers	XRF(EDS), IR, PLM	XRD, solubility, chromatography, fluorescence
Residues	XRF(EDS), IR, PLM	XRD, solubility, chromatography, fluorescence
Patinas	XRF(EDS), IR, PLM	XRD, solubility, chromatography, fluorescence
Corrosion products	XRF(EDS), IR, PLM, XRD	ICP
Unknowns	XRF(EDS), IR, PLM	XRD, solubility, chromatography, fluorescence
Inorganic		
Pigments	color, XRF(EDS), PLM, IR	XRD, microchemical tests
Gems	color, XRF(EDS)	IR, XRD
Glass	XRF(EDS)	ICP
Ceramics	XRF(EDS), PLM	IR, XRD, ICP
Stones	XRF(EDS), PLM	IR, XRD
Masonry	XRF(EDS), PLM	IR, XRD
Mortars/plasters	XRF(EDS), PLM	IR, XRD
Fillers (inorganic)	XRF(EDS), PLM, IR	PLM, XRD
Salts	XRF(EDS), PLM, XRD	IR, microchemical tests
Metals	XRF(EDS), metallography	ICP

material in a paint sample can indicate the presence of certain compounds otherwise not obvious from the IR spectrum. Additionally, in an IR spectrum of a multiple-component sample, knowledge of one material simplifies the identification of the remainder. Thus, the complementary use of other analysis techniques (such as scanning electron microscopy [SEM], XRF, X-ray diffraction [XRD], optical microscopy, etc.) in conjunction with IR is very helpful. The data resulting from each method can be fit together as pieces in a puzzle, with the addition of each piece bringing it closer to completion. For further information on the potential and strategy of combining multiple analytical techniques for the characterization of samples from works of art, see Masschelein-Kleiner, Heylen, and Tricot-Marckx (1968), Schreiner and Grasserbauer (1985), Roelofs (1989), Karreman (1989), and Erhardt and coworkers (1988).

Sampling location

When the removal of samples is permitted, the value of the object and its state of deterioration often dictate very specific restrictions on the number, size, and location of the samples removed. The object's history, previous analyses, treatments, revarnishing, and retouching are considered in the sampling design. In view of the complexity and variety of possible components in a sample from an art object, the first step for its analysis is to obtain all the background information possible. This measure can substantially reduce the amount of time necessary for sampling and for the later spectral interpretation, as well as minimize the chances for erroneous conclusions.

Questions, such as the following, should be asked before the sample is taken:

- What is the background of the object?
- What is the reason for analysis?
- What information is needed?
- What is the desired format for results?
- Can a sample be removed? If so, how much? Where?
- Does the sample have a single component or is it multicomponent?
- Are all components original to the piece?
- Has any previous analysis been done?

After background information has been obtained, the object is thoroughly examined in visible and UV light to evaluate its condition and homogeneity, as well as to inspect potential sampling sites. All information is recorded. Then, when possible, all involved personnel—conservator, curator, and scientist—discuss the sampling plan and make a joint decision as to sample numbers, locations, and amounts.

In the simplest sampling decision, an unknown contains a single homogeneous component or matrix. Examples include a new sheet of paper, a uniform varnish layer, an adhesive from one location, or a modern (unaged) sculpture out of a single metal or polymeric resin. In these situations, a sample can be taken from any location, and the analytical results will be the same. However, a homogeneous, single-matrix situation is more often found in manufactured objects than in works of art.

For an inhomogeneous matrix or set of objects, the ideal sampling protocol is to assign an identifier to each object or sample location (such as with a grid) and to select randomly the objects or sites for analysis. This method will provide a statistical basis for sample selection. If cost, time, or value of an object limits the number of samples that can be taken, then visual classification of the object by color, texture, structure, UV fluorescence, and so on, can be used to justify minimizing the sample number and locations.

It is rare that a random and truly representative sample is removed from a valuable object. More often "convenience" samples are selected because of their accessibility or because they can be taken from regions of previous damage. For example, a typical sampling procedure for a surface finish from a museum furniture object is to remove one or two barely visible samples ($< 50\ \mu g$) from arbitrarily chosen, obscure areas near or under metal mounts, in the rear of the carcass, or on the legs. In such circumstances, the sample may not be truly representative of the object. However, with care and knowledge of the situation, good data can still be obtained and justified as valid when combined with background information and visual examination results.

Sampling Implementation

Sampling implementation is the collection and preparation of the sample for analysis. Once the analyst has removed or received the sample, additional steps may be taken to prepare it for a specific analytical method. As it is the sample that determines the quality and utility of the results, care is taken in each step of its removal, storage, preparation, and analysis to prevent contamination or loss. Appropriate sampling tools and containers for storage are necessary to keep the sample from changing prior to analysis.

Tools

Several types of sampling tools are useful for the mechanical removal of a sample (Fig. 3.1). Fine-point forceps are appropriate for samples that are visible to the naked eye. Smaller particles may be manipulated with a tungsten needle with an extremely fine ($1\ \mu m$) point. A tungsten needle is prepared by heating the tip of a tungsten wire (3 cm of 24 or 26 gage wire attached to a wire holder) to red hot in a burner, then quickly drawing it through sodium nitrite (McCrone and Delly 1973; Teetsov 1977). Another fine-point probe for manipulating small samples can be made from a cat hair or a thick human eyelash that has been adhered to a wooden stick (Reid 1972). Also helpful in retrieving small particles is a fine artists' brush (no. 2 or no. 3) or a brush modified to contain only a few bristles. The static electricity in a small sliver of cured silicone, freshly cut from a flexible mold, makes it extremely useful for picking up and transferring small particles. Fine-tip paper points, available from

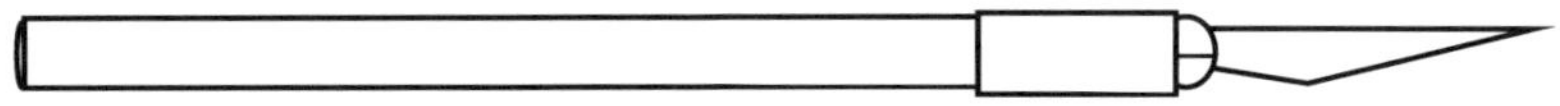

Disposable **scalpel blade** in universal metal handle; both pointed- and curved-end blades are useful for sample collection.

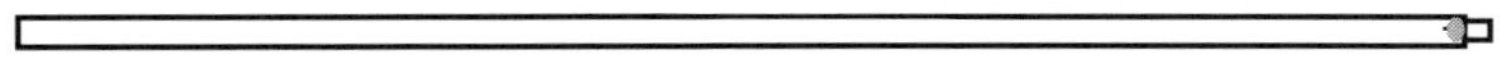

Disposable **blade** prepared from a cut razor blade and a wooden stick.

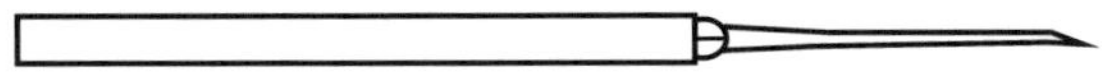

Disposable **eye blade** in a 3 inch (7.62 cm) Beaver blade handle; this blade is useful for removing cross sections.

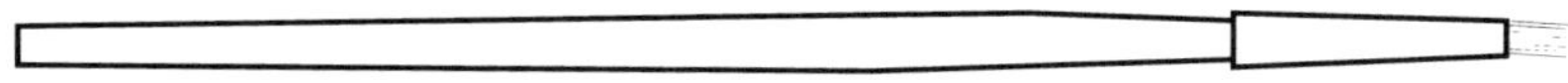

An inexpensive **artists' brush** with all but a few bristles removed; the brush fibers are useful for picking up groups of particles.

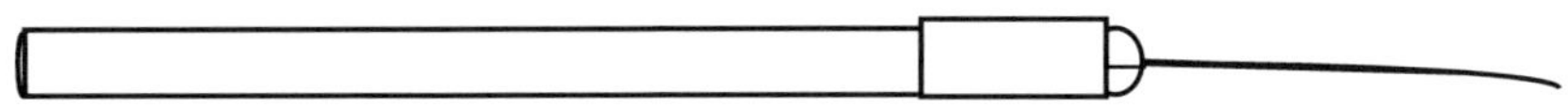

A **tungsten fine-point needle** in a universal metal handle; a fine-point needle is used for separating and transferring single particles.

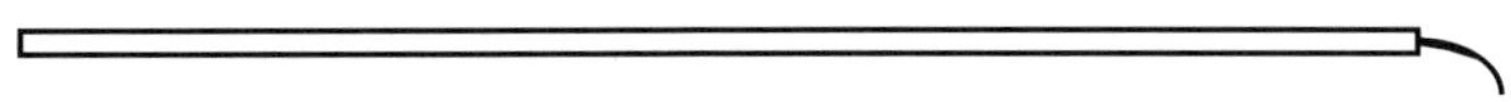

An **eyelash brush** prepared by adhering an eyelash or cat hair to a wooden stick; this tool is useful for handling fragile particles or thin sections.

Figure 3.1

Several types of sampling tools.

microscopy supply companies, may also be used for picking up tiny particles; dampening a point with a microdrop of distilled water creates a tool that will pick up the most obstinate particle; however, this treatment subsequently makes it difficult to release the particle into a container.

Scalpels are used to remove barely visible particles as well as multiple-layer cross section samples. Microsurgical scalpels have sharp, thin blades that work well; of particular note are eye blades. Another option is a microscalpel made by adhering 1–2 mm broken from a razor blade to a sharpened wooden applicator stick (Hill 1989). With any tool, it is important that the cutting or sampling end is immovably secured to the handle.

Sample removal and subsequent sample preparation require a steady hand. The probe or scalpel should be held at a low angle to the working surface (McCrone 1982). Optimally, the working hand and arm should be supported on a vibration-free surface. This is rarely possible during the sample removal step, unless an exterior bridge (boom arm, stool, scaffolding) is available, since the object itself should not be used as a support. The sampler's other hand can be used to steady the working hand, provided it is not required for holding the sample container.

Sample documentation and storage

Documentation of samples is crucial for understanding and interpreting the analytical results and relating the sampled area to the conservation problem. Proper documentation starts with a picture of the object (e.g., a photocopy or a Polaroid) on which the explicit sampling areas and distinguishing features can be marked. Corresponding labels are placed on sample containers, with the date of sampling and the initials of the sampler noted. All pertinent information is recorded in a sampling notebook. This information includes a complete description of the object, the sampling area, and the sample, along with the reason for sampling and the potential types of analyses.

Each sample is placed in its own well-labeled container. Containers can introduce their own sets of problems. Plastic containers (BEEM capsules, Ziploc bags, etc.) often produce static electricity that hinders the addition and removal of samples. Also, additives in and on plastic containers (such as slip agents added to plastic bag surfaces to keep them from sticking together) readily contaminate samples. Samples in solution should never be placed in plastic containers because of the potential for leaching of plasticizers or dissolution of containers. Gelatin capsules should also not be used for sample storage, as they may hamper the determination of the presence of proteinaceous materials within the sample. Organic samples are best contained in clean glass containers, such as a depression slide sandwich (described below) or a vial with a Teflon-lined lid. Aluminum foil is an acceptable alternative.

A glass depression slide is ideal for small samples. A standard microscope slide can be placed on top of the depression slide to serve as a lid; tape hinges can be added on one side and a tape latch attached to the other (Fig. 3.2). The cover slide provides protection as well as a suitable flat area for the required labeling or coding. The primary advantage of the glass depression slide container is that samples can be examined and photographed with an optical microscope multiple times without the container being opened, so that opportunities for contamination or loss are minimized. When the cover is removed, the sample is readily accessible to scalpels, tungsten needles, cured silicone rubber slivers, or other probes

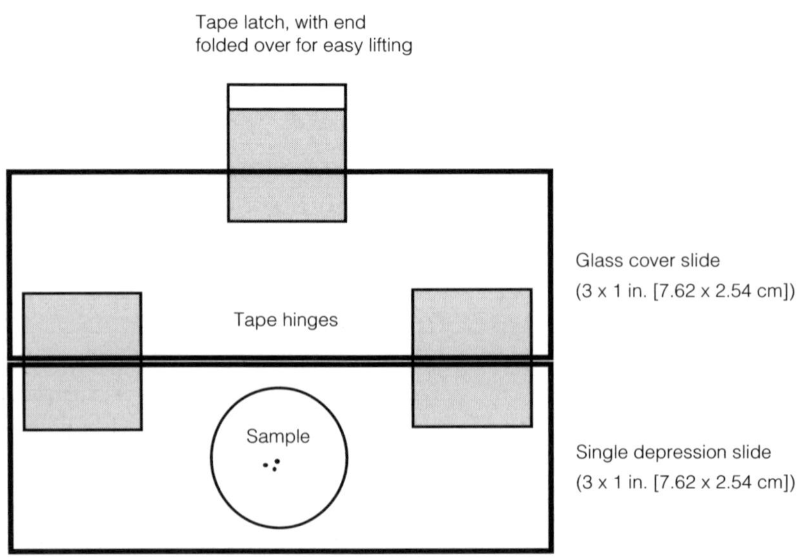

Figure 3.2
A sample holder made from a single-depression glass slide with a standard glass microscope slide as a cover. Single-sided transparent adhesive tape is used to make hinges and a latch. The tape latch should be placed on the cover slide and not on the depression slide to minimize the chance that the sample might become attached to the adhesive.

used to select and transfer a portion for analysis. The slide sandwich also allows the samples to be carried and stored in microscope slide trays.

Avoidance of contamination

Sample purity is always a primary concern, especially in microanalysis, where contaminants, such as dust particles, can be as large as or larger than the sample. There are numerous means by which a sample may become contaminated, and it is always best to analyze a sample blank as a check. A sample blank should be exposed to all environments and come into contact with the same materials and solvents as the sample itself.

The most common sources of contamination are unclean sampling tools, storage containers, and analytical support materials, such as IR windows. Remnants from previous samples can be incorporated into a new sample, producing a déjà vu spectrum. Storage containers—such as plastic vials, bags, and lid liners—can pollute the sample with particulates, plasticizers, and processing oils. Glass slides, even though marked "precleaned," may be covered with a thin layer of formic acid, calcium carbonate, sodium sulfate, or silicates (Sommer and Katon 1988). Whenever solvents are used, chromatographic or spectroscopic grades should be selected to minimize contamination from solvent impurities. Residual impurities in solvents, as well as contaminants, such as silicones, from capillary pipette holders, are readily detected by the examination of residue left after evaporation of a solvent drop on a glass slide or IR window.

Environmental substances can taint a sample before or after it is removed from the object. Fibers (natural, synthetic, and hair) are the most commonly recognized contaminants in nonfiber samples. Inorganic particulates, such as quartz, clay, gypsum, and calcite, are generally soil related; it is difficult, if not impossible, to tell whether the soil source is current contaminant (nonsignificant) or from the weathered crust of the object (potentially significant). Organic materials may also be distributed as airborne matter. Activated carbon particles may come from the air-conditioning and filtration system. Plants, insects, animals, and humans are the sources of oils and protein bits that may produce potentially ambiguous results in the analysis of art materials containing natural products. Work areas and containers should be clean, and sample handling with bare fingers and powdered gloves (containing talc, starch, etc.) should be avoided. Optical surfaces should not be sprayed with compressed air from canisters, as the spraying may leave a fine residue of oil on the surfaces (Leyshon and Roberts 1983).

Lang and coworkers have identified a number of common contaminants that make IR microanalysis very difficult (Lang, Katon, and Bonanno 1988). An excellent chapter on the detection and identification of contaminants commonly found in laboratory samples has been published by Aldrich and Smith (1995). Launer provides an extensive list of spurious IR absorption bands that may be due to contaminants (Launer 1962). Of particular interest on the list are the bands that are attributed to internal impurities in halide windows (see Table 4.2).

Sample Collection and Preparation Procedures

Many, if not most, samples will be analyzed by more than one method. Thus, unless additional samples can be obtained easily, the entire sample must never be used during any single sample preparation step. If a small, irreplaceable sample must be used in totality, it should first be analyzed by a noncontaminating, nondestructive method, such as optical microscopy or IR microspectroscopic analysis—after which it can be retrieved and stored for further testing.

Good laboratory practices are important. Sample documentation, which began during the collection phase, is continued, with all preparation steps recorded in a lab notebook. Photomicrographs are obtained on the sample whenever possible. If not, drawings and visual observations are recorded. This procedure helps to distinguish the physical characteristics of the sample. Subsamples, such as solubility fractions or isolated inhomogeneous particles, are always labeled accordingly, along with identification of the parent sample.

The following describes a possible sample preparation procedure for a multilayer paint sample after its arrival in the lab. The sample is examined under a stereomicroscope, and a drawing or photograph—with colors and relative sizes noted—is made of the visually discernible layers. Particles are removed for further characterization and identification by polarized light microscopy, if this has not already been done. Then a small portion of the sample (no more than half) is embedded. After polymerization of the mounting medium, the embedded sample is microtomed. This method achieves two goals: First, thin sections are produced for analysis by IR transmitted light. Second, a flat, even surface is obtained on the remainder of the embedded cross section, so that the effort required for polishing the surface is minimized. At this point, photomicrographs, in visible and UV light, are taken of the cut surface, preferably adjacent to the thin section layer removed for IR analysis. Later, if desired, the embedded residual sample can be used for IR reflectance mapping studies, SEM, and fluorescent staining. The compiled data set from these analyses should give a clear image of the structural composition of the paint sample. The unembedded portion of the sample is retained for later chromatographic studies or for additional embedments, if problems arise from the first sample.

In the interrelated sampling scheme, the steps required for sample collection and preparation depend not only on the analysis methods but also on the type of sample (Table 3.3). The following sections discuss a variety of sample collection and preparation procedures that apply to IR analysis, as well as to other analytical methods. The IR-specific techniques mentioned in this section are described in chapter 4.

Gases

IR analysis is rarely used to monitor pollutant levels within museums and display cases because these are relatively clean environments, with low-level contaminants that are better measured by quantitative techniques such as chromatography. IR analysis can, however, be used to perform

Sample type	Examples	Potential collection methods	Potential preparation methods	Potential analysis methods
In situ nondestructive analysis	surface studies	none	none (possible precleaning)	Photography (IR, UV, etc.); XRF
Gas	pollution measurements; humidity studies	none; syringe; flow-through cells; pump; sorbent tube	none; desorption; derivitization; preconcentration	IR; GC; HPLC; colorimetry; visual indicator
Liquid	residue; exudates; prepolymer components	syringe; capillary tubes; swab; sorbent	none; solvent evaporation; derivitization	IR; GC; HPLC; TLC
Solid Solvent-soluble	coating; adhesive	solvent-soaked swab	extraction; solvent evaporation; derivitization	IR; GC; HPLC; TLC
Solid Particles	corrosion layer; stone; pigment; coating; alloy; adhesive	scraping; scalpel; forceps; probe; silicone sliver	mounting; physical separation; grinding; flattening; dissolution; derivitization	IR; GC; HPLC; TLC; SEM/EDS; PLM; fluorescent microscopy; chemical microscopy; XRD
Solid Cross section	painting; alloy; weathered crust	scalpel; needle; drill	mounting; physical separation; grinding; embedding; microtoming; derivitization	IR; GC; HPLC; TLC; SEM/EDS; PLM; fluorescent microscopy; chemical microscopy; XRD
Solid Fibrous	textile; contaminant; basket	forceps; probe	none; mounting; embedding; flattening; cutting; extract dye	IR; PLM; SEM/EDS; UV/Vis

Table 3.3

Sample types and related sample collection, preparation, and analysis methods (GC = gas chromatography; HPLC = high performance liquid chromatography; PLM = polarized light microscopy; SEM/EDS = scanning electron microscopy with energy dispersive spectroscopy; TLC = thin-layer chromatography; UV/Vis = ultraviolet/visible spectroscopy; XRD = X-ray diffraction; XRF = X-ray fluorescence).

comparative studies of environments, evaluate changes in atmospheres, and identify some offgassing materials. Koestler demonstrated that IR spectroscopy could be used to detect the by-products of insect activity in an inert atmosphere (Koestler 1993). A later example in this book shows the effectiveness of gas-phase IR measurements in the evaluation of the interaction of a reactive gas (Vikane) with different types of materials (chap. 6, case study 7).

When a gaseous environment is to be analyzed, a sample may be collected by (1) opening an evacuated gas container in the area of interest, (2) pulling an air sample into a gas syringe, or (3) using a pump to make the air flow over a sorbent that will specifically trap the components to be analyzed. Due to its volatility, a collected gas sample should be stored at cool temperatures and analyzed as soon as possible.

Alternatively, the IR gas cell may be used as a collection device for an in situ experiment, where the analyte gas is produced from a sample within the gas cell via some process, such as a chemical reaction or heat. In this case, the sample is kept out of the beam path, so that only the atmosphere in the cell is measured. Outside the laboratory, portable long-pathlength IR monitors are available for the measurement of gaseous environments (see Suppliers, Midac Corp.).

Liquids

Liquid samples encountered in the analyses of art materials are usually in the form of solvents, uncured materials (coatings, adhesives, etc.), or exudates on the surface of objects. In any case, a liquid sample should be collected and stored in a well-sealed glass vial with a Teflon-lined lid.

Small droplets of a liquid can be collected with a capillary tube that is used to draw up the liquid. The capillary can be placed in a glass vial for labeling. A scalpel may occasionally be used to collect a small viscous drop and to place it in a glass depression slide. Teetsov describes the use of a small, 0.2–0.4 mm polyester filter square to collect a very small volume of nonvolatile liquid from the surface of an object (Teetsov 1995). Alternatively, a swab may be used to collect the liquid sample (see below for cautions regarding the use of swabs). However, with either of the last two procedures, it may be difficult to extract the liquid from the swab or filter later, if the liquid (or a portion of it) is insoluble in common solvents.

Liquids generally require very little sample preparation. For IR analysis, volatile liquids can be analyzed directly by use of a commercial liquid cell or simply by sandwiching a drop between two salt plates. A liquid sample that is nonvolatile or of low volatility can be analyzed by a procedure of spreading one drop on a single salt plate or other transparent surface, such as a single surface of a diamond cell (two diamonds should not be put together, or the liquid will be displaced). Alternatively, a few drops of the liquid may be dripped onto a bed of powdered potassium bromide (KBr) for analysis by diffuse reflection. Liquids may also be analyzed with an internal reflection cell. This technique works particularly well with liquids that contain water.

Sampling with swabs

Solvent-dipped swabs are used to conduct solubility studies and to collect samples when the analysis pertains to a solvent-soluble surface of an object, such as a patina or coating on a bronze sculpture. The sampling area can range from a few millimeters to a centimeter in diameter, depending on the size of the swab and the prominence of the area in question.

Commercially purchased cotton swabs or applicators should not be used for this type of sampling because the cotton is typically adhered to the stick by an adhesive, such as poly(vinyl acetate). Instead, swabs for analysis should be prepared by winding a small piece of cotton fiber, gauze, nylon, polyester, or additive-free clean-room wipe onto the end of a wooden stick (Fig. 3.3). The smaller the point of the wooden stick, the smaller the swab. Selected fibers, wipes, and sticks should be tested before sampling to ensure that they will not be a source of contamination. A sharp, pointed stick, such as an uncoated toothpick, is needed to prepare a tiny swab. The sampling end of the stick and the swab should be handled only with tweezers to prevent contamination from skin cells and oils. After preparation, the swabs may be wrapped in coating-free aluminum foil for storage.

When sampling preparations are complete, the selected area on the object is lightly swept with a sable artists' brush to remove any loose particles or dust. A drop of an appropriate solvent (water, ethanol, acetone, chloroform, hexane, etc.) is put on the swab, which is then rubbed over the small area on the object. Multiple samples taken with swabs treated with different solvents may be collected from the same sampling area. After collection, each sampling swab is placed in a clean

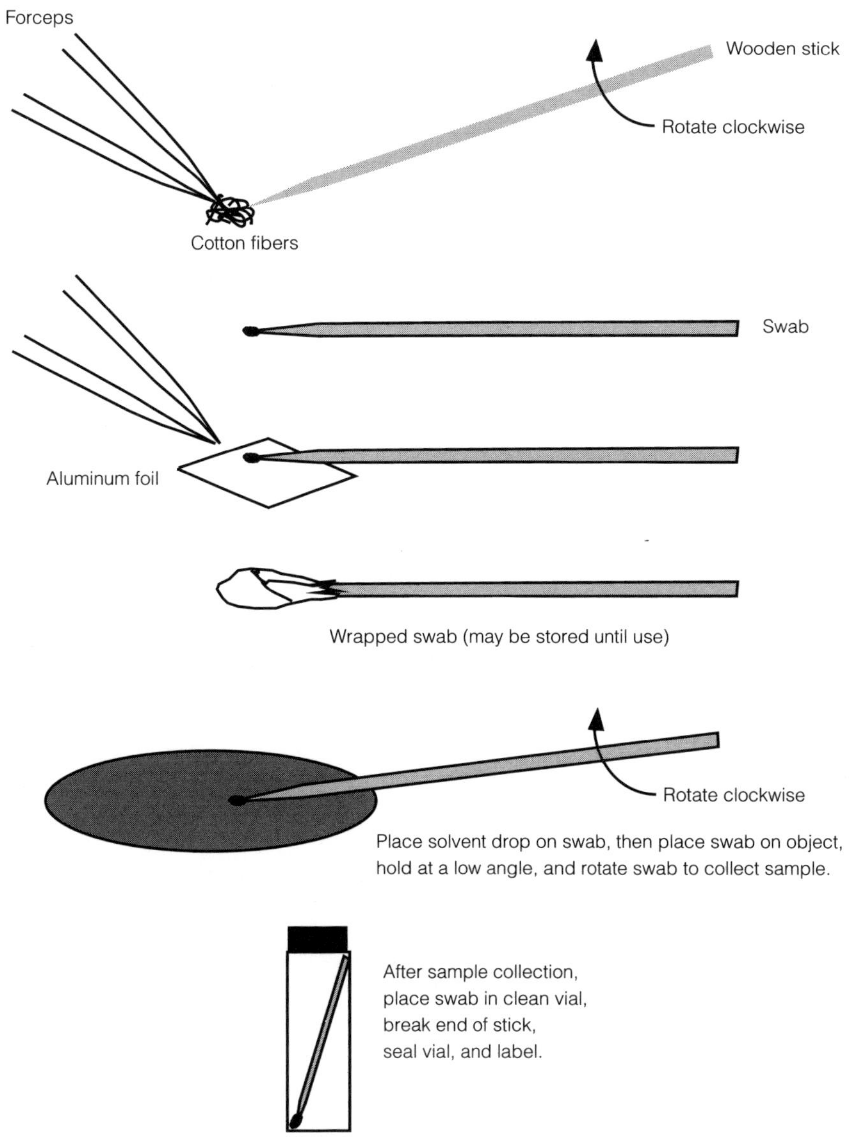

Figure 3.3
In-house fabrication of a cotton swab and collection of a solvent-soluble sample from the surface of an object. An alternative procedure is to use small squares of clean-room wipes for swabbing; they should be held with forceps for sample collection, then placed in a vial for storage.

glass vial; either the stick is removed or its end is broken off; and the vial is sealed and labeled.

Samples obtained by solvent-dipped swabs or filter squares contain the evaporation residue of the solvent-soluble portion removed from the sampled area on the object. For analysis, the residue is extracted from the swab with the same solvent that was originally used to acquire the sample. A corresponding blank should be prepared to check for contamination from solvents, capillaries, or other contact sources.

Swab sampling is not recommended when any other choice is available. While the solvent swab removes the soluble portion of a sampled area, it may be difficult to determine whether the swab sample contains only the surface coating or whether it also contains some leachate from lower layers. Additionally, particles may be removed by abrasion. Thus, this selective but nonspecific removal can create confusion in the analysis scheme. When possible, it is better to remove a scraping and perform solvent solubility tests under the microscope.

Solids, powders, and particles

Samples consisting of fine powders or crystalline particles are often collected when the analysis question pertains to pigments, stone surfaces, corrosion products, adhesives, or coatings. Loose samples can be acquired with a fine-tip brush. Single particles can be picked up with a tungsten needle or a cat-hair probe. In cases in which particles must be scraped from a surface, a scalpel is used. The scalpel blade is held at a low angle, and the blade is pulled backward or away from the cutting edge (Fig. 3.4). This technique abrades only the very top surface and minimizes the chance of accidental damage (from slicing or gouging) to the object. The particles will often cling to the scalpel; this tendency aids in the transfer of particles from the object's surface to a glass slide. When the sample surface is not horizontal, it is best to hold (or have someone else hold) the slide directly below the sample area to prevent possible loss of valuable particles that fall as they are loosened. It is advisable to wear gloves (cotton, latex, or other gloves appropriate to the object) to protect the object, sample, and future samples from contamination.

The preparation procedure selected for solid materials depends on the form and homogeneity of the sample. For IR analysis, a homogeneous sample can be ground or filed to form fine particles, then analyzed by either KBr pellet, KBr micropellet, diffuse reflection, internal reflection, diamond cell, or microscope. Although inhomogeneous samples, such as mixtures and multilayered paints, can be analyzed directly, the characterization of each component is simplified by the use of a physical or chemical preseparation step prior to analysis.

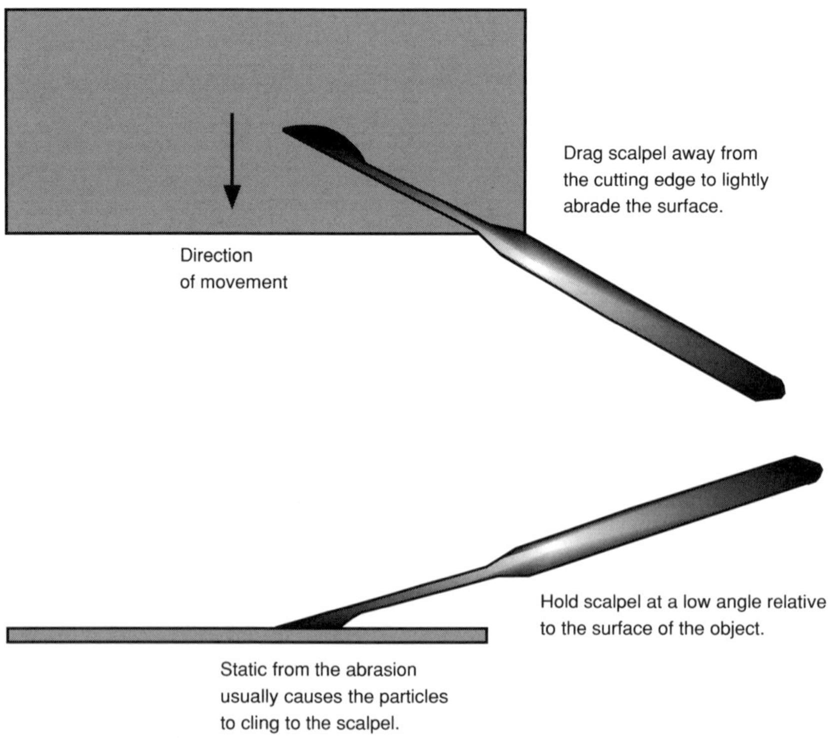

Figure 3.4
The proper use of a scalpel to remove particles from the surface of an object for analysis. Alternative procedures involve a fine brush for gathering loose particles or a tungsten needle for removing a single particle.

Mixture separation by solvent extractions

Similar to separation experiments described by Gettens, solvent extractions can be performed on multiple-component particles to identify individual components (Gettens 1959). For this procedure, a sequence of solvents is used to selectively remove one or more components in a mixture; after each extraction step, IR spectra are collected from the soluble and insoluble portions of the analyte. A secondary benefit is that the solubility information from the extraction can also aid in the identification of components. This method is especially useful when the IR spectrum of a bulk sample of a coating or consolidant is difficult to identify because of the presence of other components, particularly inorganic fillers or pigments. Figure 3.5 is a solubility schematic for many natural and synthetic materials.

For solute extraction of macrosamples (e.g., a swab sample or a particle large enough to be seen with the naked eye), the sample is placed in a micro test tube (5 × 35 mm), then covered with a few drops of solvent (about 10–100 μl). The test tube can sit for about an hour or can be briefly agitated in an ultrasonic bath. A microdrop (1–5 μl) of the extract is removed with a capillary tube or micropipette; the microdrop is then placed on an inert IR window (such as barium fluoride, BaF_2) for IR microspectroscopy analysis, or it is dripped onto KBr powder for pressing a salt pellet or for diffuse reflection analysis. The solvent is then allowed to evaporate. The drop procedure may be repeated to concentrate the amount of sample for analysis.

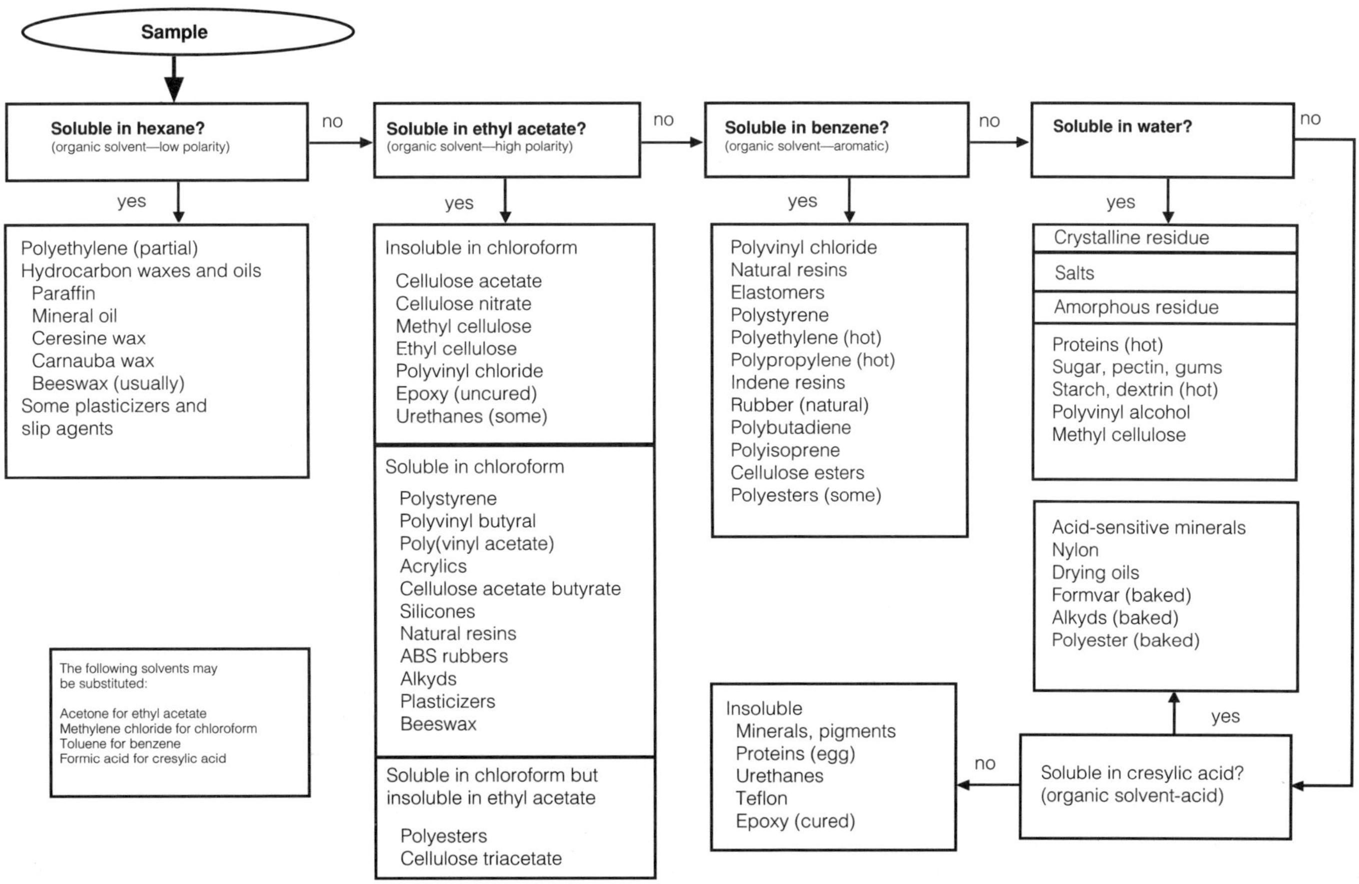

Figure 3.5

A solubility schematic for many natural and synthetic materials.

For smaller samples, a microdrop (1–5 μl) of solvent is placed directly on a sample that is on an inert window or glass microscope slide (care must be taken that the drop cover only the sample of interest and does not spread over other samples, and that the solvent chosen does not react with the substrate). Any soluble portion of the sample will dissolve and be deposited away from the insoluble material as the solvent evaporates. Depending on the solvent and the solute, the deposition may appear as a solid film, as droplets in a ring at the former edge of the solvent, or as a dried puddle (Fig. 3.6). The most highly soluble materials—such as waxes, synthetic resins, gums, and inorganic nitrates—will form a solid film that often has a more concentrated region at the exterior rim. Such analytes as nondrying oils, waxes, and natural resins tend to form a ring of droplets. "Before" and "after" photos for a mixture of natural resins extracted with chloroform are shown in Figure 3.7. In the analysis of many art materials, water-soluble proteins have been found to occur most often as dried, wrinkled puddles near or under the original sample. To identify these new puddles, it is desirable to have a good visual memory or a photograph of the sample prior to the solvent drop. Because multiple components might have been extracted and deposited in different areas, it is also important to collect IR spectra from all visually different areas after each extraction.

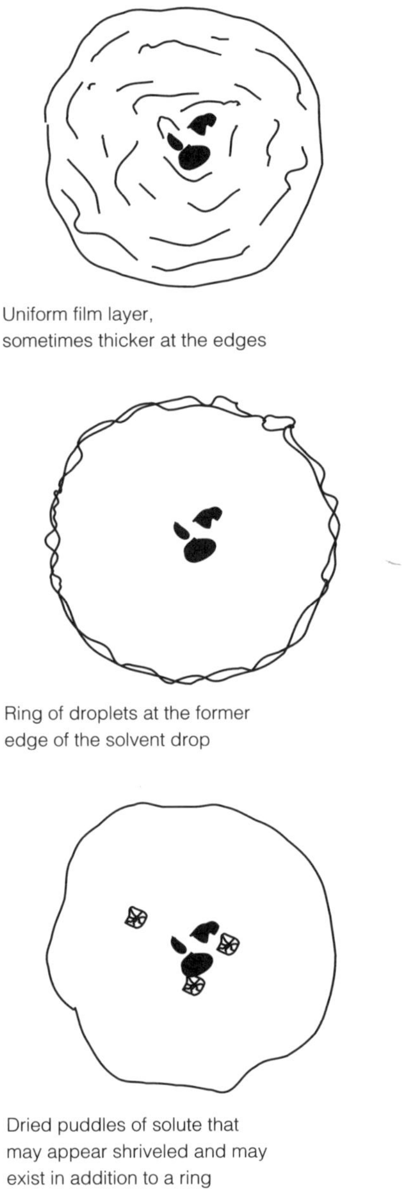

Figure 3.6
Possible deposition patterns for soluble components extracted from a sample with a drop of solvent. With IR microspectroscopy, a spectrum can be collected of the solvent-soluble and solvent-insoluble fractions.

Multiple extractions and analyses can be done sequentially on the same sample. The selection of solvents for a series of microextractions depends on the components expected to be in the sample. For a sample thought to be composed of natural products, a typical solvent series would start with a drop of hexane to extract nonpolar components such as mineral waxes. After the drop has evaporated and the spectrum collected, a second solvent is selected—typically either ethyl acetate or chloroform—to check for the presence of other waxes, natural resins, nondrying oils, and many synthetic resins. If multiple components are extracted at this point, then ethanol or acetone is used for further separation. The final solvent is usually a drop of water to check for the presence of carbohydrates or soluble proteins. After the drop of water is placed on the sample, the pellet is positioned under a warm light to heat the water; this procedure increases the solubility of many proteins while also hastening the evaporation of the droplet. After each drop dries, spectra are collected at several positions around the deposition ring. Once the solvent series is completed, the insoluble residue of the sample is analyzed. It is also important to analyze a blank for each solvent.

The same extraction procedures can also be done on a gold mirrored surface; the IR spectra are then collected by reflection measurements. Because of the nonreactive surface of gold, acids and bases may be used to further characterize the sample components. A sample with a high content of drying oil can be treated with a 5% solution of sodium hydroxide to saponify the oil and separate it from the pigment particles (Gettens 1959). Alternatively, a drop of concentrated nitric acid placed on the sample and warmed to dryness will solubilize most oil and resin media. A drop of concentrated sulfuric acid will extract many colorants from dyed fibers (Saltzman and Keay 1972). The acids and bases will

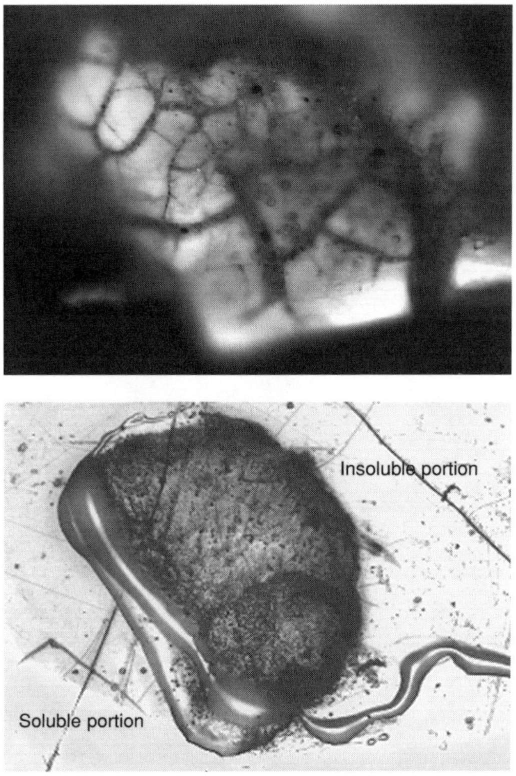

Figure 3.7
The separation of components of a carved resin sculpture by solvent extraction. The top photograph shows a sample placed on a BaF_2 window for IR microanalysis. The lower photograph shows the same sample after a microdrop of chloroform was placed on the sample, then allowed to evaporate. Two distinct components were separated. The chloroform-soluble component, later characterized as a pine resin, primarily collected in a ring at the former edge of the droplet. A chloroform-insoluble portion, later characterized as copal, was concentrated as a residue in the center. IR spectra were collected on the bulk sample, as well as from several regions of the extract and residue.

also react with some pigments and thereby provide additional information on the sample (Gettens 1959).

Disposable glass capillaries used for thin-layer chromatography can be employed to dispense the microdrops of solvent required for this procedure. However, because some capillaries have been found to be a source of contamination, it is recommended that each capillary be rinsed inside and out with a drop of solvent prior to use. An alternative to glass capillaries are the polypropylene tips for the Eppendorf GELoader (available from any scientific supply company), as recommended by Teetsov (1995). While they must also be rinsed, the tips can be bent with heat to provide an improved pipette for easier control of the drops.

Mixture separation by pyrolysis

Pyrolysis is useful for the separation of small amounts of insoluble organic materials from an inorganic matrix, such as a glass-filled epoxy resin, a low-binder oil paint, or a polyurethane-consolidated stone. For manual pyrolysis, it is best to have at least 1 mg of sample, as the procedure becomes easier as the sample size increases. Smaller samples can be analyzed manually with careful manipulation (Humecki 1995a) or with temperature-controlled, commercially available pyrolysis equipment.

To pyrolyze a sample manually, it is placed as pulverized chunks or as a pile of powder inside a glass capillary tube or disposable glass pipette with an inner diameter of approximately 1–8 mm. The sample should be at least 1 inch (2.54 cm) from the end of the tube. With the tube held horizontally with forceps or pliers, the sample area only is placed in the hot spot of a flame from a Bunsen burner or a butane lighter (Fig. 3.8). After the sample produces smoke, the tube is

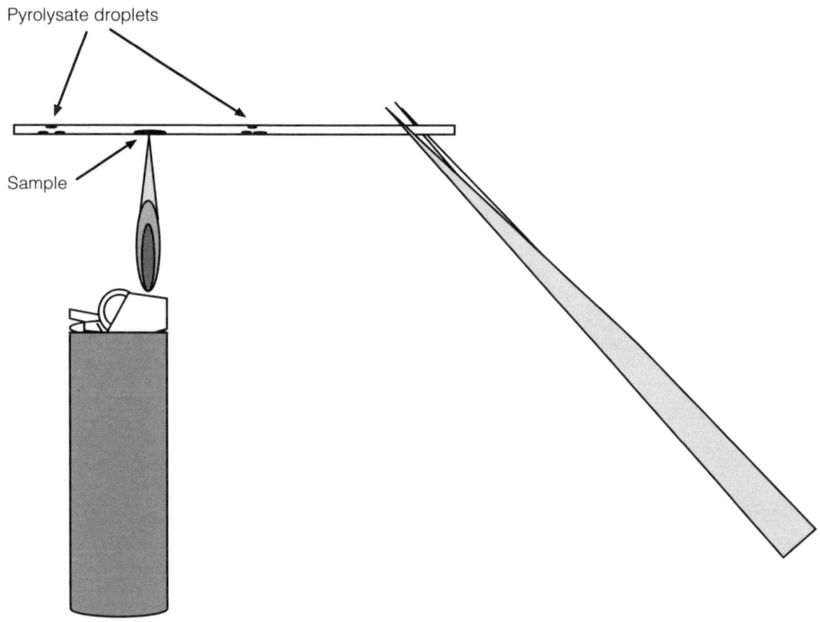

Figure 3.8
The manual pyrolysis technique useful for the separation of nonsoluble organic compounds from an inorganic matrix. A portion of the sample is compactly placed inside a glass pipette, then heated. After the sample smokes, the tube is cooled, and the pyrolysate that collects as droplets in the cooler regions of the tube is collected and analyzed.

removed from the flame and cooled. The volatiles in the smoke will primarily condense and collect as droplets in the regions of the tube not heated by the flame. These droplets can be sampled with a metal probe and placed on an IR transparent pellet for analysis. While in most cases, the spectra of pyrolysates will correspond to the parent compounds, it is best to prepare pyrolysate references for comparison.

Fibers

For microspectroscopic analysis, fiber samples may be collected from paper, textiles, and ethnographic objects with fine-tip forceps. A single fiber can be grasped and pulled loose from its thread or mat. Some small bristles or lint particles embedded in or attached to painted or coated surfaces may require a scalpel for removal. When a larger fiber sample is needed for macro IR analysis, a sharp scalpel can be used to cut the thread while it is held in position with forceps. Since this maneuver often requires three hands, this job is best performed by two people. Some forensic examiners have used adhesive tape to collect loose fibers and particles from a surface (Ryland 1995). The adhesive from the tape can, however, contaminate the surface of the sample.

Fiber, hair, and brush samples require little sample preparation. They may sometimes even be analyzed "as is" by taping the ends of a single fiber to hold it across a hole in a metal disk, or by simply laying it on an IR window (Tungol, Bartick, and Montaser 1995). However, the fiber is often too thick and absorbs too strongly for direct transmission. In such a case, one alternative is to use a reflection method, such as diffuse reflection or internal reflection. When a large sample, such as a length of yarn or a piece of textile, is available, internal reflection is a very good nondestructive method.

A second alternative is to make the fiber thinner. The advantages to flattening a fiber are twofold. First, the horizontal area of the fiber is increased. Second, the surface is now planar. The use of the diamond cell is a quick and easy method for flattening fibers. Another method is to press the sample between two IR transparent windows in a compression cell. A final option is to use a scalpel to slice a thin section from the end or side of the fiber. This method is sometimes preferred for the examination of the fiber structure, since flattening destroys its physical shape and may change its crystalline orientation.

Cross sections

When the surface of an object, such as a painting or a folk art carving, has a complex structure with multiple layers, a cross section sample is taken that incorporates this stratigraphy to facilitate the study of the sample. Paint cross sections have been used for the examination of paintings for over eighty years (Laurie 1914; Gettens 1932). The technique of embedding paint cross sections was revived in the 1950s; it quickly became a standard method for the study of painting techniques (Plesters 1956).

With the use of a stereomicroscope, a cross section sample is removed by cutting with a scalpel from the top down to the bottom substrate. When possible, a sample is taken from the edge of the painting or near a preexisting crack, provided that the sampling area has not undergone previous restoration (Plesters 1954). Alternate methods for cross section removal have been tried but have not proved to be as successful as skillful extraction with a scalpel. Laurie used a hypodermic needle, and Gettens constructed a workable, although complex, apparatus (Laurie 1914; Gettens 1936). While these tools are useful with some paintings, crumbling can occur with brittle paints.

Once removed, the resultant cross section sample is often too small to be picked up with forceps and may not have enough static charge to cling to the scalpel. In these cases, an artists' brush modified to contain only a few bristles is used to pick up the particle and transfer it to a glass depression slide. Occasionally, for a few obstinate samples, a short breath of air on the brush (directed away from the sample) will supply enough moisture to the bristles to adhere the particle temporarily for transfer. However, when a very sensitive analytical technique, such as gas or liquid chromatography, is used, it is important to minimize the potential for protein contamination. In these cases, the brush may instead be moistened with a mist of distilled water (Johnson and Packard 1971).

Before IR microspectroscopy was commonly used as an analytical technique in conservation, the two usual IR sampling techniques for the analysis of paint cross sections consisted either of examining the entire chip or of selectively removing visually different sections for analysis. Both methods can misrepresent the composition of the paint. In the first method, when an entire multilayer chip is ground into a homogeneous mixture, usually with KBr salt to make a pellet, any stratigraphic or layering information about the sample is lost. In the second method, physical separation of layers is time-consuming and difficult unless the layer is large. Small layers may be missed altogether.

To obtain meaningful analyses of multilayer media, it is important to work with a high-quality cross section of the sample. IR microspectroscopic analysis can be done on cross sections with reflected or transmitted light. Cross section samples from paintings contain many materials, such as binders and pigments, with various particle sizes. These variations can produce a reflectance spectrum that, even after mathematical corrections, looks different from the spectra collected by transmission from the same components because of the dissimilar reflection and absorption characteristics of each material. For the analysis of binders, which generally absorb IR radiation better than they reflect it, transmission FT-IR is usually more successful. To produce good-quality IR transmission spectra, most paint samples need to be 1–10 μm thick. Microtoming is normally used to prepare a thin section of a multiple-layer sample, and an embedding medium is needed for the support of small and fragile paint samples during microtoming. Audrey Glauert provides an excellent reference on embedding and microtoming procedures (Glauert 1972).

Embedding media and procedures

Polyester resins have been commonly used in art conservation since the 1950s for embedding paint cross sections prior to microscopic and analytical studies (Plesters 1954). Indeed, polyester resins have many desirable properties for embedding resins, and they are nearly ideal embedding media for most samples found in fine-art paintings (Table 3.4). Polyester resins are clear and colorless and easy to section; they cure at room temperature and do not react with most samples. A polyester resin may, however, dissolve some compounds, such as some waxes on furniture-finish cross sections (Godla 1990), organic dyes on inorganic carriers in modern (post-1850) pigments (Stodulski 1994), and fresh natural-resin layers (Derrick et al. 1992). Additionally, Derrick found that polyester resins infiltrate porous samples (Derrick, Souza et al. 1994).

Embedding with Polyester Resin. To embed a typical paint sample in polyester media (brand names are noted in Table 3.4; see also Suppliers), six drops of catalyst (methyl ethyl ketone ether) are mixed thoroughly with 10 ml of liquid polyester resin in styrene solvent. The transparent resin is initially light blue; it turns yellow when the catalyst is well mixed, and then it quickly becomes colorless as the reaction pro-

Table 3.4

Types and brand names of selected media tested for embedding and microtoming of paint cross sections.

Type	Brands	Source	Comments
Paraffin	Paraplast	Ladd Research Industries	Opaque; minimal shrinkage; soft; cuts well; elevated temperatures required for preparation.
Epoxy	Epon 812 LX-112 Maraglas 655 SPURR	Ted Pella, Inc. Ladd Research Industries Ladd Research Industries Ted Pella, Inc.	Generally needs elevated temperatures to cure; transparent though sometimes yellow; forms very hard block that is difficult to slice at > 1 μm thicknesses.
Acrylic	Quetol 523M LR White Butylmethyl-methacrylate Krazy Glue (cyanoacrylate)	Ted Pella, Inc. Ladd Research Industries Ladd Research Industries Borden, Inc.	Exothermic cure reactions; transparent; shrinks more than polyesters; cuts well; may infiltrate some samples; dissolves lipids; toxic.
Polyester	Caroplastic Bio-Plastic Castolite	Carolina Biological Supply Ward's Natural Science Castolite Co.	Cures at room temperature; transparent; cuts well; minimal shrinkage; may infiltrate some samples.

ceeds (Ward's Natural Science 1990). Excess catalyst, while it will speed up the curing process, will also make the final mount more brittle and difficult to microtome. A mold is initially half filled with the well-mixed embedding medium and cured at room temperature for 3–6 hours. A representative portion of the sample containing all the layers is transferred to the hardened base layer in the mold with forceps or a probe and positioned in the desired orientation. A label is placed on the opposite end of the mold, then both the label and the sample are covered slowly with freshly prepared polyester embedding medium. The embedment is cured and nontacky within 24 hours. For microtoming of the sample for IR microanalysis, the best results are obtained by allowing the embedment to set 36–48 hours before slicing. The medium continues to cure slowly over time (Demmler 1980). After one month, the microtoming becomes noticeably more difficult, and the samples tend to crumble. The bottom and top halves of the block should be prepared within a few days of each other to prevent a hardness differential between halves that interferes with microtoming. Silicone molds, ice cube trays, and peel-away molds are all useful for preparing embedments; some examples are shown in Figure 3.9. Because the presence of excess medium around the sample produces stress during microtoming, most of the plastic around the sample has to be trimmed away to reduce the area of contact with the microtome blade. Trimming time can be minimized by the use of small Pelco silicone rubber molds for embedding (see Suppliers, Ted Pella, Inc.). These molds produce embedments with trapezoidal tips that only require minimal trimming prior to slicing. While small embedments are good for microtoming, they are difficult to hold in a level position for the polishing of samples.

Infiltration. Paints are porous when the amount of binder is low enough that it does not fill the void spaces around the pigment particles (Hansen, Lowinger, and Sadoff 1994). The embedding resin can then seep into the sample, fill these spaces, and, by doing so, coat the particles. This is termed infiltration. Infiltration can occur with matte or porous paints and glue gessoes. If analysis of the binder or organic components in the sample is needed, infiltration may be undesirable, and in

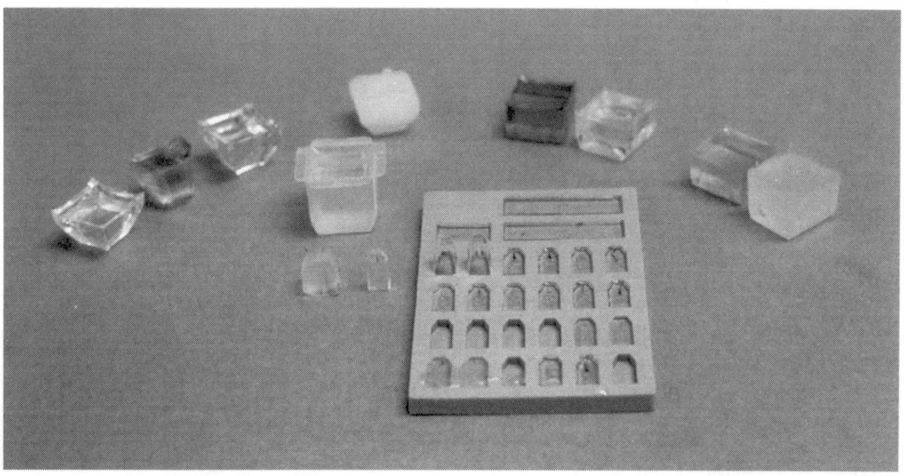

Figure 3.9
Molds and embedding media blocks used for the preparation of embedded cross sections.

some cases, steps should be taken to prevent infiltration from occurring (Derrick, Souza et al. 1994). In other cases, sample infiltration is beneficial. It consolidates the sample to produce a smooth block that is readily polished or sliced in preparation for analysis.

If several samples of a porous material are to be embedded for analytical studies of the media, it is prudent to embed only one sample initially to see if infiltration occurs and, more important, whether it interferes with analysis. Depending on the type of analysis, it is possible that infiltration will not cause a problem, but it is important that the analyst recognize that resin may be in the sample, taking that into account in any spectral interpretation. IR spectroscopy is more sensitive to the presence of the infiltrated resin than is visual or microscopic examination; thus, when IR is used, it may be important to prevent infiltration.

Visual examination of a paint cross section can often detect infiltration of the embedding resin from the discoloration or darkening of the sample. This is particularly noticeable for white paints and grounds. Samples that visually appear very white and opaque before embedding can take on a darker, transparent appearance after resin penetration. Because of the presence of embedding resin inside and outside of the sample, there is less contrast at the sample edges, and the edges may seem poorly defined. For example, two small portions of a sample from a polychrome sculpture were embedded separately—one in an acrylic medium and one in polyester (Figs. 3.10, 3.11). The sample in the polyester medium (Fig. 3.11) shows infiltration of the resin, while the sample embedded in the acrylic (Fig. 3.10) does not. The acrylic-embedded sample has very well defined edges, and the opaque white ground remains white after embedding. The polyester-embedded sample visually appears to have a more transparent, darker ground layer.

The IR spectrum of an infiltrated sample will contain absorption bands for the sample components, as well as for the polyester resin.

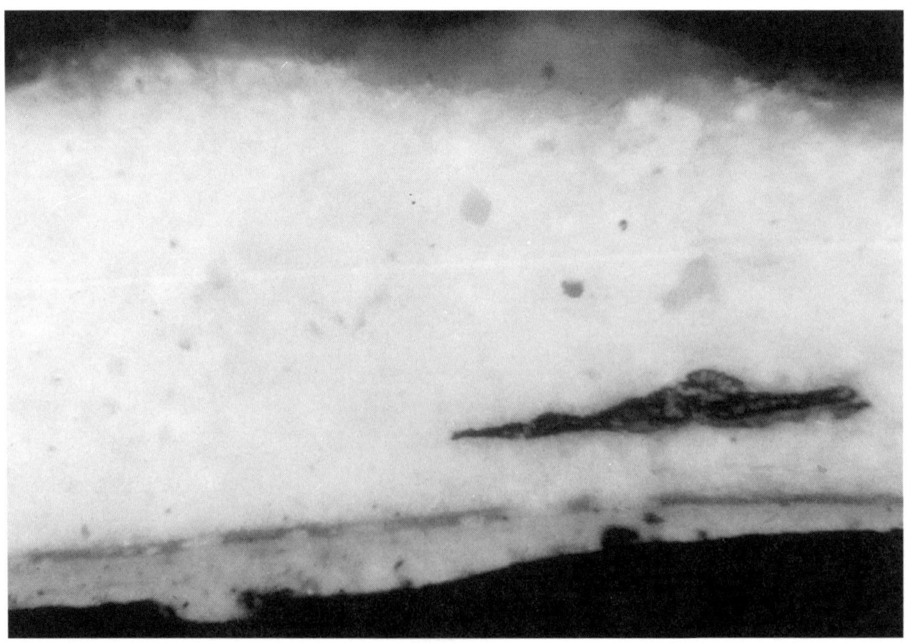

Figure 3.10

A paint cross section from a painted sculpture in a church in Minas Gerais, Brazil, embedded in acrylic medium.

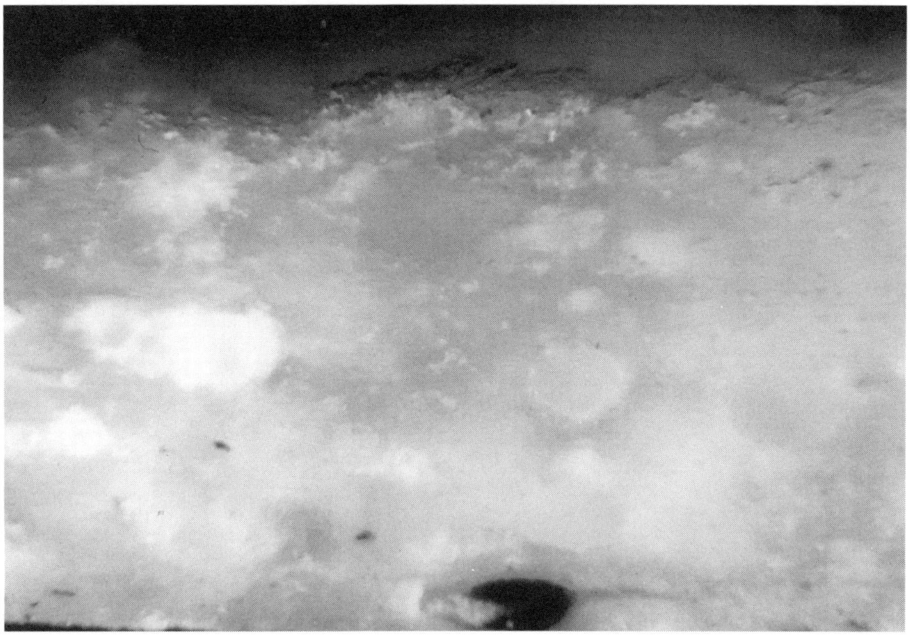

Figure 3.11
A cross section from the same paint chip as shown in Figure 3.10, embedded in polyester medium.

The polyester resin produces numerous strong absorption bands. Theoretically, the absorption bands for the polyester could be subtracted, but the subtraction process can distort the remaining bands in the spectrum, thus limiting the detection of other components. Many binding media are present in low concentrations and have absorption bands at wavenumbers similar to those of polyester, making the binder especially difficult to detect. Since infiltration of polymer severely inhibits the IR analysis of the sample, several methods for preventing this occurrence were examined (Derrick, Souza et al. 1994). The most successful method discovered for inhibiting the infiltration is to precoat the sample with a thin layer of acrylic emulsion (Rhoplex AC-33; see Suppliers, Conservation Materials) thickened with fumed silica to form a gel. A thin layer of this gel dries quickly to encapsulate very porous samples, even plaster, thereby preventing polyester infiltration without inhibiting the optimal slicing properties of the polyester.

Microtoming procedures
Rotary microtomes were first developed in the 1950s. They are available with manual and motorized cutting arms. The cutting arm moves the clamped sample past a sharpened knife or blade to produce a thin section of the sample. Detailed information on microtomes and microtoming techniques is available in Reid (1972).

Four types of knives (steel, tungsten carbide, glass, and diamond) are readily available for use with any type of rotary microtome (Figs. 3.12, 3.13). The steel blade is an all-purpose, inexpensive microtome blade that can routinely provide slices 15–30 μm thick. However, depending on the absorption of the sample, this may be too thick for IR analysis. Both the tungsten and the glass knives routinely produce good, clean slices 1–10 μm thick. The diamond knife is typically used for ultramicrotomy of sections, since it should only be used to cut sections

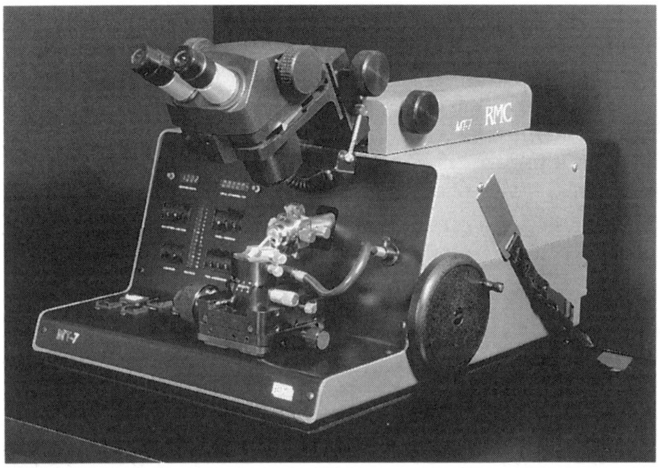

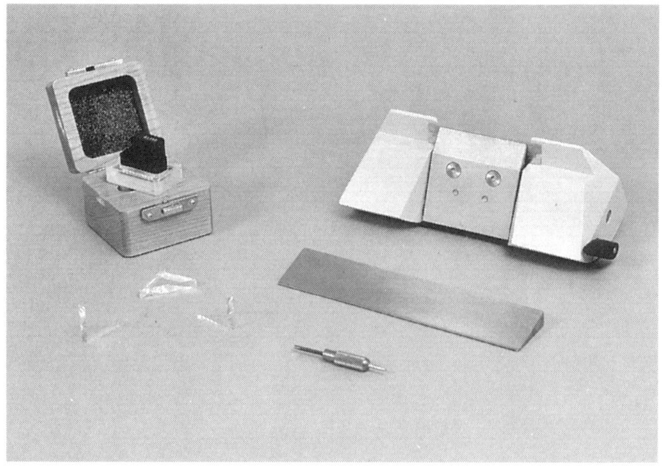

Figure 3.12
A motorized rotary microtome with a stereobinocular microscope for viewing and positioning the sample.

Figure 3.13
A selection of microtoming blades: diamond (in the box at the upper left), glass (lower left), tungsten (lower right), and stainless steel (in the blade holder at the upper right).

thinner than 1 μm. A general dictum is that harder samples require harder knives.

Of the four blades, glass knives are the easiest to use. Glass knives are made by scoring and breaking a 2.54 cm (1 inch) square of glass to create two triangular pieces, each with a very sharp edge (Figs. 3.14, 3.15). This procedure produces a 45° angle on the blade. The angle may be increased for cutting harder materials and decreased for cutting softer materials (Malis and Steele 1990). The glass knives section best on polyester-embedded paint samples containing finely ground pigments; in such instances they produce transparent, cellophane-like thin sections.

Small samples are easier to microtome because smaller cutting-surface areas require less force. Cutting-surface areas that are too large may produce sections that curl or cause layers to debond. Thus, to reduce the size of the cutting surface, self-supporting, strong, hard samples—such as polymers, laminates, and metals—should be trimmed to a sharp point. Harder samples require finer points. The facet point for a self-supporting polymer should be less than 0.5 mm; for aluminum it

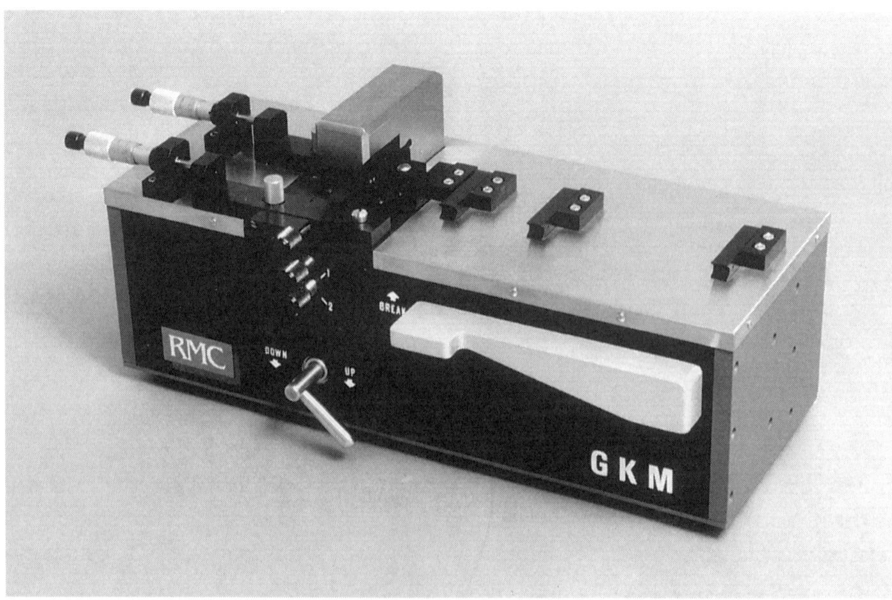

Figure 3.14
A glass knife maker for the accurate scoring and breaking of microtome blades.

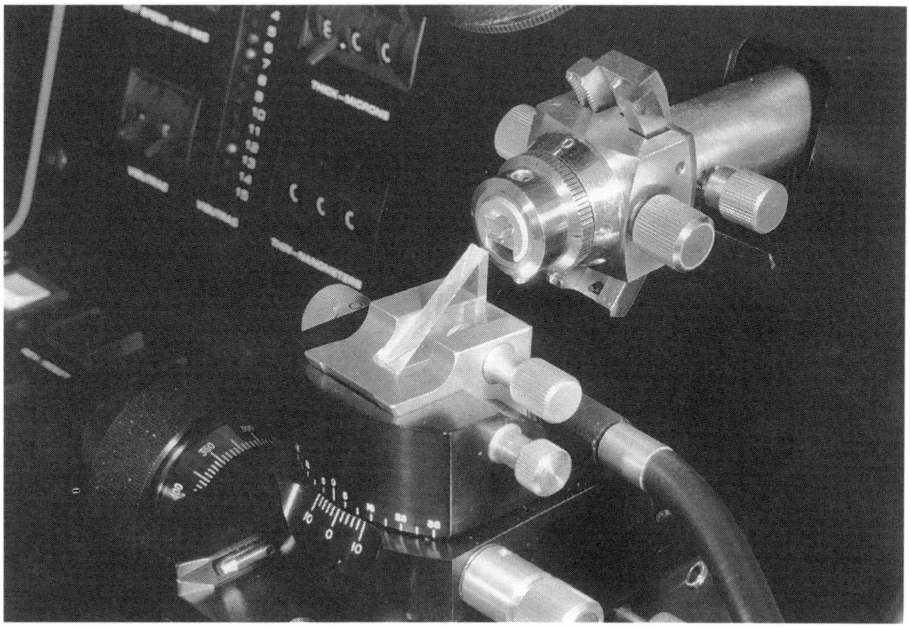

Figure 3.15
A glass knife positioned in the rotary microtome. The sample is held in the chuck and slowly moved past the stationary knife.

should be less than 0.1 mm; and for ceramics, the facet should be only a few microns (Malis and Steele 1990). Delicate, small, or soft samples, such as paints, waxes, papers, or textiles, should be cut into a pie shape and positioned in a mold for embedding, such that the point is cut first (Fig. 3.16). An embedded sample should have all excess medium trimmed from the embedment so that only a trapezoidal facet with 1–2 mm of medium outside the sample at the cutting surface is left (Fig. 3.17). A trapezoidal shape of plastic around the sample helps in later microtoming by providing a pointed region to initiate and end the cutting stroke.

Unexpected thickness variations and jagged cutting surfaces are often due to vibrations or movement of the sample block during microtoming. Thus, the embedment or sample block should be fastened tightly in the microtome chuck, and the fastening should be checked regularly, since some materials, such as polymers, can creep under pressure. Moreover, positioning the majority of the sample block within the clamp, such that only the cutting tip extends, will increase stability.

Orientation of the sample to the knife edge can affect the quality of the section. In the following description, a multilayer sample is considered as a series of parallel lines that can be placed either vertically or horizontally in a rotatable vise of a microtome with a horizontal knife edge (Fig. 3.18). For most samples, the initial cut is best made with the sample rotated 10° from vertical, with the most important layer closest to the knife. Additionally, a corner of the trapezoid, rather than a flat edge, should be cut first; this order minimizes stress on the block. Exact vertical orientation of the layers can increase section curling and particle loss, while exact horizontal orientation can result in the compression of the sample. Depending on the sample and how it behaves during cutting, the orientation may need to be changed.

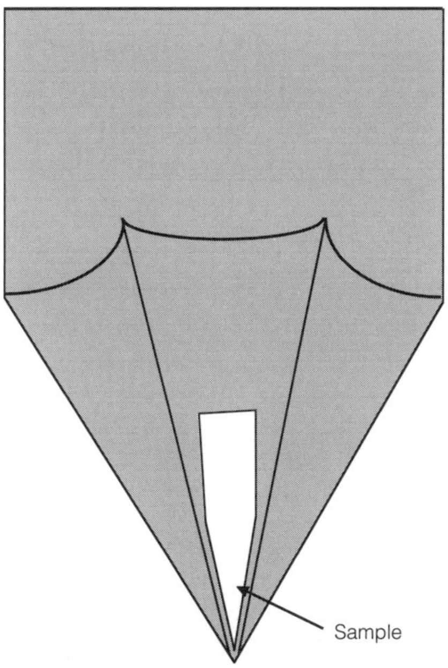

Figure 3.16
An optimal embedment for microtoming, with a small triangular sample positioned at the tip of an embedment cone.

The standard procedure is to cut the multilayer sample with the layers orthogonal to the knife edge. However, other orientations are

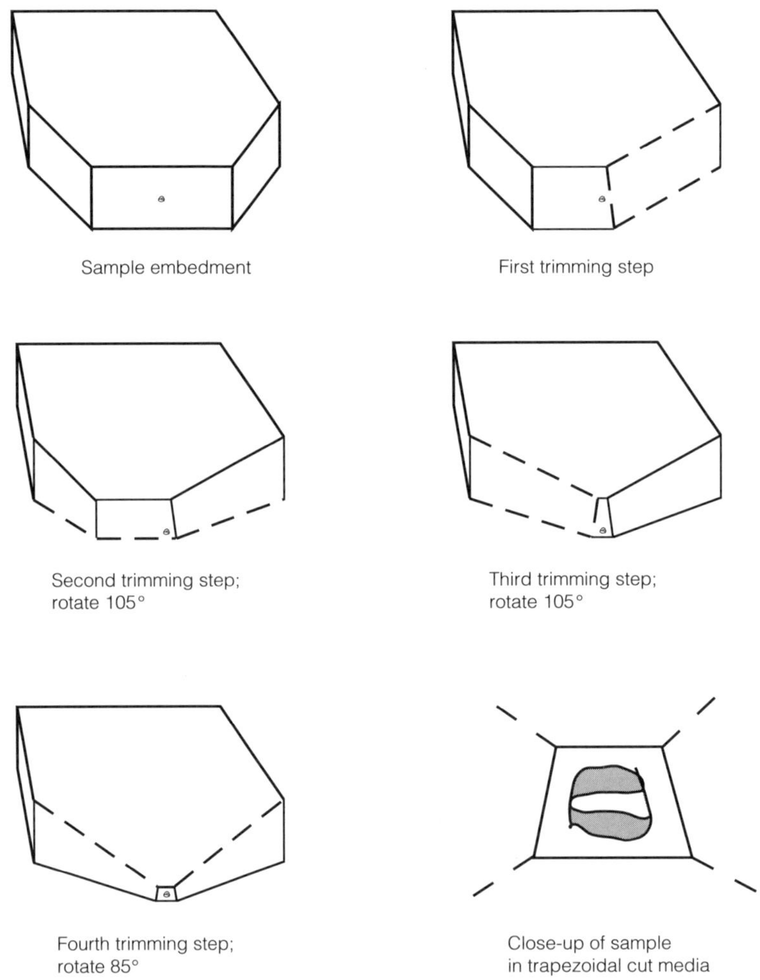

Figure 3.17
The precise trimming angles required to produce a trapezoidal tip for optimal microtoming. The trapezoidal shape is placed in the microtome vise, with the parallel sides vertical (as shown) or rotated slightly to the left, so that the cutting blade, when striking the block from the bottom, will first cut into a small corner, then progress to the rest of the sample. This procedure minimizes stress on the knife as it cuts into the hard polymer block.

possible. Cartwright and coworkers placed the sample with the layers parallel to the cut edge and produced slices of individual layers that were then placed in a micro diamond cell for analysis (Cartwright, Cartwright, and Rodgers 1977). Oblique sectioning is another method used in the case of thin layers to increase their analysis width effectively (Reffner and Martoglio 1995). In this method, the sample is oriented such that the layers are at an angle to the cutting edge (Fig. 3.19). As shown in Reffner and Martoglio, an angle of 10° will increase the free width of a 10 μm layer to nearly 53 μm. Figure 3.20 shows a sample from the varnish layer of an eighteenth-century gilded table. The analysis question was to determine the adhesive under the gilding. Since this layer was

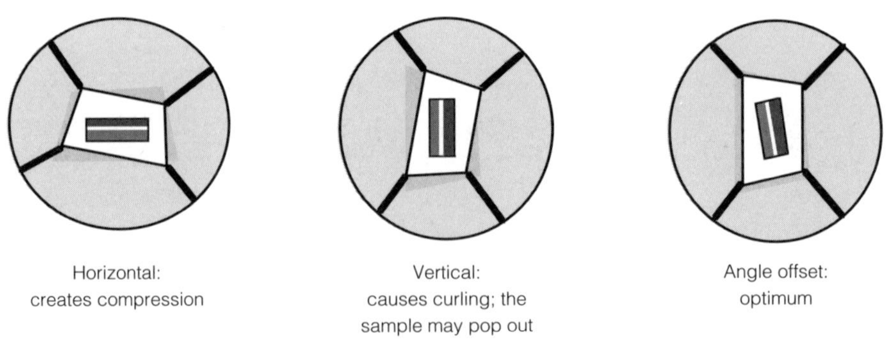

Figure 3.18
The orientation of sample layers and the trapezoidal-trimmed tip for optimal microtoming. The cut direction is from the bottom to the top, with the imaginary knife edge aligned parallel to the bottom edge of the page.

only about 10 μm thick, the cross section was cut at an oblique angle to increase the width for analysis.

The optimal sample thickness for IR transmission analysis is 1–10 μm (Derrick, Landry, and Stulik 1991). Sections thicker than 15 μm usually absorb IR radiation too strongly to allow for analysis. Moreover, an attempt to cut thick sections can result in a sample popping out of the slice, in chattering of the knife, or in damaged knife edges. When a microtoming problem occurs, it can often be solved by slicing thinner sections (1 μm), then returning to slice a thicker section (5–10 μm). All cuts should be smooth, slow, and executed with a motorized microtome, if available, set at speeds of 0.1–3.0 mm/sec. Cutting speeds that are variable or too fast can cause particle loss and curling of the thin section.

The optimum environment for microtoming is a small room with no foot traffic, drafts, or temperature and humidity fluctuations. The microtome should sit on a sturdy, vibration-free table. When cutting problems occur, all sources of heat are suspect and should be modified when possible. These heat sources include lighting, hand temperature, and the human breath. If a motorized microtome is used, the microtome and its bench should not be touched during the cutting stroke.

As the section is being cut, it can be encouraged to cling to the knife. This technique can be performed with a small, stiff bristle artists' brush (size 000) that can be used to hold the initial cut edge of the section to the knife surface without the sample region being touched. An eyelash brush or a cat-hair brush will also work well. Static charge should keep the section on the knife. After the section is cut, it can be easily and delicately picked up from the glass surface with the paintbrush. For IR analysis, sections are taken directly from the microtome, placed on an appropriate IR window, and transferred to the sample stage of the IR microspectrometer. If necessary, folds or wrinkles in the section can be removed by the placement of another transparent window on top of the section. Alternatively, ethanol or heat may be used to relax the polymer, provided that the sample will not be adversely affected.

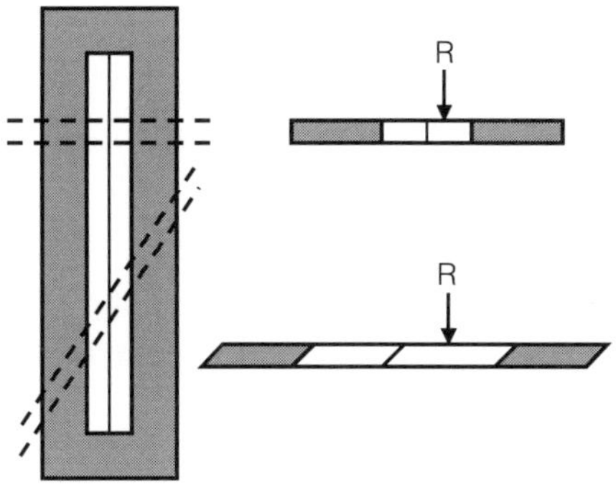

Figure 3.19
Two possible cuts for a multilayered sample. Cutting a sample at an oblique angle will increase the effective analysis width of the layer (R = IR beam direction).

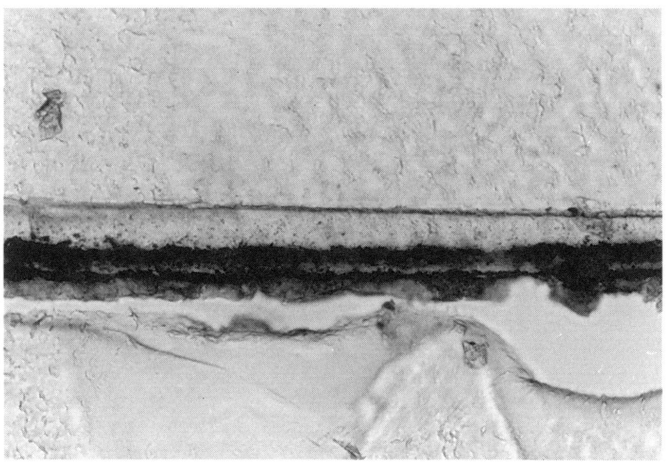

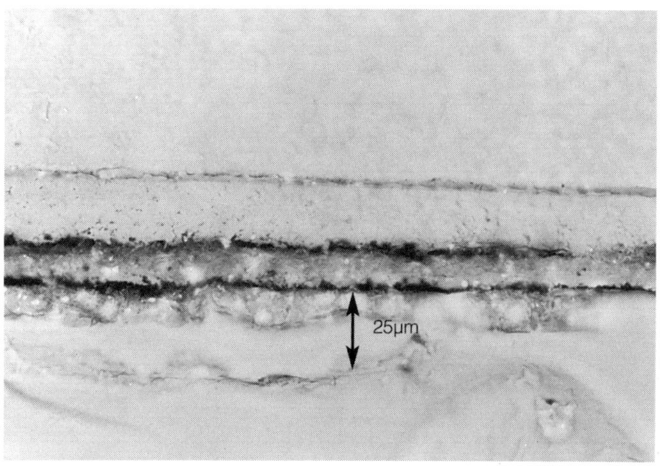

Figure 3.20
A finish cross section from a gilded table sample. The photograph on the left shows a thin section that was microtomed with the cut edge perpendicular to the top surface of the sample. The photograph on the right shows a thin section from the same sample that was cut at an oblique angle (45°). The width of the top coating layer was increased significantly.

Summary

The ultimate goal of any analysis is to produce useful and reliable information about the sample. Critical steps toward that goal are the selection, collection, and preparation of the sample. Errors at any point in this process—such as selection of a sample that poorly represents the object or contamination of the sample through placement in a poor storage container—can produce analytical results that are, at best, recognized as meaningless or, at worse, misdiagnose the object and its condition and lead to inappropiate treatment decisions.

Additional Reading

Chalmers, J. M., and M. W. Mackenzie
1988. Solid sampling techniques. In *Advances in Applied Fourier Infrared Spectroscopy*, ed. M. W. Mackenzie. New York: John Wiley and Sons.

Humecki, H., ed.
1995. *Practical Guide to Infrared Microspectroscopy*. New York: Marcel Dekker.

Krishnan, K., and S. L. Hill
1990. FT-IR microsampling techniques. In *Practical Fourier Transform Infrared Spectroscopy*, ed. J. Ferraro and K. Krishnan. New York: Academic Press.

Perkins, W. D.
1993. Sample handling in infrared spectroscopy—An overview. In *Practical Sampling Techniques for Infrared Analysis*. Boca Raton, Fla.: CRC Press.

Chapter 4
Infrared Analysis Methods

Proper instrumentation—that is, a well-functioning spectrometer—is the first step toward producing good IR spectra. The design, operation, and maintenance of IR spectrometers, however, is beyond the scope of this book. Resources for information on IR instrumentation are listed in the additional readings at the end of this chapter. In particular, Low and Baer clearly describe the interferometer design, along with the advantages Fourier transform infrared (FT-IR) spectrometers have over dispersive instruments (Low and Baer 1977).

The second step toward the production of good spectra is the selection of a suitable technique for presenting the sample to the spectrometer. There are a wide variety of methods using accessories that fit into the sample compartments of most spectrometers. The selection of an appropriate analysis method depends on the type, form, and amount of sample to be analyzed. No one method can meet the analysis needs of all samples.

Theoretically, one material should give one, unique IR spectrum that relates to its chemical composition. However, the physical state of the sample, its preparation, and the analysis method chosen have an effect on the resulting spectrum and can shift band positions, as well as change band shapes and relative intensities. Figure 4.1 shows spectra for Acryloid B-72 (methyl acrylate–ethyl methacrylate copolymer) obtained by four different analysis methods. Slight, but recognizable, spectral variations are seen in these four spectra, because of their different analysis techniques. Thus, while spectra are reproducible within each method, caution must be used when spectra acquired from one technique are compared to reference spectra collected through a different method.

This chapter focuses on the equipment, sample requirements, spectral variations, and limitations of several of the most common analysis techniques. Transmission methods—in which the beam passes through the sample—are used for the analysis of gases, liquids, and solids. Reflection methods measure changes in the IR beam as it is reflected by the sample; these methods are generally used for solid samples. Microspectroscopy, often the best method for the analysis of small samples taken from works of art, will be discussed in detail. Additionally, microscale versions of each method will be presented for use with limited amounts of sample when an IR microspectrophotometer is not available. Table 4.1 presents analysis methods described in this chapter and their relationship

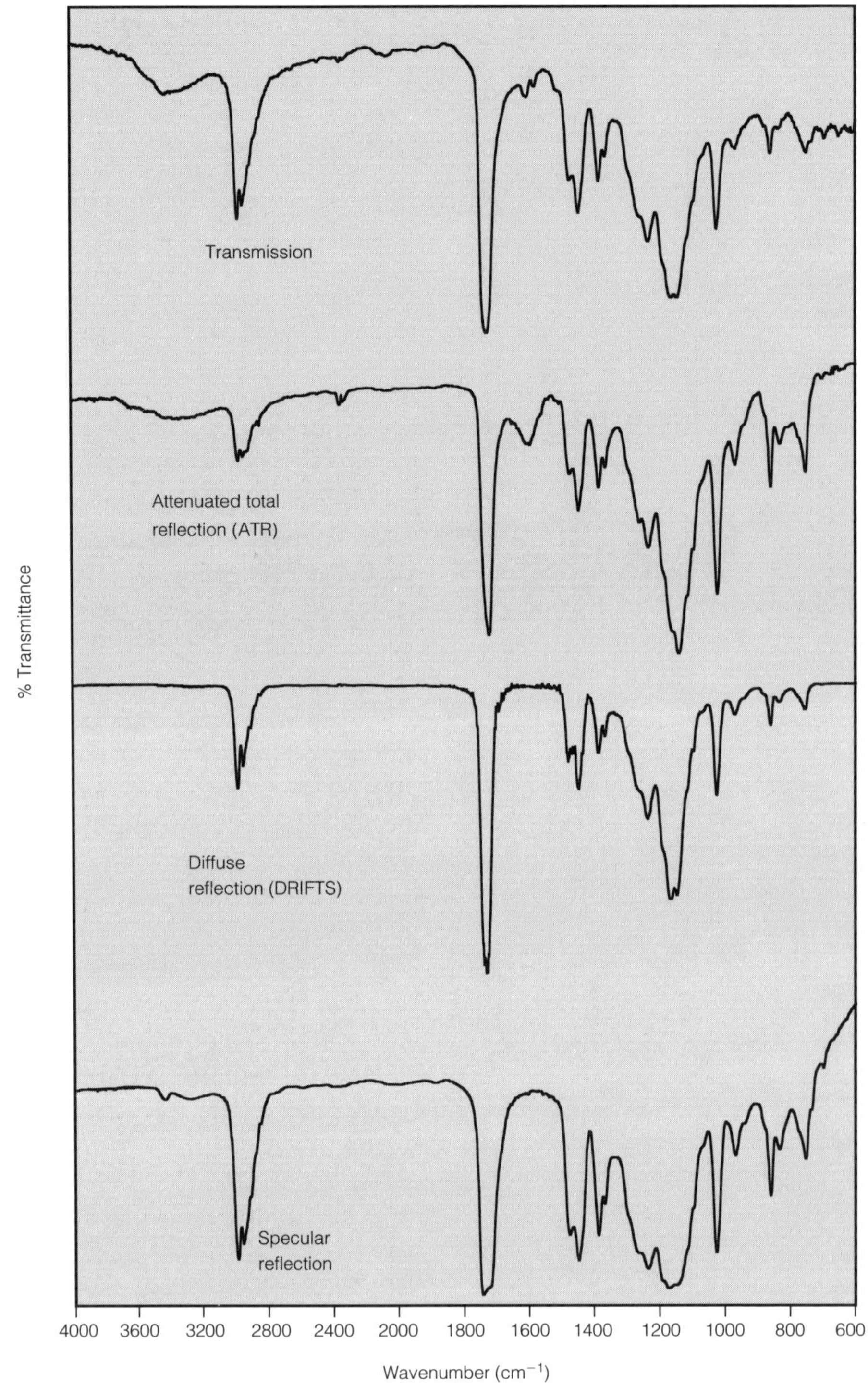

Figure 4.1

IR spectra of Acryloid B-72 collected by four different analysis methods. The overall spectral pattern is similar, but slight variations in band position and intensity can be related to the analysis method.

to different sample types. In-depth information on the techniques presented here, as well as on other techniques (chromatographic, thermal, photoacoustic, and emission), can be found in the Additional Reading section.

Sample form	Examples	Potential analysis methods
Gases		
Nonreactive gas, gas mixture, trace atmospheric components	pollutant, fumigant	transmission gas cell
Liquids		
Viscous	oil, paste, gel, curing adhesive, plasticizer	film on transparent window or diamond cell, horizontal reflection-absorption, liquid internal reflectance cell
Nonviscous	solvent	liquid cell/liquid internal reflectance cell, film on transparent window, horizontal reflection-absorption
Solution	emulsion, varnish, paint, consolidant (before curing)	diffuse reflectance, pressed pellet (dried residue), cast film on window or reflective substrate, diamond cell, internal reflectance
Solids		
Hard (smooth)	plastic, glass	external reflection, grind for KBr pellet, section for transmission or reflection-absorption, film on transparent window or diamond cell
Hard (rough)	ceramic, stone, corroded metal	diffuse reflectance, grind for KBr pellet, section for transmission or reflection-absorption, film on transparent window or diamond cell
Soft	flexible polymer, adhesive, paper, parchment	dissolve in solvent and recast as thin film, internal reflectance, film on transparent window or diamond cell, reflection-absorption on mirrored substrate
Thin film	plastic film	dissolve in solvent and recast as thin film, internal reflectance, film on transparent window or diamond cell, reflection-absorption on mirrored substrate
Coating	coated paper, coated metal, adhesive	dissolve in solvent and recast as thin film internal reflectance, film on transparent window or diamond cell, external reflectance
Fiber	fabric, fur, carpet, polymer, wood	diffuse reflectance, internal reflectance, film on transparent window or diamond cell
Powder	mineral, colorant, clay, stone, corrosion product	diffuse reflectance, grind for KBr pellet, internal reflectance, film on transparent window or diamond cell

Table 4.1

IR analysis methods for various types of samples. Each technique may also be performed on microsamples.

Infrared Transmission Measurements

Until thirty-five years ago, transmission measurement was the only method of IR analysis. Because of its versatility, IR transmission was used for both qualitative and quantitative analysis of gas, liquid, and solid samples. Large libraries of spectra from IR transmission analyses were produced. Thus, IR transmission spectra became a standard that is still used today and against which spectra generated from other analysis methods are compared.

One advantage of transmission analysis is that the total sample absorption is proportional to the thickness of the sample (pathlength) and its concentration. This correlation establishes a simple relationship for quantitative calculations. Also, because of high energy throughput, transmission analysis methods have greater sensitivity than reflection analysis.

The character of a transmission spectrum is dependent on sample preparation, particle size, and absorptivity. For example, spectral distortions, such as band broadening, occur when the transmitted beam is partially reflected or refracted (Fig. 4.2). Additionally, whenever another material is added to the sample, either to dilute it or to support

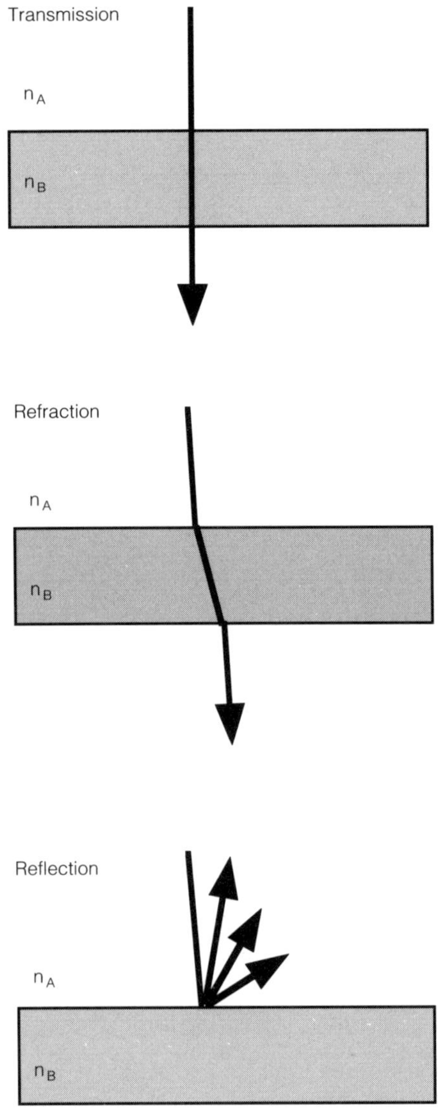

Figure 4.2

A radiant beam path when the beam is transmitted, refracted, and reflected by a sample and/or substrate (n_A = refractive index of air; n_B = refractive index of the material).

it, there is an added risk of contamination, which can result in spurious absorption bands. Some common spectral anomalies and solvent bands seen with transmission spectra are listed in Table 4.2. Many of the diluents mentioned in this table are no longer common in IR analysis labs; however, they may still be encountered in published reference spectra because of their previous high incidence of use.

Infrared window materials

For transmission analysis, the sample can be placed freestanding in the IR beam, or it may be held in place by a supporting material. Ideally, the supporting material would not, in itself, absorb any of the IR radiation. While many materials come close to this ideal, each has its own limitations regarding frequency range and chemical compatibility. The selection of the support material is dictated by the character of the sample and analysis method. Table 4.3 lists the common types of support materials used, their useful frequency ranges, refractive indices, and chemical compatibilities.

While glass is a good support for the visual examination of samples, it absorbs strongly over most of the mid-IR region. Figure 4.3 shows the IR spectrum of a standard glass microscope slide, as well as the characteristic spectra of two materials, potassium bromide (KBr) and barium fluoride (BaF_2), commonly used as IR windows in the mid-IR region. The glass slide essentially absorbs all the IR radiation from 2200 to 400 cm^{-1}. Since the main property desired for IR window material is its transparency to the IR radiation emitted by the source, glass is not used as a support material for IR analysis.

Many IR transparent materials are available for use as IR window materials; each one has certain advantages and disadvantages, and it is up to the analyst to select the material most suited for a particular experiment. The relatively inexpensive KBr and sodium chloride (NaCl) windows are commonly used. The KBr and NaCl windows are useful over the entire mid-IR region, but are moisture sensitive and will dissolve in water. Thus, the use of aqueous solutions must be avoided. Moreover, KBr and NaCl pellets, if left in humid air, will fog and become pitted. These windows are, therefore, typically kept in a desiccator. Also, NaCl and KBr windows are easily scratched and gouged, so that the requirement for routine cleaning and polishing makes it hard to maintain the original level of transparency. Poor window surface quality and transparency are particular problems in microspectroscopy—both during the process of aperture positioning as well as at the photodocumentation step.

BaF_2 windows are not attacked by water. Therefore, they can be stored in the open, and aqueous solutions can come in contact with the material without dissolution. Also, because the windows are very clear, they provide high-quality imaging for light microscopy and photodocumentation. However, BaF_2 windows are roughly four times as expensive as KBr windows. While they can be readily cleaned and polished for reuse, they are also very fragile and will break when dropped even a few inches. Finally, these windows have a cutoff at 700 cm^{-1} and

Sample preparation method	Problems and causes	Remedies
Free film	interference fringes from shiny sample surfaces	abrade the sample surface or analyze by another method
	bands totally absorbing from a too-thick sample	make sample thinner by flattening or cutting
Film, liquid, or particles placed on a support	bands totally absorbing from a too-thick sample	make sample thinner by flattening or cutting
	curved or slanting baseline due to nonflat sample surface	flatten sample or place in a compression cell; avoid particle edges
Dissolved in carbon tetrachloride (CCl_4)	interfering bands at 700–800, 980, 1250, 1550 cm^{-1} due to solvent	use only for 4000–1350 cm^{-1} range
Dissolved in carbon disulfide (CS_2)	interfering bands at 1400–1650, 2150, 2300 cm^{-1} due to solvent	use only for 1350–650 cm^{-1} range
Diluted with Nujol (mineral oil)	interfering C-H bands (2850–2960, 1460, 1380, 720 cm^{-1}) due to diluent	use only for 1300–200 cm^{-1} range
Diluted with Fluorolube (fluorocarbon oil)	interfering bands at 1080–1200, 965, 900 cm^{-1} due to diluent	use only for 4000–1300 cm^{-1} range
Ground and diluted with KBr, CsI, AgCl, then pressed into pellet	salts are hygroscopic; may absorb water with interfering bands at 3400, 1640 cm^{-1}	grind and mix sample and salt in dry atmosphere and/or under heat lamp
	may contain sharp NO_3 = band at 1384 cm^{-1}	use only fresh salt supply and store in dark container

Table 4.2

Possible spectral anomalies due to the sample preparation method.

thus may not be appropriate for the identification of some inorganic materials with characteristic absorptions that occur at lower energies.

Silver chloride (AgCl) windows are also water insoluble. They have good transmission properties from 4000–400 cm^{-1}, are relatively inexpensive (only twice the cost of KBr), and are fairly durable. These windows, however, are sensitive to light and will darken upon exposure to short wavelength visible light—therefore, they must be kept in the dark. For IR microspectroscopic analyses, their relatively high refractive index ($n = 2.00$) may affect focusing and sample isolation.

While the most commonly used window size for an IR salt plate is a 13 mm diameter disk of 2 mm thickness, some instruments have sample holders specifically designed for small (6 × 1 mm) or large (32 × 2 mm) salt plates. Windows can be found in a 4 mm thickness, but these tend to decrease the available energy in the low-wavenumber end of the usable window region and may actually cut off 50–100 cm^{-1} higher than the 2 mm window. Windows of 1 mm thickness can be specially ordered. These will have better transmission properties than the 2 mm thick variety and therefore may be desirable for some situations. They are likely to be more expensive, though, and are substantially more fragile than the 2 mm thick windows.

In normal use, IR windows may become fogged or scratched or accumulate insoluble deposits. When this happens, the surface can be reconditioned by being polished with fine-grain emery paper. The coarseness of grit selected for the initial polishing depends on the extent of the damage. Polishing kits from many of the IR supply companies have a

Material	Range (cm^{-1})	Refractive index	Compatibility
Glass	55,000–3000	1.45	water insoluble, attacked by hydrofluoric acid
Quartz	40,000–2500	1.54	water insoluble, attacked by hydrofluoric acid
Sapphire	20,000–1780	1.7	good strength, no cleavage
Diamond	40,000–200 except 2500–1800	2.42	insoluble, inert, hard, expensive
Calcium fluoride	70,000–1110	1.43	water insoluble, does not fog, incompatible with ammonium salt solutions
Barium fluoride	65,000–700	1.47	water insoluble, does not fog, breaks easily, incompatible with ammonium salt solutions
Zinc selenide (Irtran IV)	10,000–550	2.49	water insoluble, resistant to most solvents, brittle, slightly soluble in acids
Sodium chloride	40,000–625	1.54	hygroscopic, fogs slowly, easy to polish, low cost, water soluble
Silver chloride	25,000–400	2.07	water insoluble, very soft, sensitive to light
Potassium chloride	40,000–500	1.49	hygroscopic, fogs slowly, easy to polish, water soluble
Potassium bromide	40,000–400	1.56	hygroscopic, fogs with moisture slightly faster than NaCl; softer than NaCl
KRS-5 (Thallium iodide-bromide, Tl_2IBr)	15,000–250	2.38	water insoluble, toxic, soft, soluble in bases
Cesium bromide	10,000–250	1.66	hygroscopic, soft
Cesium iodide	10,000–200	1.79	hygroscopic, soft, fogs slowly
Teflon (thin film)	5000–500, except 1200–900	1.51	inert, unaffected by any solvent, available as disposable cards
Polyethylene (thin film)	4000–30, except 3000–2800, 1460–1380 and 730–720	1.53	water insoluble, resistant to most solvents, very soft, difficult to clean, available as disposable cards

Table 4.3

IR transmission characteristics of support materials.

good selection of fine-grain papers, generally ranging from 400 to 1200 grit. The grinding process is performed with emery paper wetted with alcohol and moved in a figure eight motion. The figure eight is necessary to prevent the flat crystal from becoming convex. It is usually necessary to use only one or two strokes with each grade of paper. For final polishing, a cloth-lap procedure is used (Chicago Society for Paint Technology 1980). For this technique, a soft, lint-free cloth is stretched tightly and fastened over a thick flat surface, such as a piece of glass. A small amount of very fine polishing agent (0.5 μm or less) is placed on one area of the cloth and wetted with a little solvent (or, for salt plates, with alcohol). The crystal to be polished is swept in the figure eight motion until it appears clear and smooth. Then the crystal is placed on a clean, dry area of the cloth and slid across in a straight motion to gently dry and clean its surface. To determine whether the crystal is still flat, it may be placed on an optically flat piece of glass and illuminated with a sodium vapor lamp to be examined for fringes: fewer fringes indicate a flatter crystal. Plastic gloves should be worn during the polishing process to keep finger oils and moisture from affecting the window and to protect the polisher's hands from hazardous materials. (For more polishing methods, see Smith 1979:115–17; for polishing methods for microsize KBr pellets, see Teetsov 1995.)

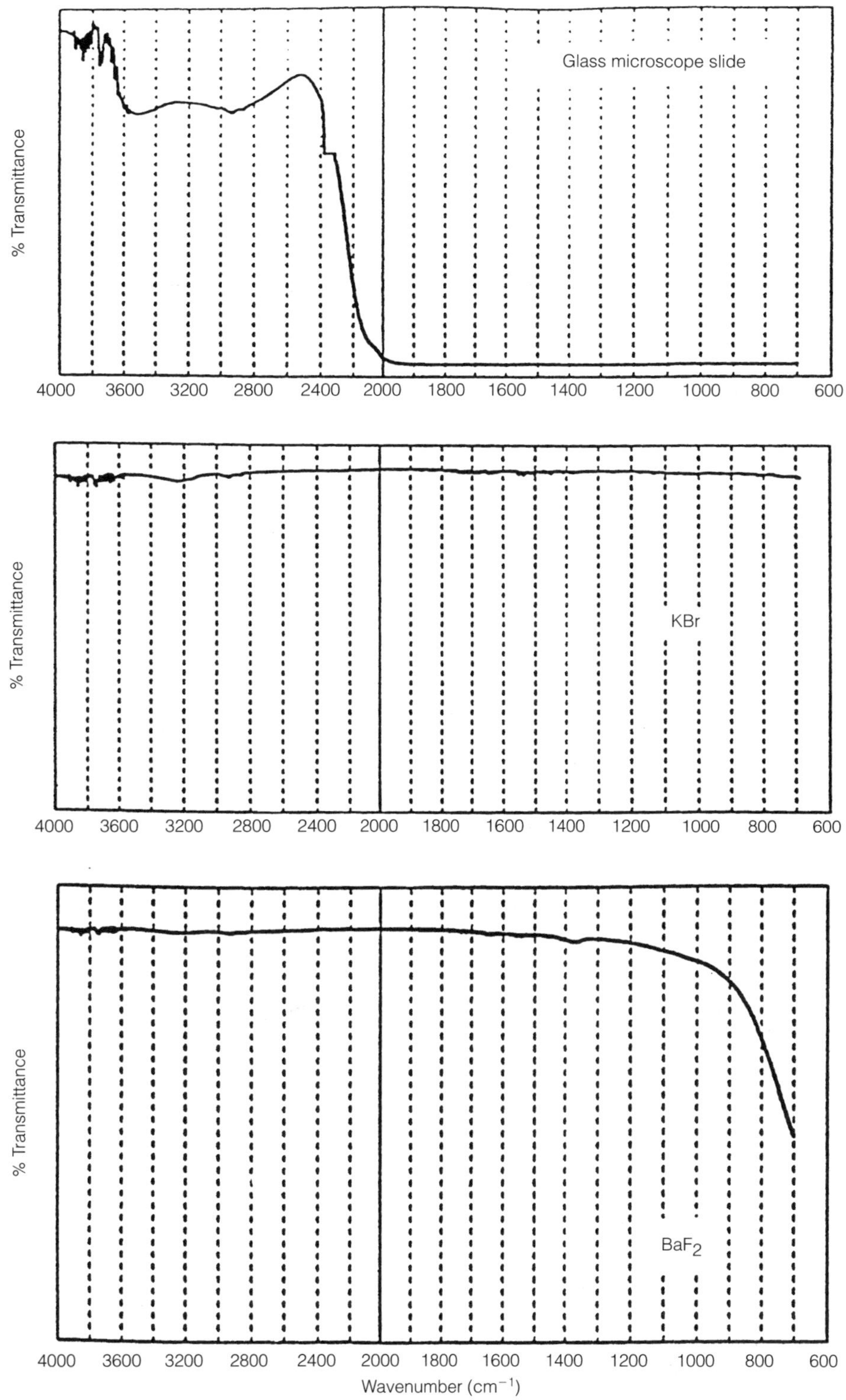

Figure 4.3

IR spectra of glass (1 mm thick), potassium bromide, and barium fluoride substrate materials. The IR transmission range is shown.

Most IR windows should be stored in a dry atmosphere to protect them from moisture. Desiccators and low-temperature ovens are available for this purpose. When limited amounts of sample are available, it is sometimes prudent to save the sample on or in a pellet for later reanalysis. Sample holders can be used to label and store pellets for later use.

Transmission analysis of gases

Materials that are in the gas phase at room temperature may be analyzed as a gas or condensed in a cold trap with liquid nitrogen and analyzed as a liquid or solid. Gas-phase molecules rotate freely, producing very sharp rotational absorption bands that are not seen in the liquid or solid-phase spectra for the same sample (as shown in Fig. 2.4). These bands are very characteristic, though, and are often used for the identification and quantitation of components in a gaseous mixture. For quantitative analysis of a gas, its absorbance is proportional not only to its concentration and to the IR pathlength but also to its pressure. High pressures can, however, result in absorption band broadening.

A gas cell consists of a glass or metal cylinder fitted at each end with a transparent, nonreactive IR window (Fig. 4.4). The cell has two stopcocks for evacuation and filling. The windows must be well sealed during assembly to prevent leakage. The typical cell is single pass and 10 cm long, though other pathlengths are available. A multiple-pass cell contains mirrors on its ends to reflect the beam back and forth through the sample several times before passing it on to the detector. This increases the effective optical pathlength and thus increases the absorbance and sensitivity of the measurement, allowing the detection of concentrations in the parts-per-million range. Gas cells with reflections of 40–100 times, producing pathlengths up to 120 m, are available.

Miniature, variable-temperature, flow-through gas cells can be used to identify the small amounts of a gas-phase material obtained from the effluent of a gas chromatograph (GC) or thermal analyzer. Alternatively, a unique GC-IR interface, in which the gas-phase effluent from the GC is frozen onto an IR transparent window cooled to liquid nitrogen temperatures, was developed by Fuoco, Shafer, and Griffiths (1986). The window is then rotated into the IR beam to obtain spectra of the condensate fractions.

Transmission analysis of liquids

Solution spectroscopy—that is, spectroscopy of liquids or solids in solution—has the advantage of being highly reproducible. Problems and spec-

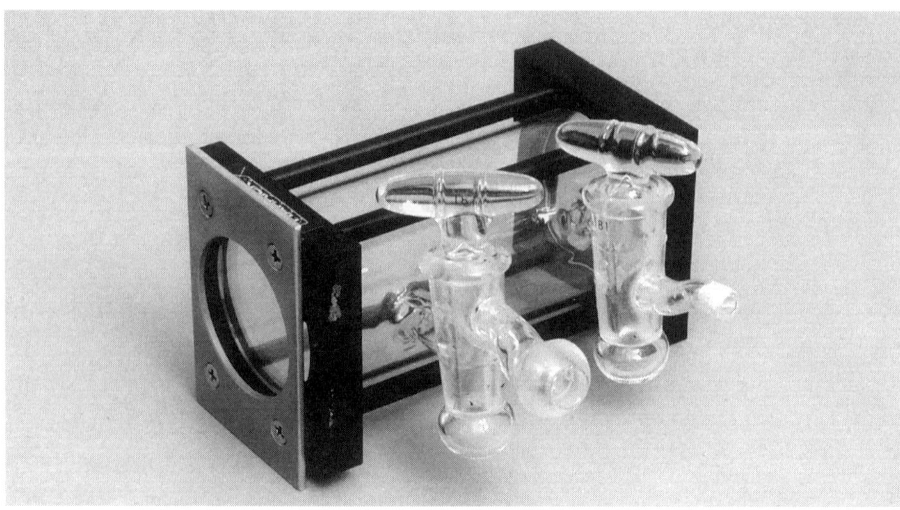

Figure 4.4
A single-pass, IR gas cell (10 cm). IR transparent windows at each end of the cell allow the passage of IR radiation beams.

tral artifacts due to sample preparation (heat, pressure, and particle size) are rarely encountered.

Liquids generally require very little sample preparation. A nonvolatile liquid sample can be spread as one drop on a single salt plate (BaF_2 is nonreactive to most liquids) or other transparent surface, such as a single surface of a diamond cell. However, unless they are applied to the support as very thin films, most liquids absorb too strongly; the result is spectra with some absorption bands having less than 10% transmittance. If this is the case, the sample must be diluted with a solvent. Carbon tetrachloride and carbon disulfide solvents were commonly used as diluents in IR spectroscopy because they themselves have few interfering IR absorptions. These diluents are rarely used today because of their toxicity and because of the availability of alternate IR analysis methods.

Volatile liquids can be analyzed by use of a commercial liquid cell or simply by sandwiching a drop between two salt plates. Commercial liquid cells consist of two IR transparent windows wedged together in a metal frame. The cavity for holding the solution is either a drilled space in one of the windows or a space created by a gasket of known thickness between the two windows. The spacers are typically made of Teflon or lead; the spacer thickness controls the pathlength and capacity. For quantitative analysis of a liquid, absorbance is proportional to both concentration and cell pathlength.

Sealed and unsealed liquid cells are available (Fig. 4.5). For cleaning, a permanently sealed liquid cell is flushed thoroughly with solvent, then dried with a stream of nitrogen or clean, dry air. Nonsealed, or demountable, liquid cells can be cleaned after disassembly. The demountable cells provide flexibility in changing the IR windows as they become clouded or scratched. However, since these demountable cells have the potential to leak, they are best used for viscous liquids or mulls. Micro liquid cells are available for the analysis of small sample volumes (0.5–50 μl). Other specialty cells, such as variable-temperature, variable-thickness, and flow-through cells, are also sold. Liquids can also be analyzed by some reflection methods, such as diffuse reflection (DRIFTS) and internal reflection (IRS).

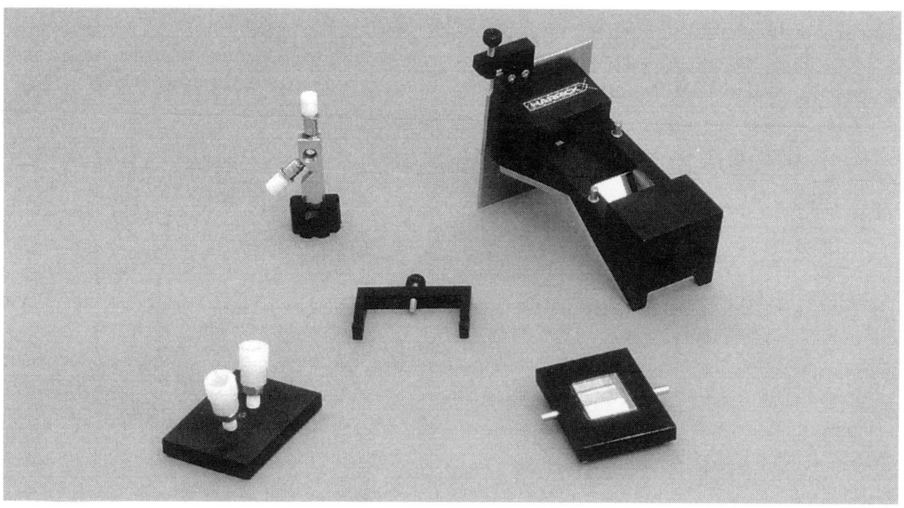

Figure 4.5

Single-reflection and micro liquid IR cells. For both types of holder, the cell is sealed, and then the liquid is injected with a syringe. The cells can be disassembled for cleaning.

Transmission analysis of solids

Several methods can be used for the IR transmission analysis of solid samples. In the most commonly known method, the samples are powdered, then mixed with an IR transparent material, such as KBr, and pressed into a clear pellet for analysis. In other methods, samples are analyzed as free, or unsupported, films or as thin films flattened or deposited on a transparent window. The deposition of films on a support is a highly versatile technique that has become the primary method used for samples analyzed with IR microspectrophotometers. Even hard and tough materials can be compressed by use of a diamond cell for transmission studies. Examples of materials that may be analyzed by these IR methods are viscous or nonvolatile liquids (oils, adhesives), films dried from a solution (varnishes, paints, emulsions), and flattened, solid samples (particles, pigments, fibers, scrapings). A disadvantage to the transmission analysis of solid materials is that grinding, dissolution, melting, or flattening is usually required. Any of these procedures can produce changes in sample orientation or in crystalline structure; the result is a slightly altered spectrum.

Pellets and micropellets

The pressed-pellet technique for IR analysis was first introduced in 1952, and it was used as the primary method for the analysis of solids for over two decades (Stimson and O'Donnell 1952). It is still a commonly used technique because it requires no expensive instrument accessories, it allows for control of analyte concentration, and it results in a sample that may be conveniently stored for later reanalysis. Butz's text on IR absorption analysis of powdered samples is recommended as a good basic reference, since it gives detailed information on the preparation and problems of preparing salt pellets (Butz 1960). The analysis of small samples with KBr micropellets has been successfully applied to early microphotographic emulsions (Newman and Stevens 1977), artist paint materials (Newman 1980; Meilunas, Bentsen, and Steinberg 1990), and furniture finishes (Derrick 1989).

The standard pellet size for commercial dies is 13 mm. Micropellet dies can produce disks of 0.5 or 1.5 mm in diameter. For analysis, a sample pellet is placed in a pellet holder and inserted directly into the sample compartment of the IR spectrometer. A background scan is usually taken with no pellet in the IR beam path. Micropellets require a beam condenser unit that focuses the energy of the beam to the correct size and position of the sample. Figures 4.6–4.9 show pellet dies and holders for both the normal and the micropellets, along with a ×4 beam condenser.

Beam condensers allow micrograms of samples to be analyzed, because they focus the radiation to a smaller diameter, thereby maximizing the amount of the energy in a minimum size area. The beam condenser must be aligned when placed in the sample chamber with both the pellet holder and empty pellet in place, to ensure that the beam will pass through the pellet area. Most instruments have a white light source that can be used for the rough alignment of the mirrors in sampling accessories when needed. Turning out the room lights increases the visibility of the alignment beam. A white card or piece of paper can be inserted at each position to locate the beam and to ensure that the focal point occurs at the pellet position. Final alignment is done by optimizing the IR beam intensity at the detector. The pellet and its holder must be

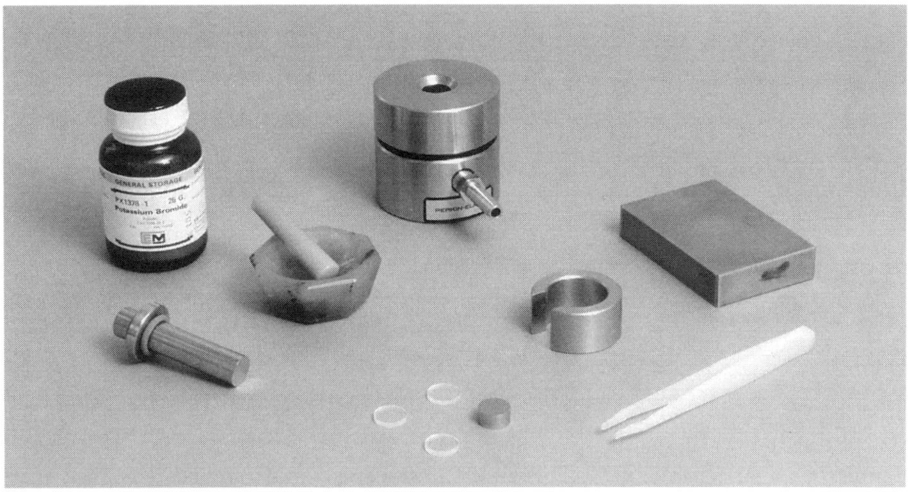

Figure 4.6
A 13 mm pellet die (top center) and holders. A sample is ground into fine particles and mixed with an alkali halide salt in the mortar and pestle, then transferred to the pellet die and pressed into a transparent pellet.

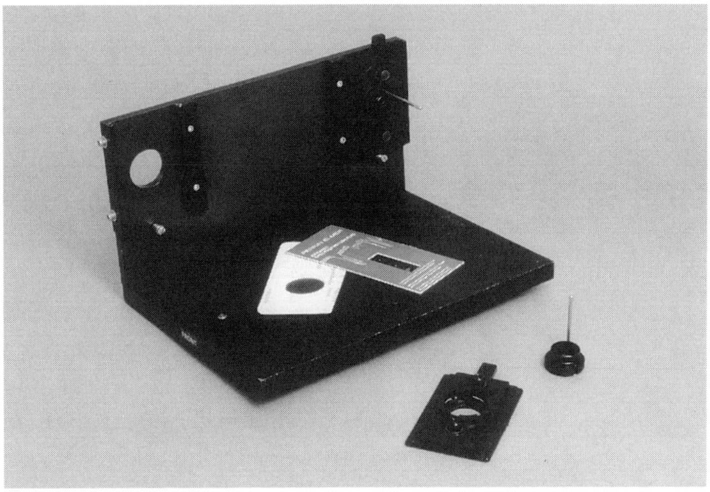

Figure 4.7
A standard sample holder with accessories for 13 mm pellets and thin films. The standard holder fits into the sample compartment of the IR spectrometer.

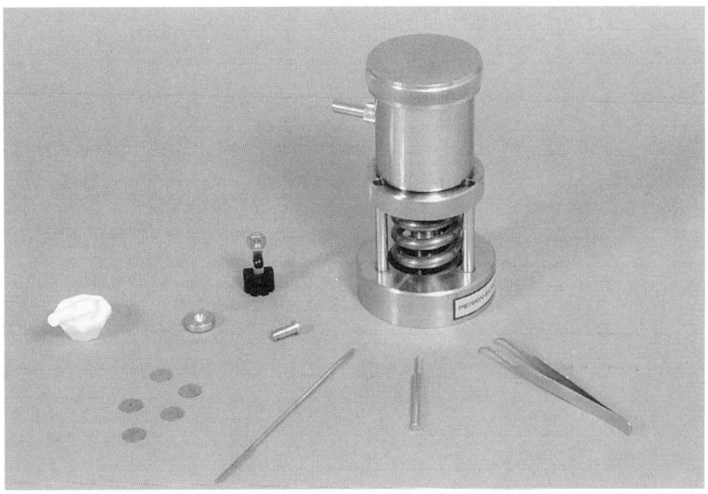

Figure 4.8
A micropellet die (1.5 mm) and holder. Sample preparation for micropellet analysis requires small tools and an extra-small mortar and pestle.

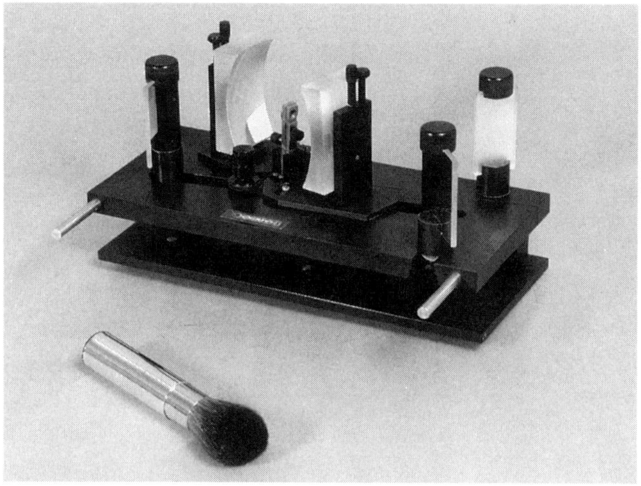

Figure 4.9
A beam condenser (×4) shown with a micropellet at the center focal point. The beam condenser may also be used with the internal reflection holder and the micro liquid cell.

repositioned in exactly the same manner each time, with the entrance face always oriented in the same direction. It is important to recheck alignment with the sample in place, since the sample itself may slightly change the focal length of the beam. If a spectrum with a low signal-to-noise ratio is obtained, the first thing to check is the position of the beam relative to the pellet.

To prepare a pellet, a small amount of sample is first placed in a clean agate mortar and ground with a pestle to produce particles that are smaller than the wavelength of IR radiation (5 μm). To determine when the sample is ground finely enough, use the pestle to form a smear on the mortar; the sample should feel slippery, should have no grit, and should spread out in a waxy film. For pulverizing large numbers of samples, a stainless steel ball mill or vibrator-grinder (see Suppliers, Wig-L-Bug, Crescent Dental Manufacturing Co.) can minimize the tedium. Heat and pressure produced by excess grinding, however, can cause changes in some samples. As with the mortar and pestle, all grinding containers should be thoroughly cleaned between samples to prevent cross contamination. Some tough, hard samples—such as epoxy resins, paper, parchment, and rubber—are very difficult to grind. Adding a few drops of liquid nitrogen to the sample will freeze it before crushing, thus making it brittle and easier to grind.

After the sample is ground, it is uniformly mixed with a powdered matrix, such as an alkali halide salt, that has a broad window of transparency in the mid-IR region. While potassium chloride (KCl), KBr, cesium bromide, cesium iodide, AgCl, and polyethylene have all been used to press pellets, KBr is the most commonly used. When KBr is used, it is important to mix quickly, as KBr is hygroscopic and will pick up water for the atmosphere that will appear in the resultant spectrum as broad bands at 3450 and 1650 cm^{-1}. Preparing the samples under a strong incandescent light or heat lamp can slow water absorption.

The mixed matrix is transferred to a pellet die and spread evenly. Then the plunger portion of the die is inserted, twisted, and lightly tapped against the sample to ensure that the sample is distributed evenly within the die. An uneven distribution of the salt matrix will produce a pellet with cloudy areas. The surfaces of the stainless steel dies are highly polished mirrors that produce transparent faces on the pellet and thereby minimize scattering effects. It is important that the die surfaces are clean and not scratched. They should not be touched with fingers and should be cleaned thoroughly between samples and before storage to minimize chances for corrosion.

After loading with the sample, the die is usually attached to a vacuum line and evacuated for a few minutes with a high vacuum pump. This procedure removes water, solvents, and air from the mixture and results in a clearer and longer-lasting pellet. The die is then compressed under several thousand pounds of pressure, either by hand or by a hydraulic press. After the pressure is released, the die is opened, and the sample pellet is carefully removed with plastic forceps. The sample should be immediately transferred to a pellet holder in the instrument for analysis. For analysis at a later time, the pellet must be stored in a water-free environment, such as a desiccator, a low-temperature oven, or a purged IR bench.

A pellet may also be prepared from a sample dissolved in a volatile solvent: a few drops of the solution can be placed on some powdered salt; once the solvent has evaporated, the solute is mixed with the salt powder, then pressed into a pellet. Any residual solvent, however, will hinder the production of a clear pellet. Microsamples dissolved in a solvent can be drawn up and concentrated into the tip of a KBr cone commercially available as a Wick-Stick (see Suppliers, Perkin-Elmer Corp.). The tip of the cone is broken off, ground, and pressed into a micropellet. A solvent blank pellet should be run alongside any sample to check for potential contamination and impurities.

Micropellets containing only a few micrograms of sample may also be prepared. With small samples, extraneous absorptions due to contaminants can appear as strong as absorptions from the sample. Therefore, it is critical to ensure that clean techniques are used. One method for preparing tiny, irreplaceable samples is to do all grinding, mixing, and transferring inside a glove bag that is slowly purged with dry air or nitrogen. This method will minimize moisture pickup, as well as other sources of environmental contamination.

Most solid samples can be analyzed as pellets, but some problems may result. Nonhomogeneous samples, such as a paint cross section, will be homogeneous after grinding. Since individual components of multicomponent mixtures can be difficult to identify from an IR spectrum, a preseparation method should be used when feasible. Grinding some samples can cause changes in crystallinity. When particles are not ground finely enough in a pellet, some scattering of the beam may occur, resulting in energy loss and a broadening of the absorption bands. The sample spectrum should always be examined for potential absorption bands due to the pellet matrix, such as water absorption, contaminants, or even reaction products (for example, some organic acids can react with alkali halides).

Unsupported films

Self-supporting materials, such as films, fibers, and tapes, can sometimes be mounted directly in the path of the IR beam without the aid of a supporting material; in this way problems of contamination and interfering bands are avoided. The only sample preparation is mounting the sample. A holder should maintain the sample at the focal point of the beam without blocking the beam. Commercially available magnetic film holders easily hold a large sample. Other sample mounts, such as pellet holders, can often be converted for use by stretching the sample across the empty pellet opening and holding it in place with tape.

While easily prepared and analyzed, freestanding samples can be difficult to analyze. A sample that is sturdy enough to be self-supporting is often too thick and absorbs too strongly, producing saturated spectral bands. Samples that are not flat, such as fibers, can cause the radiant beam to scatter. Scattering can be reduced by applying mineral oil to the sample (if the hydrocarbon band region is not critical) or by flattening the sample.

Free film samples can produce interference fringes. Interference fringes are the result of internal reflections of the IR beam inside the sample. Interference fringes are seen as a sinusoidal wave superimposed on the sample's molecular absorption pattern (Fig. 4.10). The intensity of the wave is a function of the refractive index of the sample and the glossiness of its surface. The frequency of the wave is dependent on the wavelength of the radiation, as well as on the thickness of the sample film. In fact, counting the fringes is an accurate method for the calculation of film thickness; the equation shown in Figure 4.10 is used (Harrick 1971). The fringes can be minimized by abrading the sample surface, by using polarized incident light along with an off-axis entry angle of the beam, by sandwiching the sample between two pellets or diamond surfaces, by coating the sample with an oil, or simply by analyzing the sample by another method, such as internal reflection.

Films on transparent supports

Most solid samples can be cast, melted, pressed, or spread into thin films for transmission analysis, but they require a supporting surface in order to place and maintain them in the path of the IR beam. The support is typically an IR-transparent material or window, such as a halide salt pellet. Another type of support, disposable IR cards (see Suppliers, Spectra-Tech, Inc., 3M films), has recently been introduced. These cards contain a thin polymer film (either Teflon or polyethylene) stretched on a cardboard holder.

The analysis of a thin film of solid material on an IR window is one common method used for IR microanalysis. This method has the advantage that the sample can be prepared in a horizontal orientation on the support material and can then be moved onto the sample stage without further disruption or transfer. Films may be prepared from samples in many forms—that is, solutions, powders, or chunks.

Solutions of solid samples in volatile solvents (e.g., emulsions, varnishes, and paints), as well as some viscous liquids (e.g., oils),

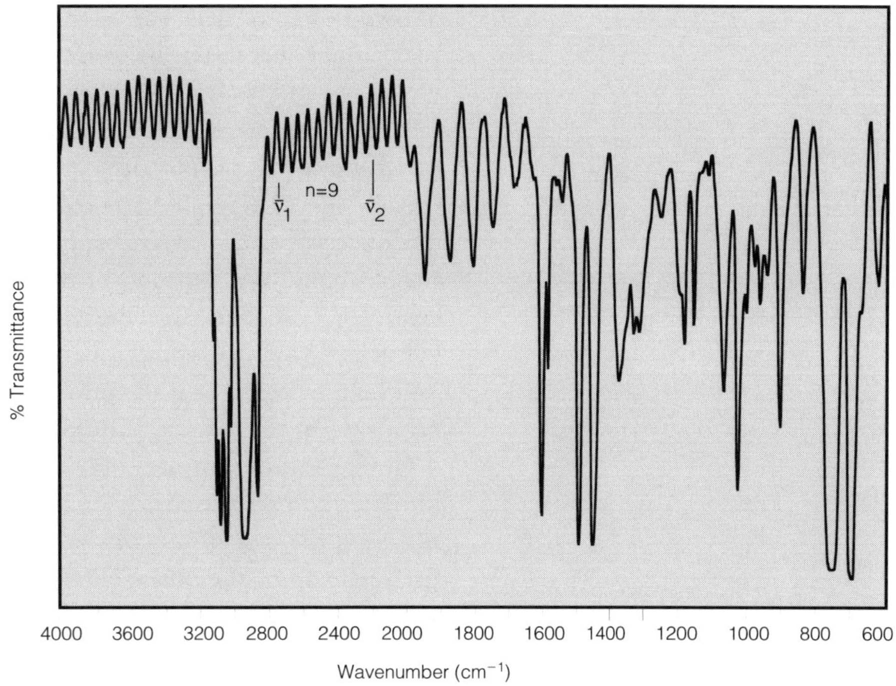

Figure 4.10
An interference fringe pattern obtained from a free film of polystyrene. The equation

$$t = \frac{10n}{2RI\,(\nu_1 - \nu_2)}$$

may be used to calculate the thickness of the film (n = number of oscillations; ν_1 = starting wavenumber [cm^{-1}]; ν_2 = ending wavenumber [cm^{-1}]; t = film thickness in mm; and RI = refractive index).

can be cast or spread onto the surface of a transparent window and dried into a thin film. The liquid is drawn up into a capillary pipette and dripped onto the surface of the window. Thick liquids may need to be spread thinly and evenly over the surface by use of the pipette horizontally as a draw-down edge or by pressing another pellet on the sample and compressing. The ideal sample should have a uniform thickness over the analysis area. For macrosamples, the area of analysis may cover the entire salt pellet, while for samples to be analyzed on a microspectrometer, the area of analysis may be only 1 × 1 mm or less. Thus, when a 13 mm window for microanalysis is used, multiple samples may be placed on one window. It is important, however, that the liquid samples not overlap and mix. The viscosity and surface tension (flow) of each liquid sample may be checked by observing a drop of the sample on a microscope slide prior to placing a microdrop on the sample window.

Dispersion of particles in an oil mull is a method that has been used for years in IR analysis. Particles are ground finely as for pellets, then mixed with either a mineral oil (Nujol) or a fluorinated oil (Fluorolube). Since these oils absorb in separate portions of the IR spectral regions, a sample is usually split into two portions, and half is mixed with each type of oil. The mull is spread as a thin film on a salt pellet for analysis. The mineral oil mixture gives a spectrum of the fingerprint region, while the fluorinated oil mixture provides a spectrum for the functional group region. Because this method is time-consuming and does not have any clear advantage, it is now rarely used, and alternate methods of analysis are preferentially selected.

Some polymers and waxes may be melted and then spread as thin films on IR windows. The sample is placed on a glass slide, metal plate, or tip of a metal spatula, then heated gently on a hot plate. The melted sample is spread quickly on the window in a thin, even film and allowed to cool. Another method melts and cools the sample on a metal plate, then analyzes it by reflection-absorption (Fischer and Bader 1994).

Particles and fibers may be flattened or pressed to form a thin film. The sample is placed on a hard surface, such as a reflective metal or IR window. It is compressed with a flat surface, a metal probe, or a metal roller. A sample flattened directly on a transparent window can be analyzed by transmission, and a sample flattened on a metal surface can be analyzed by reflection-absorption. Soft samples flatten readily and retain their shape. More difficult samples—such as fibers, adhesives, or elastomers—may be placed between two transparent windows, flattened, and held in place by a compression cell. The presence of windows on each side of the sample will decrease any scattering effects resulting from a nonflat sample surface. However, an increase in the total thickness of an IR window (caused by the two windows) will decrease the energy reaching the detector. If this is a problem, two 1 mm thick windows can used instead of the normal 2 mm windows. An alternative procedure is to mix the sample with a small amount of a nonabsorbing matrix material (such as KBr) and then to flatten the mixture as described above (Reffner and Martoglio 1995). This procedure essentially forms a miniature pellet that increases the sampling area, minimizes scattering, and makes the sample easier to transfer with a metal probe or razor blade.

Films on diamond cells

Salt windows can break and scratch when the samples are hard or noncompressible (e.g., minerals, epoxies). In 1959 the diamond anvil cell was first introduced by Weir and coworkers (1960). The high pressure, hard surface, and inertness of the diamond cell allowed reactive, opaque, hard, or rubbery samples to be converted to thin films for IR analysis. It also eliminated the potential for contamination due to an added matrix and allowed the sample to be retrieved for additional analyses. The usefulness of the diamond cell for the analysis of small art and archaeological samples was demonstrated in the 1970s by scientists at the Canadian Conservation Institute (McCawley 1975; Laver and Williams 1978). Forensic spectroscopists have also applied diamond cells to the analysis of tiny paint chips (Rodgers et al. 1976).

The diamond anvil cell consists of two industrial-grade diamonds, each with a flat, polished face, that are carefully aligned in a steel compression cell. The diamonds serve as windows, on which the sample is placed and through which the IR beam travels. The sample, preferably a homogeneous particle, is placed on the clean diamond facet, the opposing diamond is positioned on top, the cell is closed, and pressure is applied to flatten the sample. The original high-pressure diamond anvil cell is generally used with a ×4 or ×6 beam condenser in which the whole device—diamonds with compressed sample—is placed in the sample compartment for IR analysis.

The newer, low-pressure, miniature diamond anvil cell (see Suppliers, High Pressure Diamond Optics, Inc.; Fig. 4.11) and the micro diamond compression cell (see Suppliers, Spectra-Tech, Inc.) are microsampling devices that can either be placed directly on the stage of

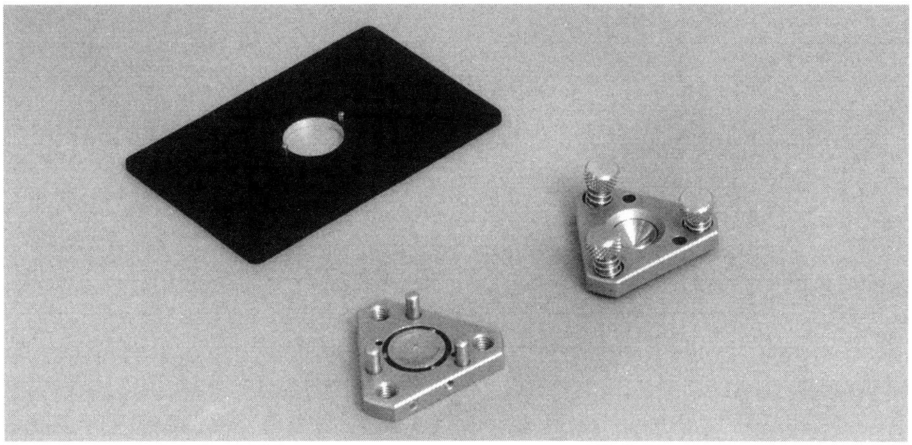

Figure 4.11
A miniature diamond compression cell. Samples are compressed between the 1.5 mm diamond surfaces. Samples may be analyzed with the cell halves together or separate.

an IR microspectrophotometer or used with a beam condenser. Commonly, the two halves of the cell are separated after sample compression, and one facet containing the flattened sample is placed directly on the analysis stage. Soft or thin samples can be placed directly on a single facet and analyzed with or without the compression step. Liquids may be analyzed on a single surface of the cell but should not be compressed, as they are readily displaced.

After analysis, the sample may be removed from the diamond facet with a cotton swab or even with a scalpel, since there is no danger of scratching the diamond. However, solvents should not be used to clean the diamond, since the diamonds are held in place with an adhesive. Solvents may loosen the adhesive, resulting in misalignment or loss of the diamond. If misaligned, a diamond face may scratch or break its opposing surface.

As noted in Table 4.3, diamonds are not transparent over the entire mid-IR region. However, the main region of absorption is from 2500 to 1800 cm^{-1}, where few IR absorption bands occur. Also, complete absorption does not occur with the thinner diamonds used in the smaller microsample diamond cells; thus the normal ratioing of a background spectrum to the sample spectrum will eliminate the diamond absorption bands. High pressures applied to the sample with a diamond cell may induce pressure- and temperature-dependent phase transitions in the sample (Lippincott et al. 1960). However, it has been pointed out that only minimal pressures are usually needed to spread the sample evenly between the diamond faces, thus minimizing potential spectral alterations (Ferraro and Basile 1979).

Fibers

IR spectroscopy is very useful for the characterization of natural and synthetic fibers. It is a standard analysis method used in the textile industry (Wannemacher 1980; Berni and Morris 1984), as well as for the analysis of leather, wood, and cellulose (Hergert 1971; Liang 1972). Internal reflection techniques work well for the analysis of larger samples, such as fabrics (20 × 45 mm), yarns, or fibers (at least 20 mm) (Wilks and Iszard 1965). Very small samples of archaeological fibers and dyes have been analyzed by IR microspectroscopy (Lang et al. 1986;

Matsuda and Miyoshi 1989; Jakes, Katon, and Martoglio 1990; Martoglio et al. 1990; Jakes, Sibley, and Yerkes 1994). IR microspectroscopy is also standard for the identification of trace fiber evidence in forensic examinations (Cook and Paterson 1978; Bartick 1987; Tungol, Bartick, and Montaser 1991, 1995).

Fiber, hair, and brush samples require little sample preparation. Such a sample may sometimes even be analyzed "as is" by taping the ends of a single fiber to hold it across a hole in a metal disk or simply by laying it on an IR window (Tungol, Bartick, and Montaser 1995). However, the fiber is often too thick and absorbs too strongly for direct transmission. In this case, one alternative is to use a reflection method, such as diffuse reflection or internal reflection.

A second alternative is to make the fiber thinner. The advantages to flattening a fiber are twofold. First, the horizontal measurement area of the fiber is increased. Second, its surface becomes planar. The use of the diamond cell is a quick and easy method for flattening fibers. Another method is to press the sample between two IR transparent windows in a compression cell. A final option is to use a scalpel to slice a thin section from the end or side of the fiber. This method is sometimes preferred for the examination of the fiber structure, since flattening destroys its physical shape and may change its crystalline orientation.

Infrared Reflection Measurements

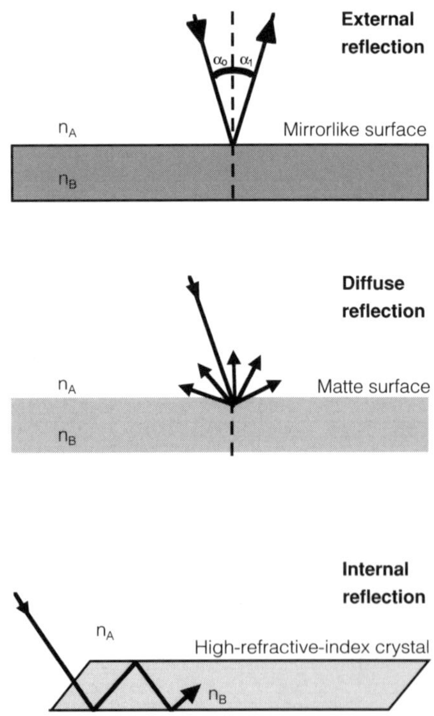

Figure 4.12
Three different reflection types (n_A = refractive index of air; n_B = refractive index of the reflecting material; α = angle of incidence).

In 1959 spectroscopists began exploring reflection techniques for IR analysis. Francis and Ellison investigated films on mirrored surfaces, but their technique of external reflectance did not become widespread until the 1970s (Francis and Ellison 1959). The technique of internal reflection spectroscopy (IRS), also called attenuated total reflection (ATR), was introduced in the early 1960s and quickly became popular, since almost any material—solid, powder, or liquid—could be easily and nondestructively analyzed. Diffuse reflection spectroscopy (DRIFTS) methods produced so little reflected IR radiation that they were not widely used until the development of FT-IRs.

Figure 4.12 illustrates the beam reflection paths used by different types of IR reflection devices. In external reflectance measurements, the angle of reflection is equal to the incident angle of the beam, as found in mirrorlike reflections. For this method, the IR beam may be reflected either by the sample surface (specular) or by a mirror surface under a thin layer of sample (reflection-absorption). DRIFTS occurs with a rough, porous, or powder sample, where the light is reflected at numerous angles that are not equal to the incident angle. IRS occurs when the incident beam is internally reflected in a high-refractive-index material placed in contact with the sample. Leyden and Murthy and Morris provide good comparisons of each of these reflection methods (Leyden and Murthy 1986; Morris 1991).

Each of these types of analysis methods can cause slight shifts in absorption band position, as well as changes in relative band intensi-

Table 4.4
Spectral alterations due to reflection analysis methods, as compared to transmission spectra.

Analysis method	Correction program	Difference from transmission spectra
Internal reflection (IRS)	none	band intensities stronger than normal at low wavenumbers, weaker than normal at high wavenumbers
Diffuse reflection (DRIFTS)	Kubelka-Munk	some slight band shifting and intensification of weak bands
		artificially flat baselines
Specular reflection	Kramers-Kronig	shifting in position of bands, derivative band shapes
Reflection-absorption (R-A)	none	may contain interference fringes and a specular component, if the surface of the sample is shiny

ties. Table 4.4 provides a summary of some of the changes that are seen in reflectance spectra as compared to spectra collected by transmission methods. Also, an expansion of the carbonyl region of the spectra from Figure 4.1 is shown in Figure 4.13, to illustrate the changes in peak shape and position.

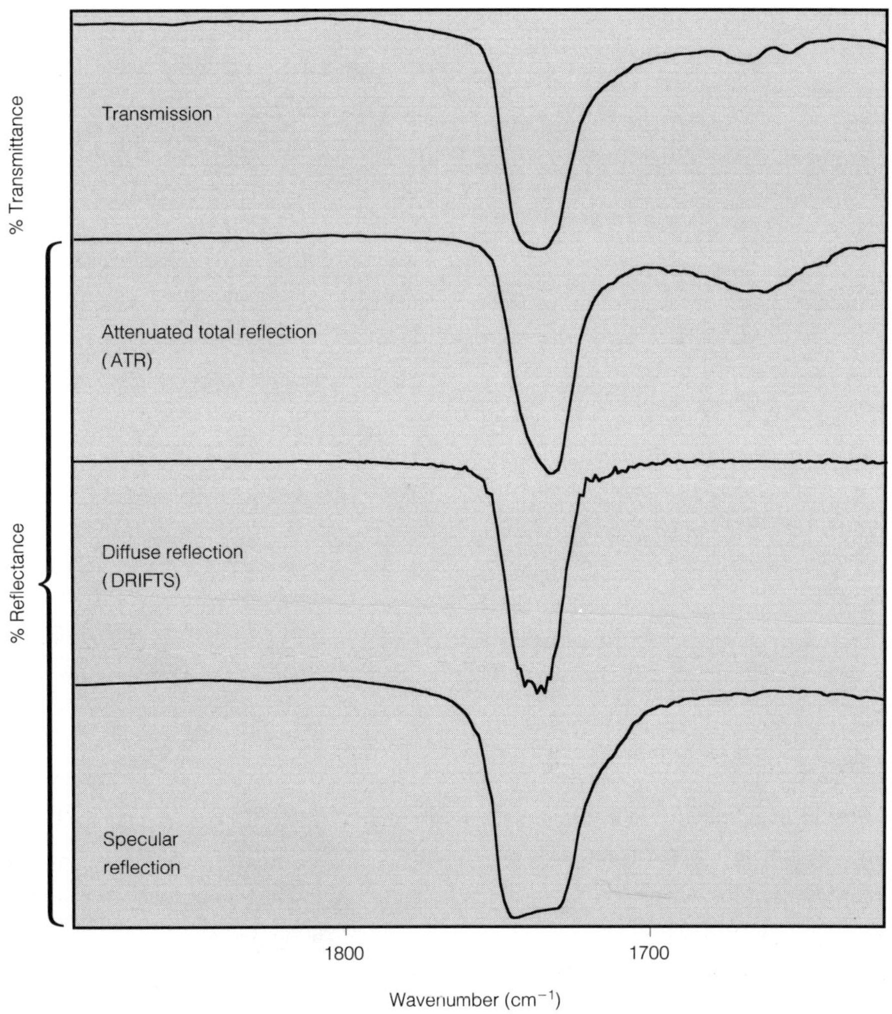

Figure 4.13
Partial IR spectra showing an expanded carbonyl region of Acryloid B-72 obtained from four analysis methods. While the overall spectral pattern is similar for the four methods, slight differences in band position and intensities warrant caution when spectra analyzed by different methods are compared. (The full spectra are shown in Fig. 4.1.)

Specular reflection

One type of external reflection, called specular reflection, occurs when the shiny surface of the sample reflects the IR beam at an angle equal to the angle of incidence. The specular reflection technique is used on samples of all sizes. Accessories for either fixed- or variable-angle reflection of the incident beam are available that fit into the IR bench sample compartment. Figure 4.14 shows a variable-angle external reflection unit. The angle of incidence is adjusted to obtain optimum beam penetration on a sample—and thus optimum sensitivity. Usually a high angle of incidence (70°–89°) is best (Harrick 1979). Larger top-loading external reflection attachments are available for the horizontal placement of samples; they can accommodate any large, flat surface placed face down.

Specular reflection spectroscopy is a nondestructive technique in that the analyzed sample does not need to be removed for analysis. However, for high-quality spectra, the sample should be highly reflective and thus may require polishing prior to analysis. In a vertical reflection holder, the sample is generally clamped in place.

The large number of polished paint cross sections already in existence in museum labs is an ideal resource for specular reflection microanalysis. Applications presented in Chapter 6 illustrate the combination of specular reflection analysis and computer-controlled stage movement to perform linear- and area-concentration maps on polished cross sections. Relative variations in paint layer composition can be determined by specular reflection, but identification of components in the samples may require supplemental analysis by other techniques.

In specular reflection, strong absorption bands appear as derivative-shaped curves because of dispersion of the radiation. The derivative-shaped bands are referred to as *Reststrahlen* (residual ray) bands. This effect is dependent on the wavelength, angle of incidence, intensity of absorption, and refractive index of the sample material. Specular reflectance spectra can be converted to transmittance-like spec-

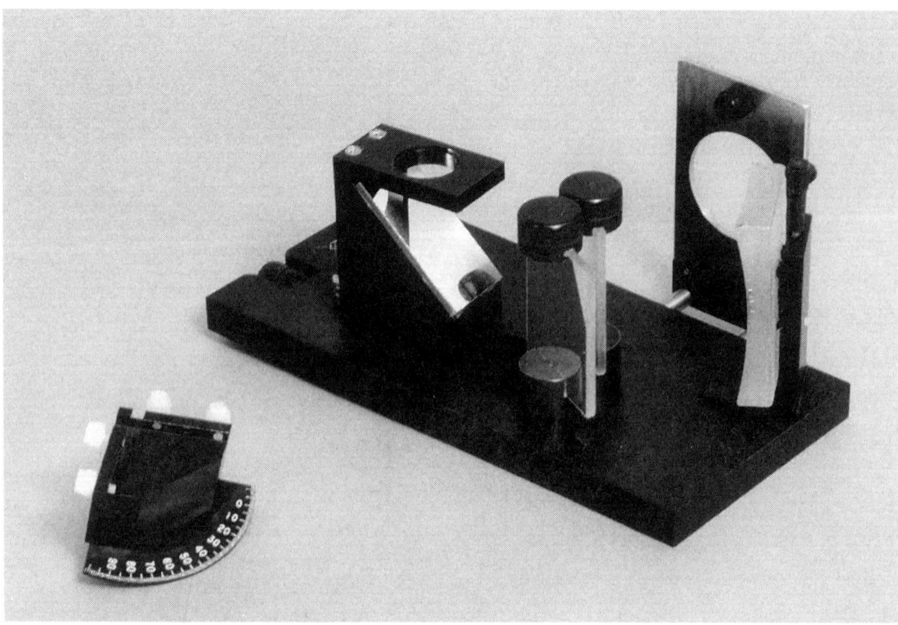

Figure 4.14

A variable-angle external reflection accessory. The sample is placed horizontally on the holder over the angled mirror. An additional sample holder (shown at left) is available to orient the sample in a vertical position.

tra with a Kramers-Kronig transformation, which calculates refractive and absorption indices from the specular data. Most IR data processing programs can perform Kramers-Kronig transformations. Such a transformation works well when the spectrum results only from specular reflection; however, other types of reflections, refractions, and scattering can occur, such that bands of different shape and position are produced. For further information on specular reflection, refer to Kortum (1969).

Reflection-absorption

A second type of external reflection, called reflection-absorption (R-A), occurs when the IR beam passes through a thin film of sample and is reflected from a nonabsorbing substrate. When the incident radiation enters the film at a near-normal angle (90°), the method is essentially a double-pass transmission experiment. Spectra generated in this manner resemble those obtained in transmission. Reflection-absorption experiments can also be conducted at a greater angle of incidence (65°–85°), thereby increasing the effective pathlength of the sample. This is called grazing-angle spectroscopy and is used to detect and analyze very thin films.

Front-surface aluminum or gold mirrors are used as reflective surfaces for thin films in a reflection-absorption experiment. The sample—liquid, polymer, particle, or fiber—is smeared, melted, or flattened onto the mirror to form a thin, smooth, and level film. The mirror with sample is then placed in a specular reflection apparatus or on a microspectrometer stage for reflection analysis.

Specular reflection is the most common spectral distortion in a reflection-absorption experiment. It occurs when the sample is shiny and some of the radiation is reflected from the surface of the sample rather than from the mirror. To eliminate the specular component, the sample surface may be abraded, or a nonabsorbing material (e.g., salt plate) may be placed on the surface of the sample (Reffner and Martoglio 1995). A salt plate must be held tightly against the sample surface to prevent interference fringes from occurring. A compression cell can be used for this purpose.

Diffuse reflection

For a non-mirrorlike surface, light is reflected diffusely, or in all directions. When all specular contributions are eliminated, DRIFTS can be related to the concentration in the sample by use of the Kubelka-Munk equation (Kubelka 1948). This conversion makes the spectra analogous to absorbance plots for transmission spectra.

DRIFTS has been used in the visible region for many years for the quantitative analysis of colorants (Billmeyer and Saltzman 1966). For IR analysis, DRIFTS was tried as early as 1913 by Coblentz, and it was later used for the examination of powdered samples (Clark 1964; White 1964; Vincent and Hunt 1968). It gained popularity in the late 1970s when an efficient collector designed for diffuse radiation was used in combination with the high-energy throughput of an FT-IR (Fuller and Griffiths 1978a, 1978b, 1980; Griffiths and Fuller 1982).

In the field of art conservation, Meilunas used DRIFTS to examine the surface of paint films and inks (Meilunas 1986). He found that while many pigments could be identified, specular reflection inhibited the identification of binders. Shearer used silicon carbide paper both as a sampling tool and as a DRIFTS substrate for the analysis of archaeological organic residues (Shearer 1987). Using DRIFTS as a nondestructive tool, Hedley and coworkers (1990) analyzed surfaces of paintings before and after solvent cleanings; Perron (1989) characterized emulsions on photographs; Poslusny and Daugherty (1988) identified the adhesives on stamps; and Faraone (1987) examined the weathering surfaces of stones. McGovern and Michel used DRIFTS in combination with other techniques to determine the composition of royal purple dye, while Dauphin applied it to the quantitative analysis of protein residues found in archaeological bones (McGovern and Michel 1990; Dauphin 1993).

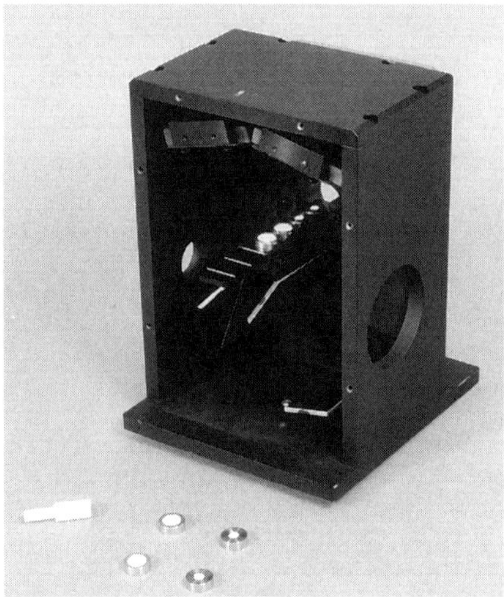

Figure 4.15

A DRIFTS unit with normal and micro (custom-machined) sample cups. Samples are held in a four-position holder that slides into position for analysis.

DRIFTS attachments are designed to fit in the standard sample compartments of most spectrometers (Fig. 4.15). The light reflected from the sample is collimated by an ellipsoidal mirror, then directed to the detector. The sample is held in a small cup on a sliding metal strip that serves to position the sample reproducibly. Some commercial DRIFTS accessories are designed to block the collection of the beam reflected at the angle equal to the incident angle—that is, the specular reflection component. This feature eliminates one problem source in DRIFTS.

Powder samples are prepared in a manner similar to that for KBr pellets. First, the sample is finely ground and mixed with KBr (or KCl) powder. Since the resolution of the spectrum is dependent on the particle size, it is critical to grind the sample as finely as possible. Distorted absorption bands will appear if particle sizes are larger than the wavelength of radiation. The mixed sample matrix is then scooped into a sample cup and packed gently with a tamper. The prepared sample must be level and uniformly packed to produce a good-quality spectrum. The IR beam penetrates only the top 0.5–2 mm of the sample. Thus, for small samples, it is possible to fill the bottom half of the cup with pure KBr powder, then put the sample matrix in the top half of the cup, to produce a more concentrated sample. Also, microsamples may be analyzed in microcup sample holders that are smaller in diameter and shallower than the regular cups.

Solutions may be analyzed by dripping a few microdrops on a prefilled analysis cup of KBr and allowing the solvent to evaporate. If the drops disrupt the surface of the salt bed, it may be possible to smooth the dried surface by gently tamping it, either directly or by first adding a few sprinkles of KBr powder on top.

For nondestructive analysis of solid samples, any matte surface, such as dyed textiles or papers, may be placed at the focal point of the IR beam in the DRIFTS apparatus (Harrick Scientific Corp. 1987). Shiny or smooth surfaces may need to be roughened with emery paper to diffusely reflect the incident beam. Also, a sample may be collected from a solid by gentle abrasion with a small circle of silicon carbide paper; the sample on the silicon carbide substrate is then mounted on a holder and placed in a DRIFTS accessory for analysis (Shearer 1987; Pretzel 1994).

In this manner, small amounts of sample can be taken from surfaces with minimal damage. The thin layer of sample on the reflective silicon carbide substrate essentially produces a reflection-absorption experiment with a spectrum similar to that obtained in transmission. The spectrum should not need the Kubelka-Munk correction used for other diffuse reflectance spectra.

Typically, all DRIFTS samples should be diluted to less than 30 w/w% concentration in a nonabsorbing matrix. On solid samples, this dilution can be done by sprinkling a thin layer of nonabsorbing material (e.g., KBr) over the surface of the sample. The powder layer minimizes band broadening and poor resolution effects resulting from saturation or specular reflection problems; additionally, the sensitivity of the measurement is improved by an increase in the depth of penetration of the beam. Spots on thin-layer chromatography plates may be successfully analyzed in this manner (Zuber et al. 1984).

The diffuse reflectance spectrum depends on sample density and refractive index, as well as on particle size and morphology. For quantitative analysis, the Kubelka-Munk function is used to relate the absorption coefficients to the material's concentration. This function tends to enhance the strong absorption bands and decrease the weaker bands in comparison to transmission spectra. Additionally, the baseline may be artificially flattened. For nonquantitative work, the diffuse reflectance spectra may be displayed without any correction function. The major problem that can occur with diffuse reflectance spectra is the inclusion of a strong specular reflection component that produces slight band distortions.

Internal reflection

Internal reflection was first noted in 1717 by Isaac Newton in his studies of total reflection at the interface of two different media. The phenomenon is observed in a clear glass of water, where one sees total reflection by viewing the water's surface from slightly below the water level (Harrick 1979). Total reflection is destroyed when another material, such as a finger, is placed in optical (air-free), or intimate, contact with the glass. In optical contact, the ridges on the finger surface can be clearly seen—but not the valleys, where no contact is made. In 1947 Goos and Hänchen described the interactions between light and the two materials as total internal reflection, and the interactions were subsequently used for recording optical spectra (Goos and Hänchen 1947). Internal reflection was applied to IR spectroscopy by Harrick (1959) and Fahrenfort (1962). While Fahrenfort coined the name attenuated total reflection (ATR), the American Society for Testing Materials has since limited the scope of the term ATR and has instead designated the overall technique as internal reflection spectroscopy (IRS) (ASTM 1981).

IRS works well with coatings, adhesives, fibers, foams, fabrics, plastics, and films, and it is the technique of choice for many opaque and intractable samples. Internal reflectance spectra of films do not exhibit any of the interference fringes that can occur in transmission spectra. In a development significant to the field of art conservation,

Paralusz did an extensive study using IRS of adhesive tapes (Paralusz 1974). This paper included studies on resin aging, delamination, and resin-plasticizer migration. An interesting paper by Wilks showed varied applications of ATR, from the analysis of skin to gas chromatography fractions (Wilks 1967). In 1985 Cain and Kalasinsky used several analytical methods to examine the degradation products on nineteenth-century paper (Cain and Kalasinsky 1985). They found that IRS was useful for the determination of gelatin size in the paper samples. More recently, IRS has been used as a nondestructive method to examine eighth-century Japanese fabrics (Matsuda and Miyoshi 1989) and to identify synthetic fabrics (Cardamone 1988).

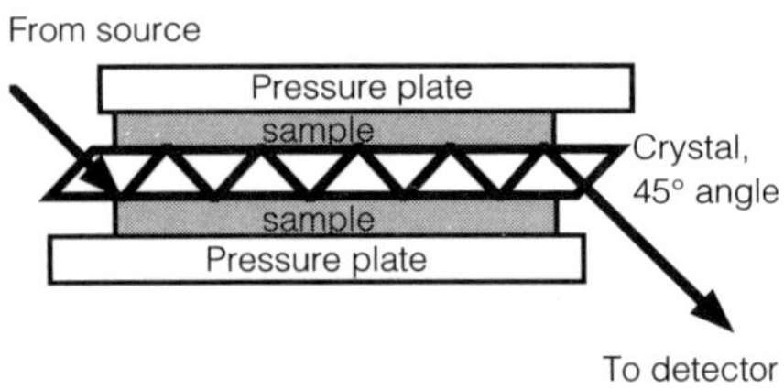

Figure 4.16
Multiple internal reflections in an IRS element. When the IRS element has a higher refractive index than the sample, the IR radiation will penetrate slightly into a sample if it is in optical contact with the element.

In IRS, the incoming radiation enters a high-refractive-index material (hence called an IRS element) and is internally reflected, usually multiple times (Fig. 4.16). Based on the fact that internally reflected IR radiation will penetrate a short distance into a lower-refractive-index medium, a sample in optical contact with the surface of the IRS element will be exposed to IR radiation. The sample can then absorb the radiation and produce a spectrum. Only the surface of the sample is analyzed, since the beam penetrates just a few micrometers into the sample. The depth of penetration is proportional to wavelength, and therefore, an increasing depth of penetration is observed at higher wavelengths (lower wavenumbers). The intensity of absorbed radiation is dependent on the amount of sample in contact with the surface of the IRS element and the number of contact points that the radiation has with the sample-element interface. Thus, increasing the number of reflections increases the intensity of the IR absorptions. Varying the angle of incidence can change the number of reflections and their position within the IRS element.

For IRS to occur, the IRS element must have a refractive index higher than the sample to be analyzed (see Table 4.3 for refractive index values). The IRS element material must also be clean, have a high surface polish, and be transparent to IR radiation. Silicon, germanium, thallium bromoiodide (KRS-5), AgCl, and zinc selenide (ZnSe) are typically used as IRS elements. Germanium and silicon are hard and brittle. They do not scratch easily; they are, however, subject to breaking. They may be cleaned in water, solvents, and dilute acids. KRS-5 elements, the most commonly used, tend to be soft and can retain the imprint of nonflat samples that are pressed into them. They can be cleaned with hydrocarbon solvents, acetone, or alcohol. AgCl, while softer and more flexible than KRS-5, is light sensitive and inexpensive enough to be discarded when damaged. ZnSe is expensive and brittle but hard and insoluble.

It is important not to scratch the optical surfaces of the IRS elements, and all handling should be minimized. Many samples, including powders, may be removed from the elements with pressure-sensitive adhesive tape. Thin films and residues may be cleaned with a wash bottle or, if necessary, an ultrasonic bath. Rubbing of the surfaces of IRS elements should be avoided. Polishing can often refurbish damaged or scratched IRS elements, but because of their toxic nature, it is often best to return the IRS elements to the manufacturer for reconditioning. Pressure plates should be wrapped in an uncoated variety of aluminum foil that may be discarded after each analysis to prevent cross contamination between samples.

Sample preparation for IRS analysis is relatively simple. The sample is placed on an IRS element (preferably on both sides), then the IRS element is positioned in its holder between two pressure plates that are uniformly compressed to place the sample in contact with the IRS element (Fig. 4.17). As the pressure is raised, the amount of sample in optical contact with the IRS element is increased, thereby enhancing the absorption intensity. For quantitative or comparative work, a torque wrench is recommended to reproducibly tighten the holder for each sample.

Samples, such as films, may need to be cut so that they do not exceed the length of the optical element and interfere with the beam path. Long fibers may be wound around the IRS element to produce an even sample surface. Fibers may also be placed on the bias to prevent orientation effects. To do this, small strips of the fiber are cut, then placed diagonally, adjacent to one another and completely covering a strip of tacky adhesive. The tape is then placed in a manner such that only the fibers, not the adhesive, touch the IRS element. Powders may be evenly spread on an IRS element, or they may be dispersed in a liquid for deposition on the IRS element. For hard samples, the pressure plates can be lined with a thin foam or thick adhesive and then covered with uncoated aluminum foil. This procedure will minimize deformation to the IRS element's surface. Liquids, gels, pastes, and even aqueous solutions can be cast onto an IRS element or run in a cylindrical internal reflection cell (CIRCLE; see Suppliers, Spectra-Tech, Inc.). A long ATR probe, originally developed by Wilks Scientific, can be immersed in liquids and solids for analysis outside of the instrument sample compartment (needle probe; see Suppliers, Spectra-Tech, Inc.).

The IRS element holder typically fits into a beam condenser unit in the sample compartment. For optimum performance, it is critical that the beam condenser be aligned to provide maximum IR beam intensity at the detector. The beam condenser should be aligned initially without a sample, after which point a background scan can be obtained. The alignment should be rechecked with every single sample that is analyzed, because the position of the element can vary with the thickness and position of the sample, as well as with the placement of the IRS element.

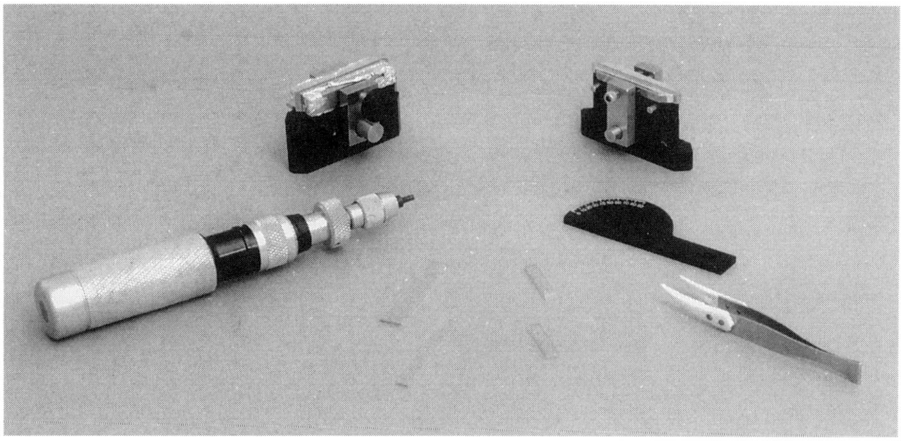

Figure 4.17
An IRS holder, elements, and torque wrench. The sample is placed adjacent to an IRS element, then positioned between the two pressure plates (covered in aluminum foil) of the holder. A torque wrench is used for even tightening of the screws that hold the pressure plates around the sample.

Otherwise, the energy losses due to poor alignment will cause significant deterioration of spectral quality.

The instrument usually contains a white light source for the rough alignment. Turning out the room lights will aid in the process. When the IRS element is properly positioned in the beam, it will glow. In fact, in a dark room, it is possible to see the points of contact of the beam with the sample-element interface. This feature is useful for the analysis of small bits of sample. These points can be marked on the top surface of the pressure plate; the holder can be removed from the instrument and disassembled; and the sample bits or fibers can be positioned in these places for optimum beam contact. Theoretically, there should be no "dead spots" for angles of incidence $\leq 45°$. In practice, however, there do seem to be regions that are more sensitive than others.

In IRS, the quality of the spectrum depends upon the angle of incidence, the depth of penetration, the number of reflections, and the relative refractive index difference between the sample and the IRS element. The spectra of thin films are virtually identical to spectra collected by transmission methods. For thicker samples, however, absorption bands at smaller wavenumbers are more intense, since the depth of penetration into the sample increases with wavelength. Poor spectra are obtained when the sample is not in intimate contact with the IRS element.

Infrared Microspectroscopy

The coupling of a microscope to the IR spectrophotometer produces a system capable of doing IR microspectroscopy. The resulting apparatus, called an IR microspectrophotometer, was first made commercially in the 1950s (Cole and Jones 1952). Although the design of the microspectrophotometer was satisfactory even by today's standards, the system proved to be costly and was limited by the low energy throughput and corresponding low signal-to-noise ratios. Therefore, the introduction of the FT-IR, with its inherent advantages, as well as advances in IR detector technology, fueled the reemergence, in 1983, of the IR microspectrophotometer. The recent success in application of this instrumentation to many areas of research (semiconductors, polymers, and pharmaceuticals), as well as forensic investigation, has established the technique of IR microspectroscopy as a powerful tool in the analysis of small samples. For more information on the history, design, and operation of IR microspectrophotometers, see Roush 1987; Messerschmidt and Harthcock 1988; Katon, Sommers, and Lang 1989–90; Katon and Sommers 1992; Reffner 1993; and Humecki 1995b.

Researchers quickly saw the applications of IR microspectroscopy to the analysis of tiny samples from works of art. The first published application of IR microspectroscopy was for the analysis of painting materials, done with a dispersive spectrometer by van't Hul-Ehrnreich (1970). Later, the attachment of microscopes to FT-IRs stimulated numerous applications for the analysis of paint layers, binding media, varnishes, waxes, dyes, and other materials (Shearer et al. 1983;

Baker et al. 1988; van Zelst, von Endt, and Baker 1988; Orna et al. 1989; Jakes, Katon, and Martoglio 1990; Martoglio et al. 1990; Derrick, Landry, and Stulik 1991; Tsang and Cunningham 1991; Derrick et al. 1992; Lang et al. 1992; Derrick 1995).

Microspectrophotometer design

The first IR microscopes were positioned in the sample compartment of the IR spectrophotometer. The microscope contained all the necessary transfer optics to direct the IR beam from the spectrometer source to the sample positioned on the stage of the microscope. In a transmission experiment, after it passed through the sample, the IR radiation was collected and directed back into the spectrophotometer and onto the detector. The optics used in IR microspectrophotometers are reflecting optics. Since both glass and quartz absorb IR light over much of the region of interest, the IR microspectrophotometers are unable to employ standard, visible-light refracting (lens) optics but rather must use reflecting (mirror) optics.

Although other detectors were used, the detector of choice was, and continues to be, the mercury-cadmium-telluride (MCT) detector. The MCT detector element is cryogenically cooled to liquid nitrogen temperature, thus providing high sensitivity and signal-to-noise values necessary for the low energy levels and small sampling areas found in IR microspectroscopy. As the microscope designs gradually changed, a major improvement in signal-to-noise levels was achieved when the MCT detector was repositioned from the spectrophotometer bench to the microscope apparatus, in order to minimize the IR beam path after interaction with the sample.

The next wave of improvements came when IR spectrophotometers were built with the ability to direct the IR source beam external to the instrument. This flexibility allowed modifications, redesign, and production of new spectrometers with optimal geometry for coupling to IR microscopes. Since the IR microscope had its own onboard detector, it was a complete system, minus the IR source and data processing/computer system. With this new design, the microscope no longer occupied the sample compartment of the spectrophotometer, so that conventional methods could still be used to analyze macro samples.

Finally, another design has produced what may be considered the first true IR microspectrophotometer. In this system, the IR spectrophotometer and microscope are no longer separate entities but are, instead, one unified body (IRμS; see Suppliers, Spectra-Tech, Inc.; Fig. 4.18). This apparatus is not capable of macrosampling but, rather, has been designed to maximize the information and sensitivity obtainable on microscale samples. Samples can be examined with polarized light and digital imaging techniques, in addition to IR microspectroscopy.

Microspectrophotometer capabilities

All IR microspectrophotometers have some type of visible-light imaging available. Indeed, this aspect is one of three that separates the IR microspectrophotometer from the IR beam condenser accessory. Imaging capabilities range from stereomicroscope viewing to that of a research-grade

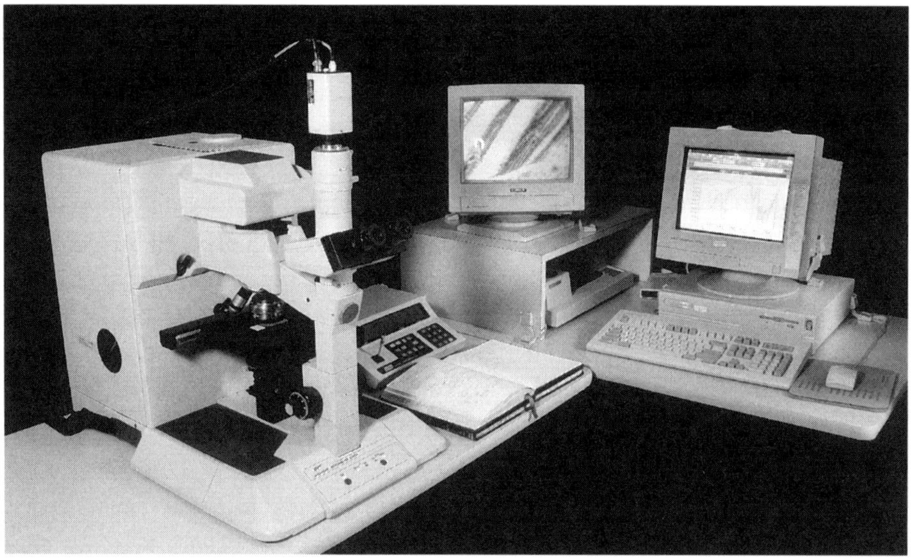

Figure 4.18
An IRμS IR microspectrophotometer (Spectra-Tech, Inc.). The instrumentation contains a source, interferometer, computer-controlled sample stage, and detector in one unified body, shown on the left. Also shown are a video camera, monitor, and data processing computer.

optical microscope. In addition, most IR microspectrometers provide photomicroscopy and/or videomicroscopy.

A second characteristic of the IR microspectrometer is the ability to isolate a particular area of the sample optically by the use of movable apertures. Apertures may be a fixed-diameter-circle or a variable-circle iris, or an adjustable, knife-edge rectangle. For heterogeneous samples, this capability means the IR microspectrometer is able to achieve both sample isolation and component identification.

Finally, all IR microspectrophotometers have the ability to collect IR reflectance spectra. Reflection IR is useful for highly absorbing samples that do not transmit IR well, as well as for providing information on the surface composition of a material. The resulting spectrum is often a complex mixture of specular and diffuse reflection components and must be interpreted with care. The ability to perform reflection analysis increases the versatility of the IR microspectrophotometer.

Particle and fiber analysis

Many of the techniques previously described in this chapter can be applied to IR microspectroscopy, and, as with macroanalysis, the selection of the technique depends on the characteristics of the analyte. Most materials are heterogeneous; however, to facilitate examination of typical analytical variables, this discussion will be limited to simple samples with homogeneous analysis areas.

The most common analyses performed by the IR microspectrophotometer involve the determination of particle and fiber composition. In the field of IR microspectroscopy, a particle is simply defined as a piece of material of suitable size for IR microspectroscopic analysis. The size suitable for analysis includes pieces as small as 10 μm in diameter, up to sizes on the order of 1–2 mm.

The initial examination and manipulation of a particle is done under a stereomicroscope, usually with the sample on a glass microscope slide. A necessary component of any IR microspectroscopic

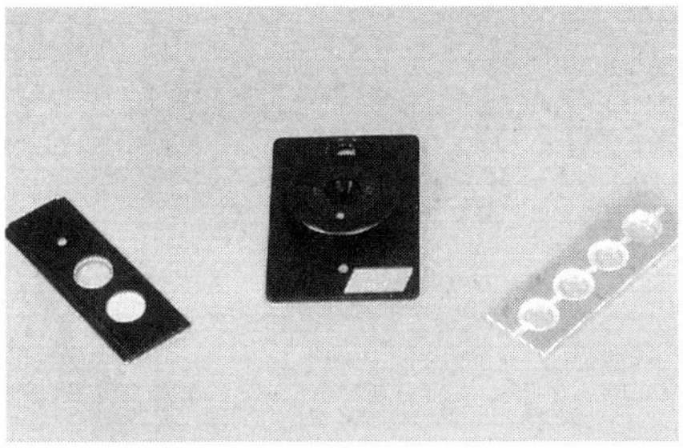

Figure 4.19
Typical IR alkali halide pellets in IR microspectrometer holders and a compression cell.

laboratory, the stereomicroscope should have variable-magnification capability, such as ×10–×50 magnification. A stereomicroscope with the head mounted to a boom arm is a valuable tool that provides flexibility for a wide variety of samples.

For transmission analysis, the first step is to place the particle on the IR window material, such as a salt plate or a diamond cell (Fig. 4.19). Corresponding procedures can also be used for reflection-absorption measurements if the particle is placed on a polished metal plate. The most important tools for particle manipulation are the probe and the forceps. Each of these tools is available in a variety of shapes and sizes from a host of suppliers. The selection of the appropriate probe or forceps is a matter of taste. Some IR microspectroscopists suggest very fine tools for the manipulation of small particles. However, if the probe is too fine, it can damage the particle. A particle can be transferred by touching it with the probe while observing it under the stereomicroscope. The probe is lifted slightly, and with the free hand, the IR window is brought into view under the stereomicroscope. The probe is lowered until it touches the surface of the window, then gently turned to release the particle. If the probe and particle are kept under the microscope, there is less chance that a valuable sample will be lost. Once the sample is on the analysis window, it is usually flattened with a probe, roller, or another window (e.g., a diamond cell).

For efficiency, multiple samples can be prepared on one salt plate for IR transmission analysis. To prevent confusion, detailed drawings of each sample's position and shape should be made before the pellet is transferred to the IR microspectrophotometer. A label can be scratched into the salt pellet next to each sample. One good method is to prepare the samples as if they were on a clock face. Another system was developed by Hill; in this method, a numbered grid is scratched into the pellet (Fig. 4.20), and a sample is placed within a square (Hill 1993). With this procedure, the samples' positions can be easily recorded in the notebook; the position numbers are readily discernible under the IR microspectrophotometer, and chaos is not introduced if the pellet is rotated when transferred to the analysis stage.

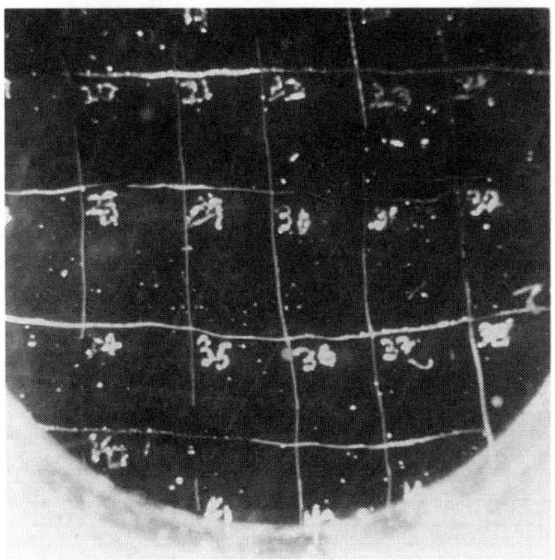

Figure 4.20
A photomicrograph (×20) of a labeled grid scratched into an NaCl pellet. This method helps locate multiple samples placed on one window.

Once prepared, the IR window with sample is placed on the analysis stage of the IR microspectrophotometer. Although the details of instrument adjustments and alignments vary depending on the manufacturer, certain operations are carried out in all IR microspectroscopic analyses. First, using the visible-light capability of the microscope, the particle is brought into focus and centered in the field of view. After this, the condenser should be checked to ensure that it is at the correct focal point for that particular sample and its substrate material. These focusing and centering steps are critical for accurate spectral results.

Apertures are used to isolate the particle, as shown in Figure 4.21. Rectangular, knife-edge apertures are the most common. When possible, the edges of the samples should not be included in the analysis area, since the curved edge of the sample can act as a lens and cause diffraction of the beam; this effect is significant in single-aperture microscopes (Messerschmidt 1988). Once isolated, the sample stage is moved

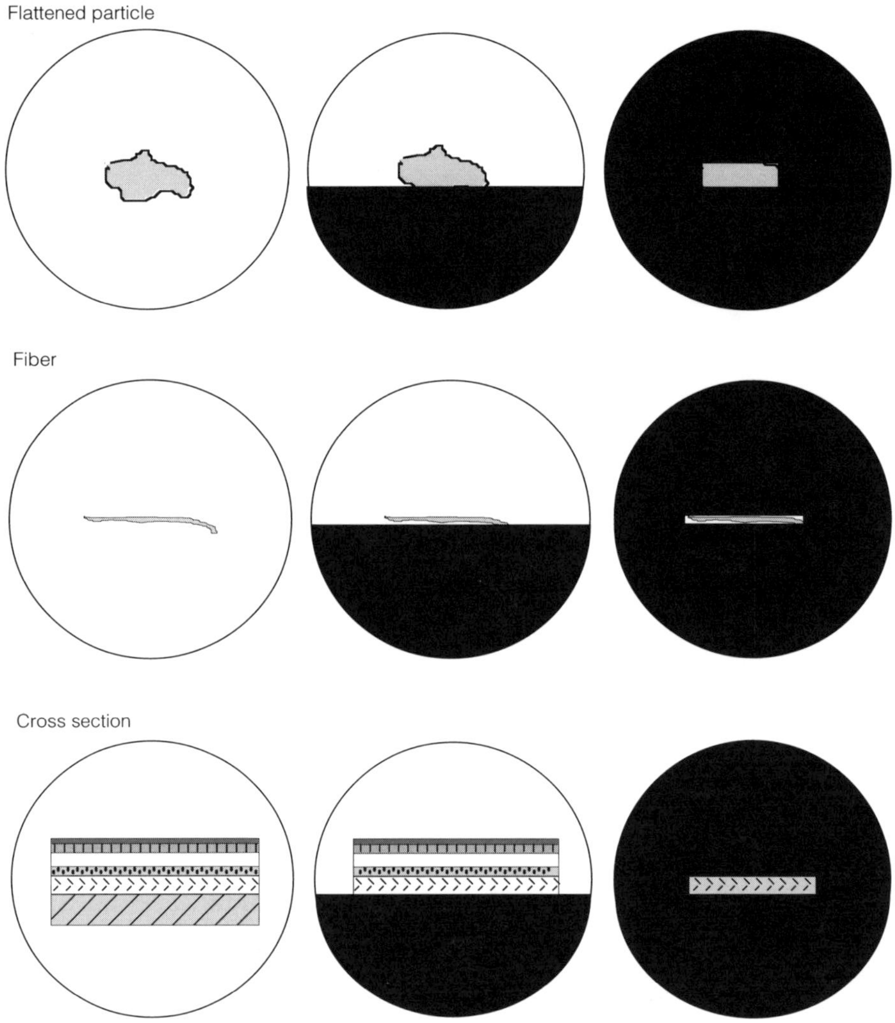

Figure 4.21
Rectangular aperture selection for a homogeneous particle, a fiber, and a cross section layer. Each of four knife-edge apertures is positioned to block the IR radiation from regions not selected for analysis.

to expose a portion of the IR window where no sample is present. A background single-beam spectrum is collected (Fig. 4.22). The stage is then moved back until the sample is brought back into view. The sample is scanned and ratioed to the background spectrum, and a transmittance (%T) versus cm^{-1} IR spectrum is produced. The isolation and analysis steps may be repeated for the analysis of various particles loaded on a single IR window. If the size of the aperture is changed from one particle to another, or if the particles are located far away from one another on the window, it is advisable to collect a new background single-beam spectrum before scanning the next particle.

Fiber analysis is handled in much the same manner. The transfer of the fiber to the window can be achieved with either the probe or forceps. Knife-edge apertures can also be used to isolate the fiber, as shown in Figure 4.21. Since many fibers are less than 40 μm in width, the aperture is usually elongated along the axis of the fiber. This elongation yields a greater sampling area, which in turn produces a better signal-to-noise ratio in the measurement.

As the size of the aperture in the sample stage (x–y) plane increases, the higher energy throughput for analysis of the sample is

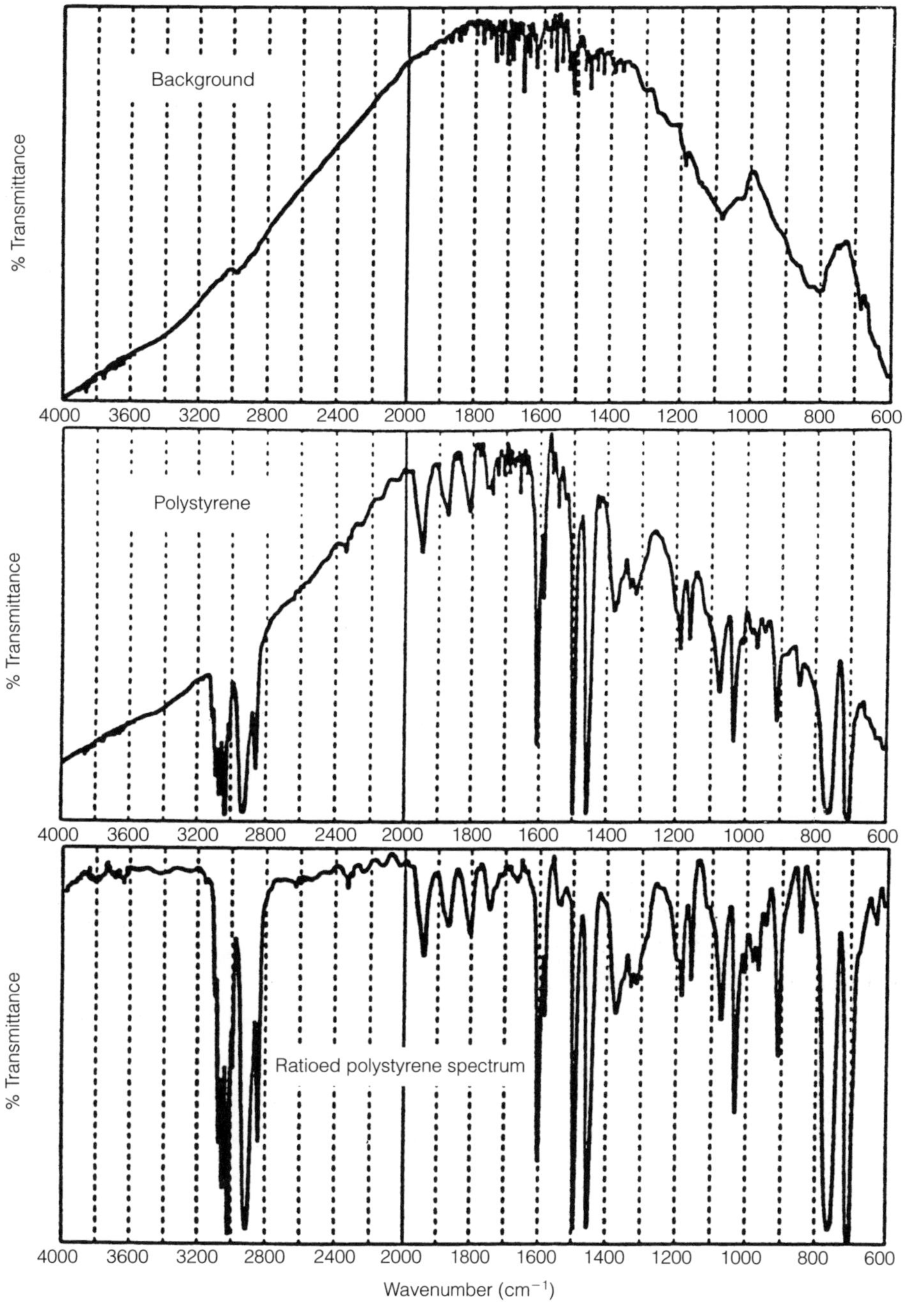

Figure 4.22

A single-beam background spectrum, a single-beam polystyrene spectrum, and the resultant ratioed polystyrene spectrum.

reflected in the magnitude of the signal seen at the detector. Figure 4.23 illustrates this effect. The top spectrum is the result of measuring a 20 μm width of a sample 100 μm in length. Below is the spectrum for the same sample analyzed with a larger aperture measuring 100 × 100 μm. A larger aperture area yields a spectrum with less noise for the same resolution and sampling time.

Diffraction effects become particularly important when apertures are employed. If light is passed through a small slit, the light rays may bend, as shown in Figure 4.24. This process is referred to as diffraction. The top drawing shows the diffraction effect for the case in which the slit size, x–y, is much greater than the wavelength of the radiation, λ.

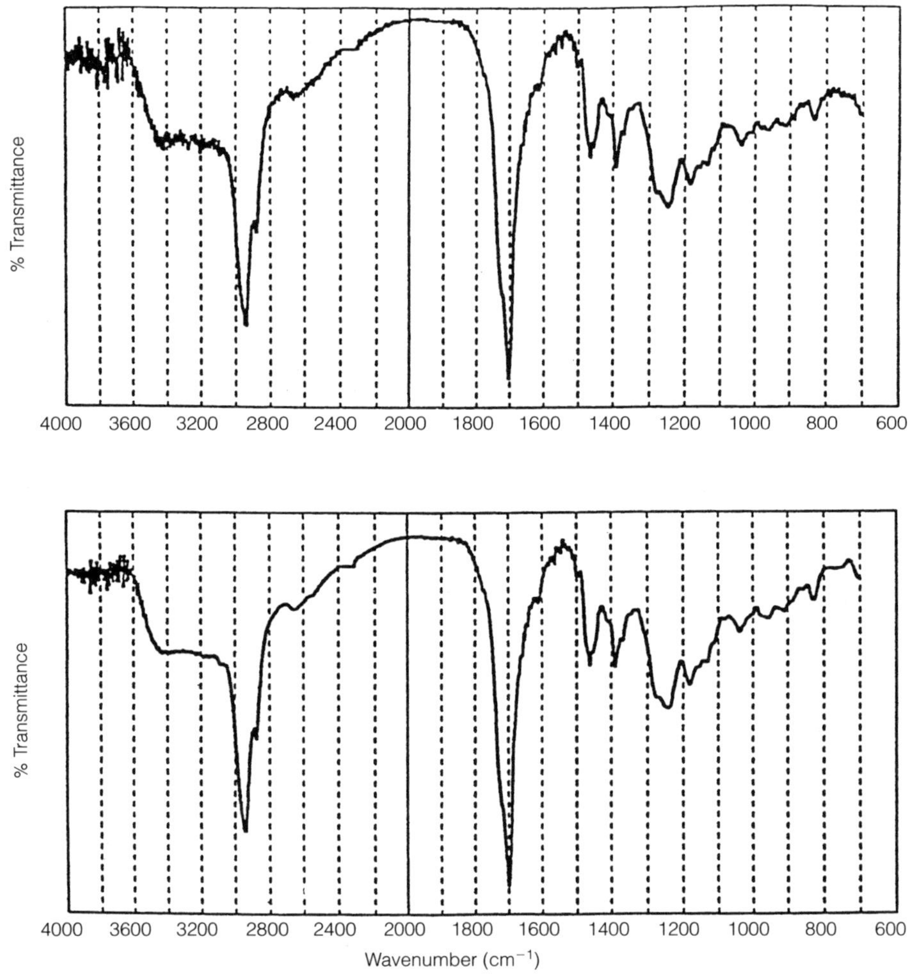

Figure 4.23
Two spectra showing the effects of a varying aperture area for the same sample, which is greater than 100 × 100 μm. The top spectrum was collected with an aperture of 20 × 100 μm. The bottom spectrum was collected with an aperture of 100 × 100 μm. Note the difference in noise. It is best to use the largest aperture that the sample and instrument will allow.

Note that the diffraction is small. But as the slit size approaches the wavelength of the radiation, as in the bottom drawing, the diffraction becomes more pronounced. Thus, diffraction is a limiting factor that relates to the minimum aperture size used in the IR microspectroscopic experiment. The longest wavelengths of IR light used in the typical mid-IR experiment are on the order of 10 μm. Therefore, this figure represents the resolution limit achievable in IR microspectroscopy. Apertures smaller than this will yield sample information from outside the region of interest, even though such areas are not visible through the aperture. Recent experiments at the Getty Conservation Institute found that a better rule of thumb is to employ apertures no smaller than 20 μm in width. This guideline ensures good quality in the resulting IR spectrum.

It is important to differentiate between the width of a sample, as described above, and its thickness. The thickness of the sample (z direction) is one factor that determines the absorbance of the material. Figure 4.25 shows two spectra of the same sample at two different thicknesses but measured with the same parameters. The top spectrum is for the sample when it is too thick. Because some bands are totally absorbing, the usefulness of the IR spectrum is reduced, and it is a poor candidate for any mathematical routine, such as spectral searching or subtraction. The bottom spectrum was collected after the sample was thinned. This time, all absorption bands are on scale. A good rule of thumb is that the %T for the most intense band should not be less than 10%T (or greater than 1.0 absorbance units). Appropriate sample thicknesses for use in IR analysis range from 1 μm to 20 μm, depending on the molar absorptivity of the compound of interest.

In terms of particle and fiber analysis, flattening of the sample may be required to obtain the desired sample thickness. Several options are available. First, the sample can be flattened with a metal probe or roller blade. Flattening can be done on a glass or metal surface prior to placement of the sample on the window; however, this procedure can make the sample fragile and difficult to transfer. The sample can also be flattened directly on the window or diamond cell. The advantage of pressing the sample directly on the window is that the sample is more likely to lay flat on the surface. For a good spectrum, it is important that the sample be flat and in focus; there should also be no air between the sample and its substrate.

A second method of flattening the sample is to place another transparent window on top. Commercially available compression cells

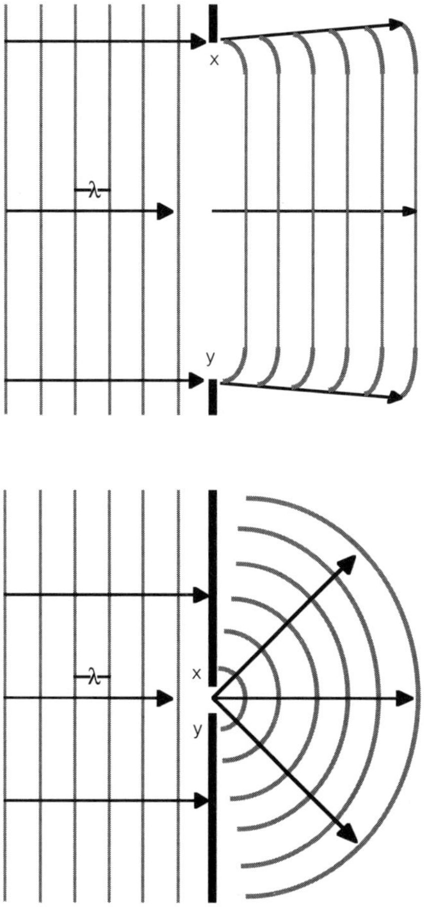

Figure 4.24
The effects of aperture slit size on diffraction. The top illustration shows that diffraction is minor when the aperture slit (x–y) is wide. The lower illustration shows that when the aperture slit (x–y) is narrow, near the wavelength of radiation (λ), diffraction is significant.

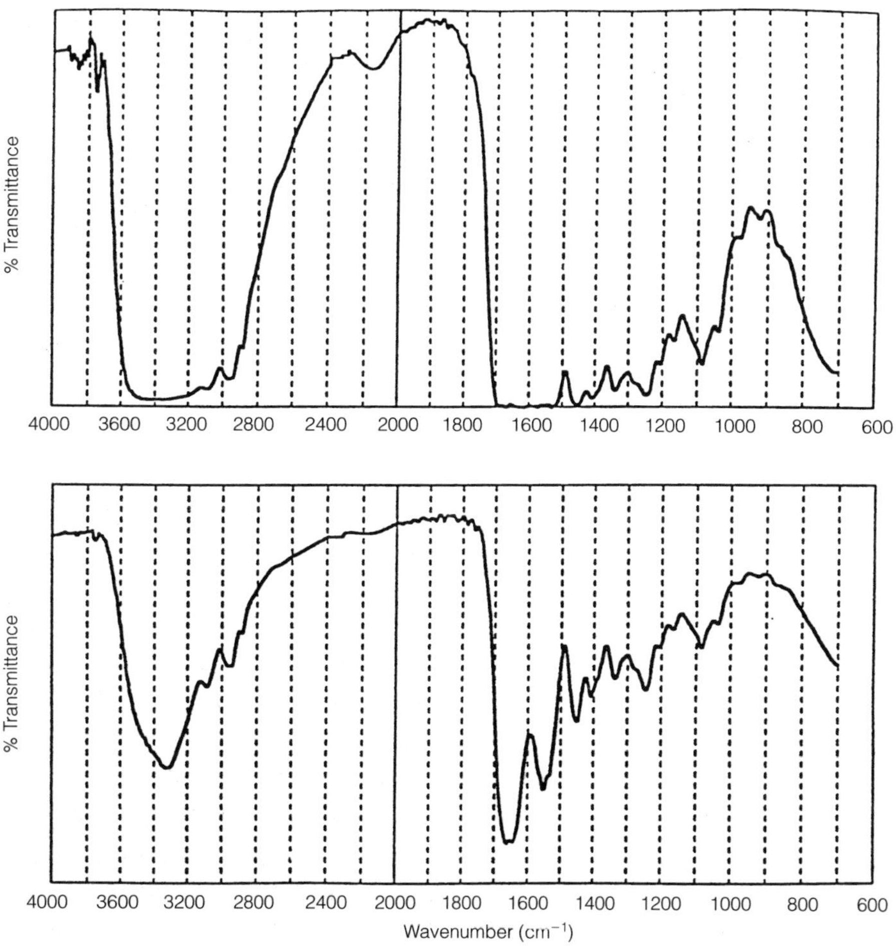

Figure 4.25
Two spectra showing the effects of sample thickness. The top spectrum is of a sample of glue that is too thick; some absorption bands are totally absorbing. The bottom spectrum was collected from a thinner area of the same sample; all absorption bands are on scale.

with alkali halide plates or diamond windows can be used to carry out this operation with good results. A nice by-product of compressing the sample is the corresponding increase in sample area. Placing a window on top of the sample is also an important step if the sample—such as a fiber—is not flat; a curved upper surface will bend the light and decrease the energy transmitted to the detector. When two IR windows are used, however, the increase in total thickness will also decrease the energy reaching the detector and may change the low-wavenumber cutoff point for the experiment. If this is a problem, two 1 mm windows, instead of the normal 2 mm windows, may be used. Alternatively, once compressed, a sample, unless it is elastomeric, will normally remain flat. Therefore, the upper compression window can be removed from the cell and the analysis done with the sample on a single window.

In either the compression cell or the diamond cell, a background spectrum is collected in an area adjacent to the sample, with the beam passing through the upper and lower windows. Because of the thickness of the sample, these windows typically have a small air space between them that can produce interference fringes in the background (Sommer and Katon 1988; Reffner and Martoglio 1995; see also Fig. 4.10). To eliminate these fringes, a small cube of salt (NaCl, KBr, BaF_2, etc.) is placed next to the sample prior to compression. The background spectrum is then collected through the pressed-salt region, and the air gap and its ensuing fringes are eliminated.

For heterogeneous particles and fibers, visually different regions of the sample can be isolated and analyzed. This approach provides a convenient separation technique without the need for additional sample preparation steps requiring time and energy.

Cross section analysis

Many samples encountered in art conservation are composed of multiple layers. The obvious example is a paint chip from a painting. One approach to handling this type of sample involves manual separation of the layers for analysis. The sample is placed under the stereomicroscope. After some observation and identification of the structure, the analyst picks off representative particles from each of the layers. The particles removed are then analyzed as described above. This technique works well with samples that have easily identifiable layers. The major drawbacks to this technique include the time and care needed to perform the separation. In addition, removal of a single particle, particularly from hard samples, may be difficult.

An alternative approach is to embed the sample in an appropriate medium and then to microtome it into thin cross sections. Once the mold, containing sample and medium, has hardened, it can be mounted in a microtome and sliced into thin sections for sequential analysis of layers under the IR microspectrophotometer.

The analysis of the thin cross section is very similar to that of particles and fibers, in that apertures are used to isolate the area of interest. In the case of a multilayer sample, the apertures are used to isolate a single layer within the sample. This procedure is illustrated in Figure

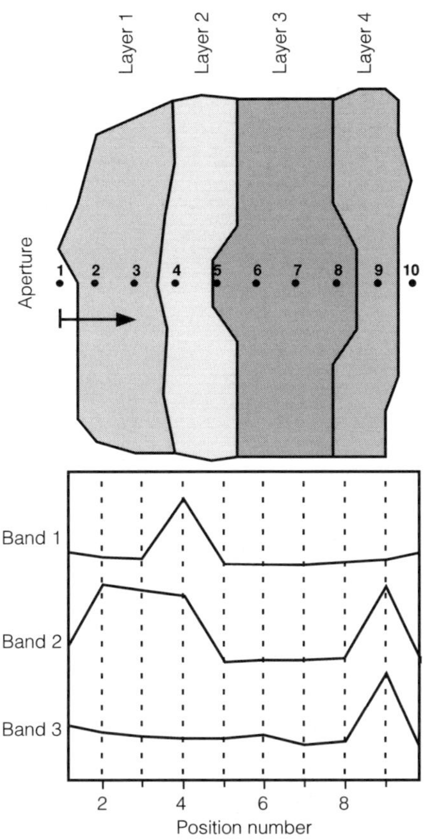

Figure 4.26

The IR linear mapping technique. The top illustration shows a set of spectral collection positions selected for the linear mapping of a multilayer material. The resolution of the map is affected by the density of the collection sites and the size of the aperture. For this example, ten spectra were collected. After the spectra are examined, a few specific absorption bands are selected, and their intensity is plotted versus their collection position on the sample to produce a line map, such as the bottom figure. Of the absorption bands plotted, only band 2 has an increased intensity from a component in the first layer. Both absorption bands 1 and 2 appear in the spectrum for the second sample layer, while none of the plotted absorption bands occur in the spectrum for layer 3. The last and fourth layer contains absorption bands 2 and 3 in its spectrum.

4.21. After the IR spectrum for one layer is collected, another layer can be isolated and its spectrum collected. In this way, the composition of each layer can be determined.

Mapping studies

Molecular mapping, also known as functional group imaging, involves the systematic study of the structure of a material of interest; molecular mapping can be done in either transmission or reflection microspectroscopy (Harthcock and Atkin 1988; Reffner 1989; Krishnan, Powell, and Hill 1995). Molecular mapping with an IR microspectrophotometer is a technique complementary to elemental mapping as achieved in a scanning electron microscope, and it may often be performed on the same sample.

In its simplest form, linear mapping is conducted as follows: (1) An aperture of an appropriate size is made, and a background spectrum is collected in a nonsample area. (2) With the sample stage, the edge of the cross section is brought into view through the aperture, and the spectrum is collected. (3) The sample stage is then moved in one direction by a given amount to reveal an adjacent area through the aperture; once again, the spectrum is obtained. (4) The process is repeated until the entire cross section has been analyzed. This procedure, known as IR linear mapping microspectroscopy, is illustrated in Figure 4.26. On many instruments, this routine can be automated by the use of a mapping stage that is operated either by a controller box or by software from the spectrometer computer. Commercially available mapping stage accessories fit many IR microspectrometers. Commercial software is available to help display the information obtained from such an experiment (see Suppliers, Galactic Industries Corp.; Spectra-Tech, Inc.; Bio-Rad Digilab).

An area mapping experiment is done to yield information about an area of interest; it may be done as a transmitted light experiment on thin sections or as a reflected light experiment on embedded, polished paint cross sections. In either case the sample is placed on the sample stage of the IR microspectrophotometer, and an analysis grid containing several hundred points (e.g., 20×30 grid $= 600$ points) is selected for the area of interest on the sample. An array of spectra is produced by the collection of a single spectrum as the aperture moves to each point on the grid (Figs. 4.27, 4.28). The effective resolution of the components in the sample is determined by the size of the analysis aperture and the density of the grid. The size of the aperture for the spectra in Figure 4.28 was 40×20 µm (x-, y-axes, respectively). The selection of appropriate step size and aperture size involves a trade-off between resolution, energy throughput, and the length of the experiment. In this case, a step size of approximately 20 µm gives an overlap of the windows in the x direction. Overlapping apertures, while not necessary for complete sample imaging, do provide an effective increase in the resolution of the components. The collection and processing of the 50 scan spectrum at each grid point requires about 1 minute. Therefore, a mapping experiment with 150 grid points, such as shown in Figure 4.27, would require about 2.5 hours.

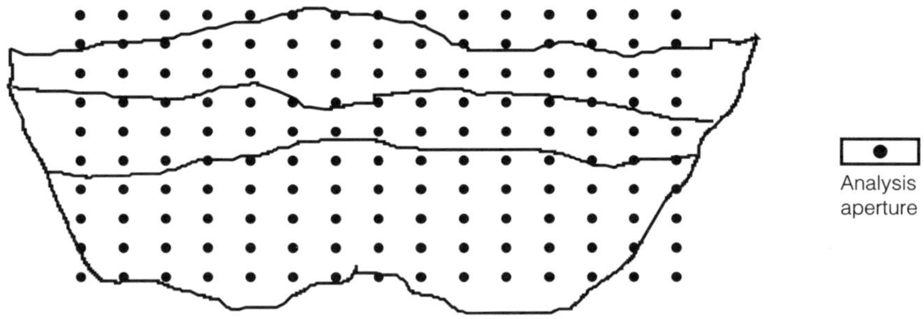

Figure 4.27
An example of grid selection for an IR area map. The dots designate x–y stage locations for the center of an analysis aperture (example on right). The resolution of the components in the sample is directly related to the size of the aperture window and to the density of the grid.

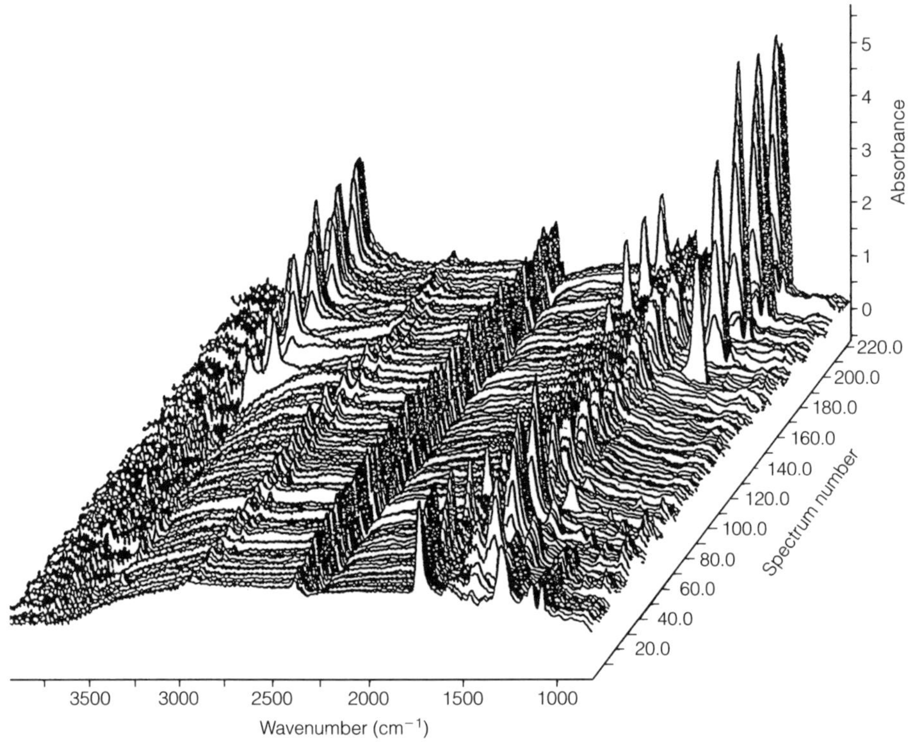

Figure 4.28
A series of spectra collected by computer from an area grid, such as shown in Figure 4.27. The computer stores the collection position of each spectrum to use later for contour plots.

From the array of spectra, contour maps may be produced that provide information on the concentration and location of compounds related to various functional groups in the sample (Fig. 4.29). This is done by selecting an absorption band of interest, such as a hydrocarbon band, and plotting its intensity versus its position in the grid where it was collected. This procedure produces an area map of the intensities of that specific band plotted as a contour map, where lines connect the areas of similar value. In these plots, line thickness has been used to represent the changes in intensity of the absorption bands. The thickest lines correspond to the areas of strongest band intensity—that is, the highest concentration in the sample—and the thinnest lines represent the lowest intensities. The intensities are relative to one another, and the background intensity may not be zero, depending on the baseline in the region.

The method of IR mapping shows potential for the determination of materials and their locations within a cross section sample. The technique is complementary to elemental mapping with an electron

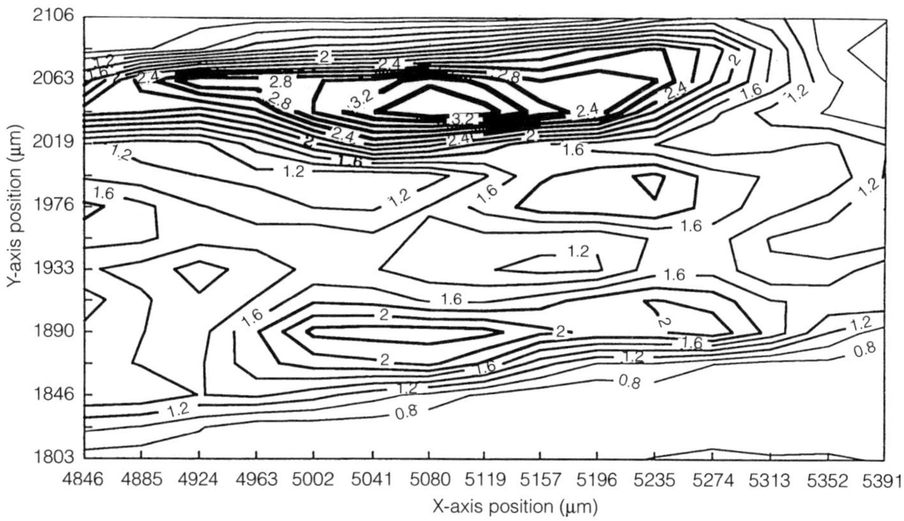

Figure 4.29

A contour map showing the intensity of the spectral carbonate band (1424 cm^{-1}) in relation to its position on the grid where it was collected. To plot the data as a contour map, areas of similar absorbance intensity are connected with a line. The heavier the line, the higher the concentration. (Paint cross section from *Allegory of Fortune* by Dosso Dossi, J. Paul Getty Museum [×67].)

microscope and may be done on the same sample. At this point, the method has two major limitations. The first is its dependence on the use of specular reflection as an analysis method; this dependence places a minimum size of 20 × 20 µm on the analysis window because of energy restrictions. Specular reflection can also result in band distortions and shifts, for which it is difficult to compensate. The second limitation is that a material cannot be reliably identified based on one IR absorption band. Thus, additional analyses are required to supplement the map and provide interpretation. Future computer programs should allow the creation of a map based on the selection of multiple absorption bands that can help identify specific compounds.

Microspectrophotometer accessories

Several accessories are available for IR microspectrometers (Reffner, Wihlborg, and Strand 1991). One is the grazing-angle objective (see Suppliers, Spectra-Tech, Inc.), which is used for low-angle reflection measurements of very thin films on polished metal surfaces, as well as for the detection of monomolecular surface films. Another useful accessory is the ATR objective (see Suppliers, Spectra-Tech, Inc.; Bio-Rad Digilab; Nicolet Instrument Corp.; Bruker Instruments; Graseby Specac; Bomem International), which may be used for internal reflection microspectroscopy to examine small areas of nonreflective materials nondestructively (Fig. 4.30). Chess provides a good paper on the use of these two objectives (Chess 1995).

ATR microspectroscopy, as with the macro version, is a surface analysis technique that requires little to no sample preparation. It has been shown to be very useful for the analysis of single fibers, hair and hair coatings, ink on paper, and paint chips (Bartick, Tungol, and Reffner 1994). Different crystals (internal reflection elements) may be selected, depending on the type of sample, its hardness, and its refractive index. ZnSe, germanium, silicon, and diamond elements are available.

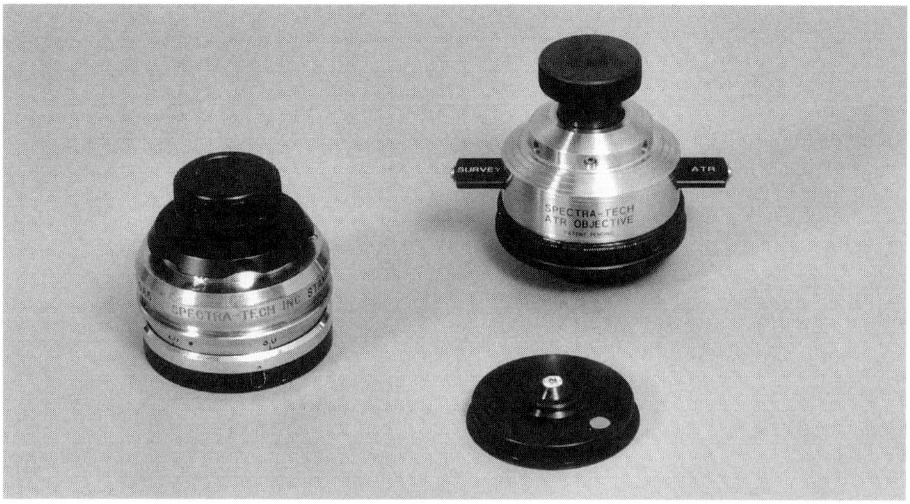

Figure 4.30
A Cassegrain reflective optics objective (left) and a micro ATR objective (right) for use on an IR microspectrophotometer.

Summary

One of the most versatile and important aspects of IR spectroscopy is that samples of any form or composition can be analyzed. Selecting the optimum method for the analysis, however, is not always easy; many sample preparation methods exist, and the analysis technique chosen often impacts the spectral results. Transmission analysis methods have become the de facto standard, because they were the earliest and most widely used techniques. Most reference spectra are generated by transmission methods.

While there is no one technique appropriate for the analysis of every sample, some methods can be used for most sample types. Many can be prepared as thin or compressed films on transparent windows, especially with the aid of a compression cell that uses alkali halide or diamond windows. The use of a beam condenser or microscope is also very beneficial to the analysis of the small samples often encountered in the art conservation field. Sample analysis techniques for IR microspectroscopy include mapping experiments used to characterize the composition and position of constituents in nonhomogeneous samples.

Additional Reading

Infrared instrumentation

Ferraro, J. R., and L. J. Basile
1978, 1979, 1982, 1985. *Fourier Transform Infrared Spectroscopy*. Vols. 1–4. New York: Academic Press.

Griffiths, P. R., and J. A. de Haseth
1986. *Fourier Transform Infrared Spectrometry*. New York: John Wiley and Sons.

Koenig, J. L.
1992. *Spectroscopy of Polymers*. Washington, D.C.: American Chemical Society.

Low, M. J. D., and N. S. Baer
1977. Application of infrared Fourier transform spectroscopy to problems in conservation. *Studies in Conservation* 22:116.

Reffner, J. A., J. P. Coates, and R. G. Messerschmidt
1987. Chemical microscopy with FTIR microspectrometry. *American Laboratory* (April):86.

Smith, A. L.
1979. *Applied Infrared Spectroscopy: Fundamentals, Techniques, and Analytical Problem Solving*. New York: John Wiley and Sons.

Willard, H. H., L. L. Merritt, Jr., and J. A. Dean
1984. *Instrumental Methods of Analysis*. New York: D. Van Nostrand Co.

Analysis methods

ASTM Committee
1979. *Manual on Practices in Molecular Spectroscopy*. Philadelphia: American Society for Testing Materials.

Chicago Society for Coating Technology
1980. *Infrared Spectroscopy Atlas for the Coatings Industry.* Philadelphia: Federation of Societies for Coating Technology.

Griffiths, P. R., and J. A. de Haseth
1986. *Fourier Transform Infrared Spectrometry.* New York: John Wiley and Sons.

Harrick, N. J.
1979. *Internal Reflection Spectroscopy.* Ossining, N.Y.: Harrick Scientific Corp.

Harrick Scientific Corp.
1987. *Optical Spectroscopy: Sampling Techniques Manual.* Ossining, N.Y.: Harrick Scientific Corp.

Humecki, H. J., ed.
1995. *Practical Guide to Infrared Microspectroscopy*. New York: Marcel Dekker.

Kortum, G.
1969. *Reflectance Spectroscopy*. New York: Springer-Verlag.

Messerschmidt, R. G., and M. A. Harthcock, eds.
1988. *Infrared Microspectroscopy: Theory and Applications*. New York: Marcel Dekker.

Roush, P. B., ed.
1987. *The Design, Sample Handling, and Applications of Infrared Microscopes*. Philadelphia: American Society for Testing Materials.

Smith, A. L.
1979. *Applied Infrared Spectroscopy: Fundamentals, Techniques, and Analytical Problem Solving*. New York: John Wiley and Sons.

Wendlandt, W. W., and H. G. Hecht
1969. *Reflectance Spectroscopy*. New York: Wiley-Interscience.

Wilks, P. A.
1965. *Internal Reflection Spectroscopy*. Norwalk, Conn.: Wilks Scientific Corp.

Chapter 5

Spectral Interpretation

Qualitative IR spectroscopy is a valuable analytical tool that allows for the identification of organic and inorganic materials. Each compound's IR spectrum contains a substantial amount of information. This information, along with some patience, skill, and knowledge about a sample's background, can be used to determine molecular structures successfully, as well as to characterize unknown materials.

This chapter covers general methods for spectral interpretation and then concentrates on specifics that apply to natural organic materials, colorants, and mixtures often found in works of art. Spectra are discussed in terms of their important features. Also included are factors that may affect the interpretation of an IR spectrum, such as its generation, presentation, and processing. Some special considerations, cautions, and limitations of IR analysis will be presented in terms of spectral interpretation.

Infrared Spectra

An IR spectrum displays detector response and is usually plotted in % transmittance (%T) or absorbance (A, which is the base 10 log of the reciprocal of T) versus IR frequency (in wavenumbers [cm^{-1}]). A frequency of radiation that interacts with the sample produces an absorption band that is characteristic of the energy required for a particular molecular group transition (usually a vibrational motion). The collective position and pattern of these absorption bands designate the combination of molecular groups found in any specific compound.

Absorption bands

The absorption bands in a recorded IR spectrum exhibit three important parameters: frequency, shape, and intensity. These band attributes are unique for each individual molecule or material.

Band frequency

The band positions, or frequencies, indicate the presence of certain functional groups in a material. Band assignment in the functional group region (4000–1500 cm^{-1}) is usually straightforward, while assigning a band to a specific functional group in the fingerprint region (1500–500 cm^{-1}) may be difficult, since many types of functional groups absorb at similar wavenumbers in this region. Identification of a material using the fingerprint region is based on the correlation between the peak pattern of the sample and the peak pattern of a standard material of known chemical composition. Many aids, some of which are discussed in this chapter, assist the spectroscopist in the determination of molecular structures based on band frequencies.

Band shape

The shapes of absorption bands provide information concerning the group functionality as well as the material purity. All single absorption bands are, by nature, symmetrical in shape, resembling a normal bell curve, with a peak maximum and equal wings on each side. Deviations in band symmetry, such as a slight shoulder or an unusual tail, indicate the presence of an overlapping band. In a complex molecule, deviations in band shape may be related to similar functional groups that exist in different molecular environments. The presence of asymmetric bands can also indicate that a sample is a mixture or has been modified, such as by oxidation. Theoretically, a normal band shape is very sharp. The width of a band can be increased by inter- and intramolecular interactions, as well as by overlapping bands. Broad bands, such as those due to hydrogen bonding or ionic functional groups, are very characteristic. Doublet bands, often produced by crystalline lattice effects, are also useful in characterizing a material.

Band intensity

The relative intensity of a band, in comparison to the other bands in the spectrum, provides information on the amount and type of a specific functional group present in a molecule. Functional groups that are responsible for a large change in the dipole moment of a molecule—carbonyl groups, for example, will produce very intense absorption bands. Of the three band attributes, band intensities are most likely to show deviations related to sample preparation.

These three attributes define each band in a spectrum. Any two identical samples prepared under identical conditions will produce identical spectra. Any deviation—in band frequency, shape, or intensity—between the two spectra indicates a difference between the two samples. While differences may be due to sample composition, they may also be due to sample preparation, spectral plotting format, or instrument parameters. While the last three factors do not change the molecular vibrations in the sample, they can cause superficial changes in the character of the spectra that need to be accounted for in the spectrum interpretation. Every good spectrum, especially a reference spectrum, should include information on conditions of analysis and instrumental parameters; this information allows the spectroscopist to make a knowledgeable comparison.

Plotting format

The interpretation of an IR spectrum is based directly on visual inspection of the spectrum, on recognition of characteristic features, and on comparison of the spectrum with reference spectra. Visual cuing is directly dependent on the plot of the spectrum. Several types of spectral presentation formats may be found in journal articles, reference books, and spectral libraries. To the novice spectroscopist, these may seem confusing and not comparable. But in fact, the spectra are consistent and reliable and, with a little practice and an understanding of the various plotting formats, they can be visually compared. Examination of the axis labels will quickly provide the mode and scales of the plot selected in any particular case.

Figure 5.1 displays the same IR spectrum of linseed oil in four commonly used plotting formats. Close examination of each plot may be needed in order to understand that, indeed, these are all plots of the same material. The most noticeable difference among these four spectra is that, in some instances, the bands or peaks descend (transmit-

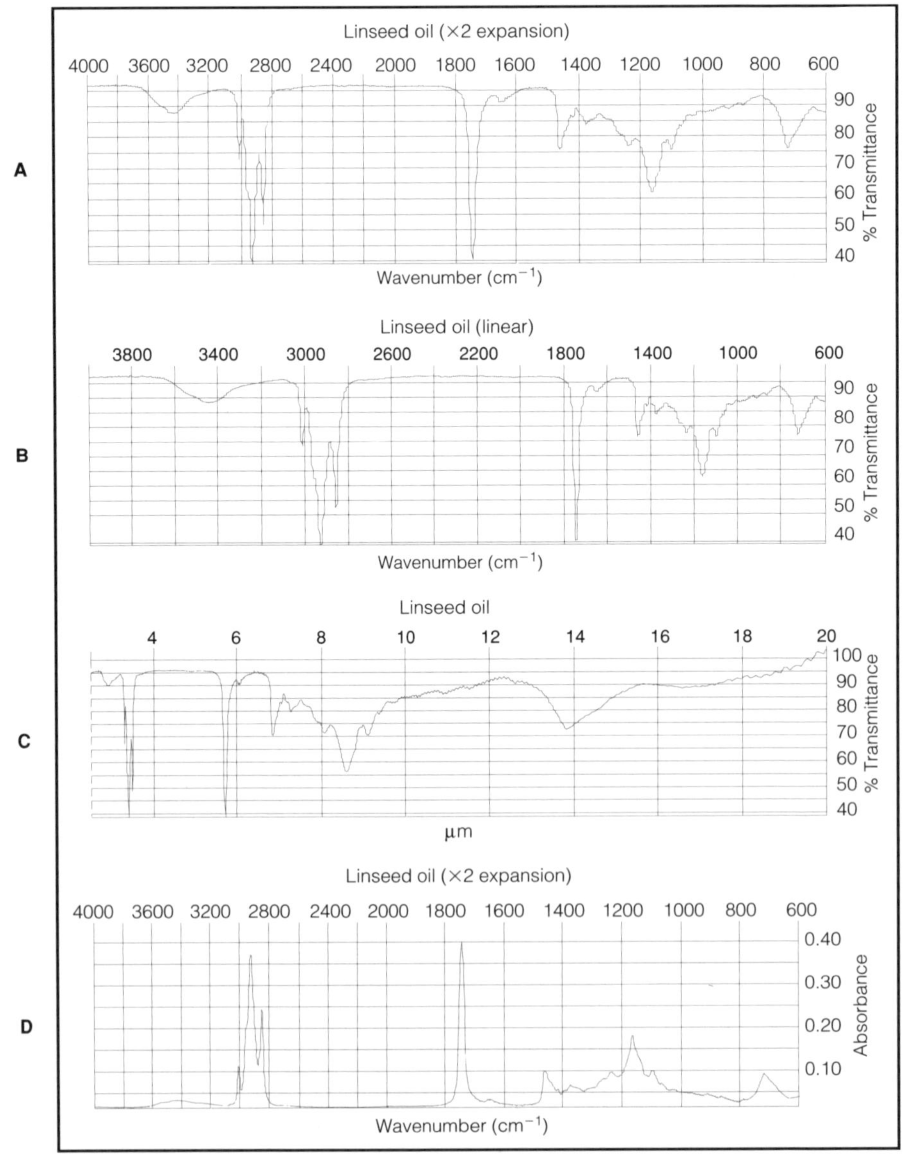

Figure 5.1
The same spectrum of linseed oil shown in four different plot formats. The format of plot A is generally considered standard. The format in plot C is no longer commonly used.

tance, Fig. 5.1, spectra A–C), and in some they ascend (absorbance, Fig. 5.1, spectrum D).

The absorbance mode is used for quantitative analysis and other mathematical processing. According to Beer's law, the intensity of each band is directly proportional to the concentration of the absorbing species—such that there is a linear relationship between band intensity and concentration. This linear relationship applies only to spectra in absorbance mode; in transmittance mode, there is a logarithmic relationship between band intensity and species concentration. Besides quantitative analysis, other computer spectral manipulation methods—such as subtractions, transformations (Kubelka-Munk, Kramers-Kronig), deconvolutions, and spectral searches—are done with the spectra in absorbance. This may be a hidden process, since computers can instantaneously switch between absorbance and transmittance. When one mode of display, such as transmittance, is selected for the spectrum, the computer will perform any conversion necessary as part of the processing and quickly return the processed spectrum to the screen in its original mode (i.e., transmittance).

Spectra plotted in transmittance are shown in Figure 5.1 (A–C). This mode has been historically used for IR spectra and is still the most commonly seen format for reference spectra. A transmittance spectrum is the ratio of the radiant power transmitted by a sample to the amount of incident radiation on the sample. Prior to the use of Fourier transform IR spectrometers (FT-IRs), this was the value read directly from the detectors and plotted. Transmittance mode has an advantage for the viewing and comparison of spectra, because the logarithmic scale amplifies the intensity of the weak bands while keeping the highly absorbing bands on scale.

With the advent of the FT-IR and its integral computer, the operator may select any mode for viewing, plotting, and calculation. The computer can quickly and easily convert from one format to another. Since it will often be necessary to compare the hard copy (printed form) of spectra in different formats, however, it is important to become familiar with the visual and mental gymnastics required to convert between absorbance and transmittance. The scales should always be checked, because there are some spectra that have been plotted in transmission but then inverted, with the bands ascending; conversely, some spectra are plotted in absorbance units, with the peaks descending. This incorrect and confusing practice occasionally appears in publications.

Varying abscissa scales are also seen in IR spectral plots. Visible light, when measured in wavelengths, is typically reported in microns (μm). For IR spectra, the corresponding use of the linear scale of wavelengths in microns (2.5–16.6 μm; see Fig. 5.1, spectrum C) is a simple representation and easily related to the other familiar spectral measurements, such as UV/Vis, that are used in spectroscopy. Older, prism-type dispersive instruments produced plots with linear wavelengths; this practice is now considered obsolete.

The surge in the generation of IR spectral libraries occurred with the grating-type dispersive spectrophotometer (without computer).

The output of this dispersive spectrophotometer was linear in wavenumbers or frequency (cm^{-1}; see Fig. 5.1, spectrum B). Wavenumbers are reciprocally related to wavelength (e.g., 1000 cm^{-1} = 10 μm) and are considered more accurate for spectral presentation, since the energy of a vibrational mode is a function of frequency and not of wavelength.

In a linear wavenumber representation, however, the fingerprint region of the spectrum (2000–500 cm^{-1}) is compressed and difficult to see. Thus, it became a standard procedure to switch the speed of the chart recorder at 2000 cm^{-1} to a slower setting; this practice resulted in a ×2 scale expansion of the fingerprint region. Thus, the plots have a linear scale between 4000 and 2000 cm^{-1} and a double-size linear scale between 2000 and 500 cm^{-1}, as shown in Figure 5.1, spectra A and D. Plots that use a ×4 scale expansion may occasionally be found.

Because of the large number of reference spectra generated in the format shown in Figure 5.1, spectrum A (i.e., most published libraries and reference books), this format has been termed standard. The spectrum is plotted in transmittance, with a linear wavenumber (cm^{-1}) scale containing a ×2 expansion in the region of 2000–500 cm^{-1}.

Instrument configuration

Several possible options for instrument configurations exist. While these options do not change molecular absorption bands, they may change the appearance of the spectra. IR mercury-cadmium-telluride (MCT) detectors are available in narrow, medium, and wide ranges. The choice of the detector affects the sensitivity and the spectral range of a given spectrometer, with the narrow-range detector providing the highest sensitivity. Thus, the scale of a spectrum may range from 4000 to 700 cm^{-1}, or from 4000 to 500 cm^{-1}, depending on the type of detector deployed.

The resolution, or number of data points per spectrum, is selectable on FT-IR instruments. A higher-resolution spectrum will contain more data points and will thereby provide better representation and separation of the true absorption bands. To collect a higher-resolution spectrum, the instrument resolution setting must be changed to a lower wavenumber, since this parameter indicates the interval for data collection. For example, an instrument set at a resolution of 1 cm^{-1} will collect a data point every 1 cm^{-1}. Thus, for a spectrum collected over the range of 4000 to 800 cm^{-1}, there will be a total of 3200 data points. An instrument set at 4 cm^{-1} will only collect one-fourth as many, or 800 data points. Therefore, a resolution of 1 cm^{-1} is higher than a resolution of 4 cm^{-1}.

A commonly used resolution is 4 cm^{-1}. However, most instruments are capable of at least 1 cm^{-1} resolution, while many digitized library spectra were collected at 8 cm^{-1}. Spectra produced at the finer resolutions (i.e., 1 cm^{-1} or below) may contain a distinct band in an area that only appears as a shoulder in a spectrum collected at a lower resolution. Interest in studying the fine spectral details is one reason for using the higher resolutions, but a possible effect on the character of the spectrum should be kept in mind when spectra recorded at different resolutions are compared. Additionally, higher-resolution spectra require more time for data collection, and since their files are larger, they also require more disk storage space. Most search routines recommend that the resolution of the search spectrum stay within a factor of 2 of the resolution of the library spectra.

Qualitative Analysis

There is no one set of procedures for the identification of unknown materials by IR spectroscopy. Nor is it possible to identify every material by IR spectroscopy alone. However, with a good-quality spectrum, it is possible to screen a sample and to obtain the general classification of a material by its major functional groups. Additionally, and sometimes most important, it is possible to rule out several types of materials based on the absence of major functional groups. Thus, it is essential to understand the analytical question framed for each sample prior to spending a significant amount of time on the interpretation of a spectrum.

In general, the interpretation of a spectrum of an unknown material requires the identification of functional groups in the spectrum, along with direct comparison to reference spectra. Several useful tools (correlation charts, flow diagrams, computer search routines, etc.) are available to speed up the process. One possible, stepwise approach for the identification of an unknown spectrum is presented in Figure 5.2. A simple spectrum may be identified within a few steps, while a complex mixture may require multiple passes through the procedure.

Spectral quality

The first step in spectral interpretation is the determination of spectral quality. Many hours can be wasted interpreting a poor-quality spectrum that misrepresents the true absorption band positions, shapes, and intensities. Table 5.1 lists some requirements for a good IR spectrum. Most

Figure 5.2
An identification scheme for the systematic interpretation of IR spectra. Many aids are available to assist in interpretation. The final step for identification is always visual comparison of the sample spectrum to known reference spectra.

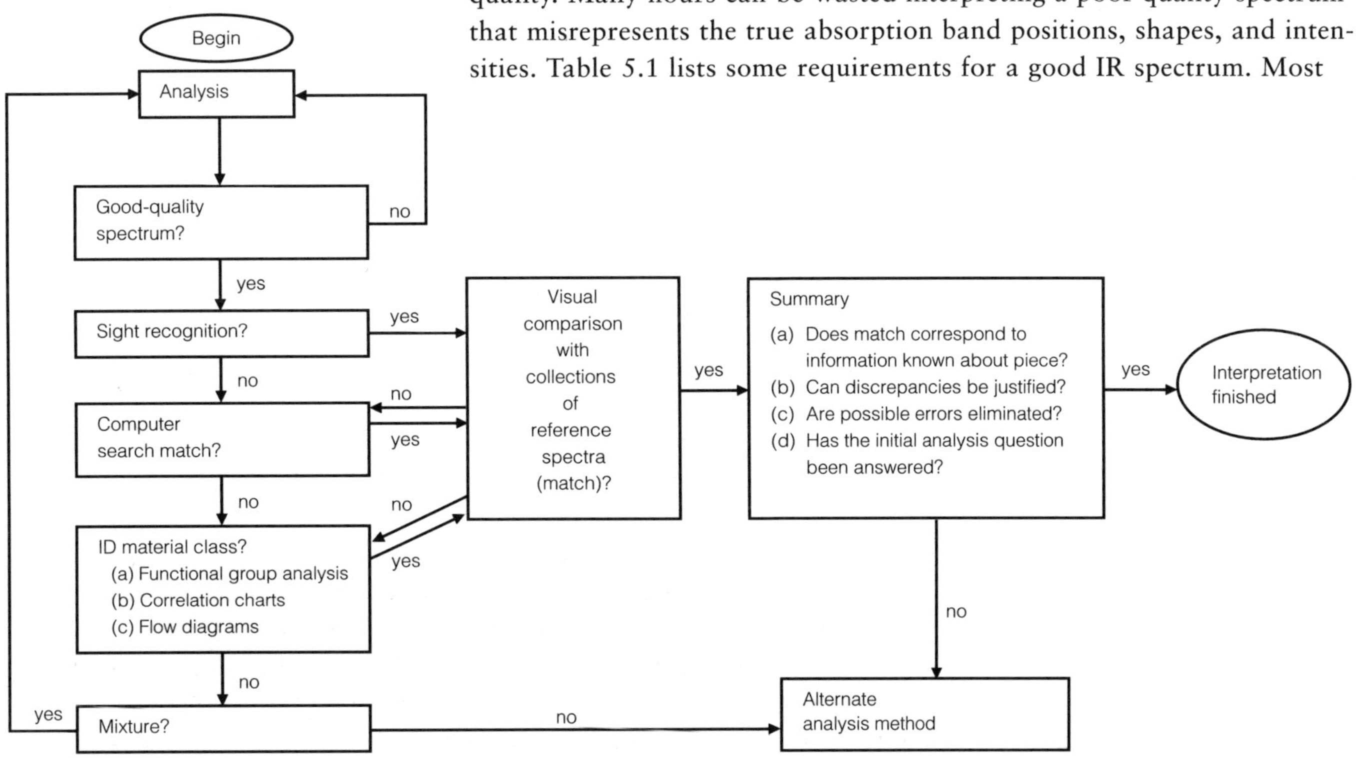

Table 5.1
Requirements for a good-quality IR spectrum.

Parameter	Specifications of a good-quality spectrum	Possible problems resulting in deviations from specifications
Baseline	flat and level, positioned near 100% transmittance (0.0 absorbance)	sample is poorly ground; sample surface is not flat; background does not match sample spectrum
Noise	maximum noise level no greater than 1% of strongest band	sample too small; insufficient number of scans collected
Intensity	signal no stronger than 10% transmittance (1.0 absorbance)	sample too concentrated; sample too thick
Extraneous bands	water vapor and carbon dioxide should be less than 2% transmittance (0.1 absorbance)	background does not match sample spectrum; instrument is poorly purged

poor spectra are the result of improper sample preparation techniques that may be corrected by preparing the sample again. Occasionally an alternate analysis method, such as reflection versus transmission, may be required to eliminate a problem. In cases in which the sample cannot be reanalyzed, computer routines may be used to manipulate spectra into a better form. Any alteration of spectra must be performed with caution—and only done as a last resort, as it has the potential to produce misleading results.

Examination of the spectrum baseline—that is, the portion that has no absorption bands—can provide information on spectral quality. One quality check for the spectrum is whether the baseline is flat, level, and positioned near 100%T (0.0 absorbance). Shifting of the baseline above or below the proper position occurs when the intensity of the background spectrum does not correspond to the sample spectrum. With IR microspectrophotometers, this shifting may occur if the aperture size for the background is different from that of the sample. The baseline can also be shifted when a sample is black, since all wavelengths will be strongly absorbed.

A sloping baseline is usually an indication that the radiation has been diffracted and/or scattered as it passed through the sample. This problem can occur if the particles in a potassium bromide (KBr) pellet are not ground finely enough or if the sample surface of a thin film on a window is not flat. Figure 5.3 shows a spectrum with a sloping baseline before and after the application of a computerized baseline correction routine. Such routines attempt to level the baseline to 100%T (0.0 absorbance). Linear and multiple-point correction curves are available. Any spectral alteration of the digitized data by baseline correction should be carefully examined for changes in the relative intensities of the absorption bands.

Examination of baseline noise level is also used as a quality check. The signal-to-noise ratio of a spectrum is calculated by comparing the intensity of the noise level at the baseline to the intensity of the

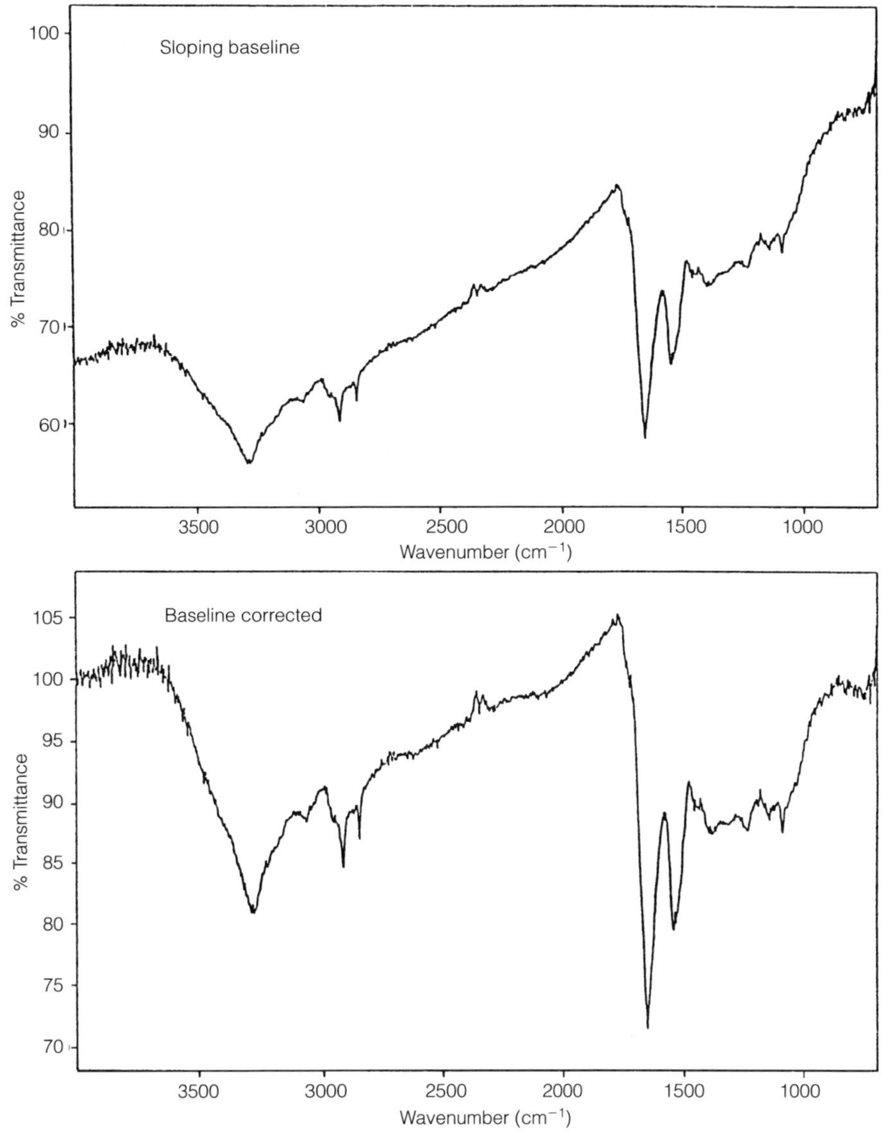

Figure 5.3
An IR transmittance spectrum for a protein before and after baseline correction. A sloped baseline is often caused by diffraction of light when it passes through the sample. Baseline correction routines attempt to level the baseline at 100% transmittance. Relative absorption band intensities may be altered by baseline correction.

strongest absorption band. This calculation is termed the signal-to-noise ratio. The higher the signal-to-noise ratio, the better the quality of the spectrum. A very small sample will often produce a spectrum with high noise levels. The best ways to minimize high noise levels are to run a larger sample or to collect and average more scans. If this is not feasible, a smoothing routine may be applied to decrease noise levels. Figure 5.4 shows an IR spectrum before and after smoothing to a 50% level. Smoothing effectively decreases the resolution of the spectrum, however, and minor absorption bands or shoulders may disappear.

The most intense absorption band in the spectrum should be between 10%T and 65%T (1.0–0.2 absorbance). The purpose of an intensity check is to ensure that the sample concentration or thickness is optimum. When a sample is too thick, the strongest absorption bands with a transmittance greater than 10%T (1.0 absorbance) will "bottom out," and information regarding band position, shape, and relative intensity will be lost. For accurate quantitative work, such as spectral

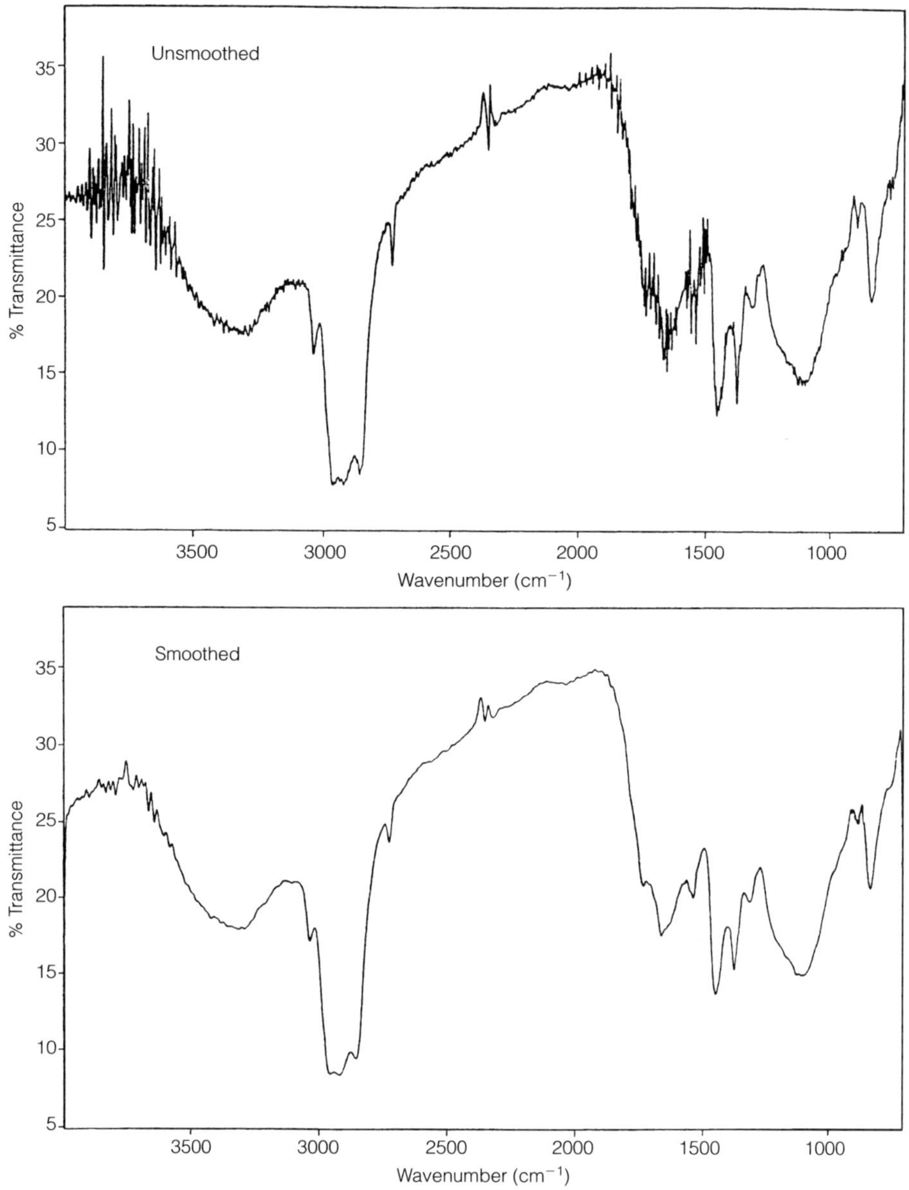

Figure 5.4

An IR transmittance spectrum before and after smoothing. The smoothing method used replaces each point with a running average for a selected number of points (n = 9). This method effectively decreases the resolution of the spectrum. While major absorption bands are not lost, minor bands or shoulders may disappear.

subtraction, the optimum absorption range is smaller, with an absorption maximum no greater than 20%T (0.7 absorbance; see "Subtraction techniques," p. 124). When the strongest absorption band is less than 65%T, the concentration of the sample is low, and some weak bands may not be easily recognized. When comparing the spectrum of a weak sample with a reference spectrum, a spectroscopist can compensate for the feeble bands, but a computer search routine may have difficulty.

Another important consideration is whether the instrument and its sample compartment are purged with dry, carbon dioxide (CO_2)–free air. Vapor-phase water (H_2O) produces small, sharp absorption bands in the regions from 4000 to 3000 and 1800 to 1600 cm^{-1}, while the predominant CO_2 absorption band occurs as a doublet at 2340 cm^{-1} (Fig. 5.5). Purged instruments will eliminate, or at least diminish, the appearance of these bands. Spectra from nonpurged instru-

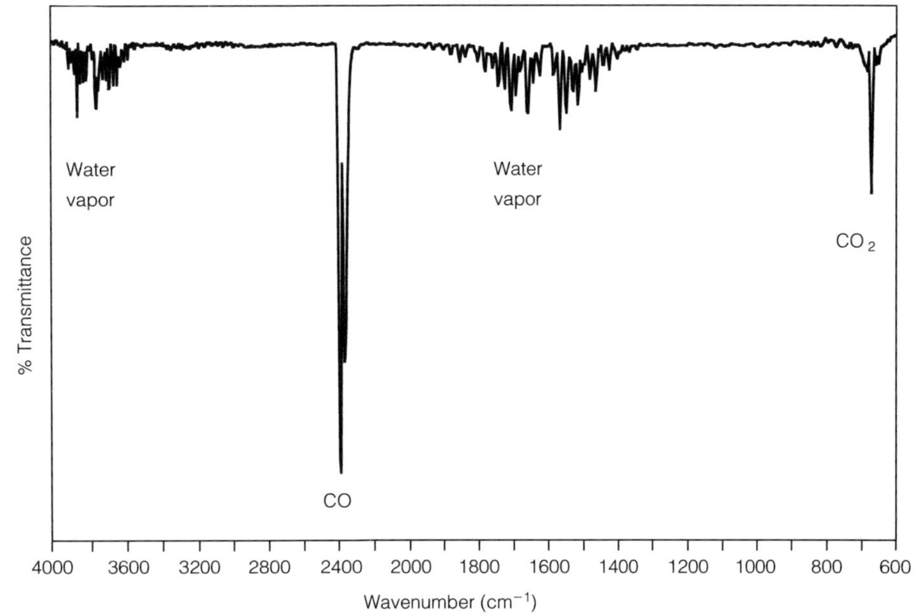

Figure 5.5
An IR transmittance spectrum of room air containing water vapor and carbon dioxide. Most instruments are purged with dry, CO_2-free air to minimize contributions due to these absorption bands.

ments will show small CO_2 and H_2O bands if the atmosphere for the background spectrum and the sample spectrum are similar. However, if the room air changes between the background and sample spectral runs because of drafts or human presence, the spectrum may exhibit stronger atmospheric bands. While usually open to the atmosphere, the sample stage area of most FT-IR microspectrophotometers can also be partially purged with the use of a removable shroud placed on the objective lens.

Visual comparison

Visual examination of the spectrum involves the direct human comparison of a sample spectrum to reference spectra of known materials. The visual pattern recognition procedure, while simple, is very important, because it provides definite identification. It is also very time-consuming, since well over 200,000 published reference spectra are available for comparison. Further, differences between spectra of similar compounds are often quite small. Time needed for interpretation can be substantially reduced by narrowing the examined number of spectra based on the sample's physical characteristics, as well as on identification of major functional groups present in the spectrum. The experienced spectroscopist will occasionally be able immediately to recognize a spectrum of a material previously encountered. In any case, the visual comparison of the unknown spectrum to a reference spectrum is *always* the last step toward verification of an unknown material.

Appendix I lists several significant books, articles, and digitized databases containing IR spectra collections. These reference collections are very comprehensive for modern materials. However, they include few spectra of natural materials, such as those used historically in artist materials. In an attempt to fill this void, more than twenty museum and conservation laboratories joined together to combine their collections of spectra into a single IR spectral library of artists' materials (see Appendix I, Infrared Users' Group 1995). Additional compilations of reference spectra prepared specifically for use in conservation are the Gettens Collection (see Appendix I, Snodgrass and Price 1993) and *Infrared Spectra of Naturally Occurring Minerals* (see Appendix I, Price and Carlson, forthcoming). Many of the compilations are available in digitized form on computer disk, for convenient addition to the digital spectral library of any FT-IR.

Computer libraries

The coupling of high-performance computers with IR spectrophotometers allows for the rapid processing of spectra together with the storage of large numbers of IR reference spectra. Many IR operating programs also provide spectrum-structure correlations, automatic spectral interpretation, and searching of large stored libraries of IR spectra. This greatly reduces the time needed for the visual comparison of an unknown spectrum to reference spectra.

Computer library search programs are typically based on calculation of the difference in peak position and intensity in the unknown spectrum versus those of known spectra. Most software library programs have an option for more than one search algorithm for distinguishing similarities between spectra. Each algorithm performs the calculations slightly differently. In order to determine the variations between these programs and the significance of the match (hit) quality, a known material may be analyzed and its spectrum searched in the library multiple times with different search options or parameters.

No matter how good the software is, it can never replace the judgment and visual skills of a well-trained analytical chemist. The list of hits from the library search is, at best, only a starting point for the visual comparison of the unknown sample's spectrum with reference spectra. Generally, a high hit index means that the spectra match well, but this is not always the case. In some programs the hit with the best match quality is 100, while for other programs it would be 0.000. For example, a hit index of 0.0716 would be an excellent match if obtained on a scale where 0.0000 is a perfect match, and a typical match of 0.2000 is considered a good fit. A computer program may find a hit that has similar bands except for one or two. If these are significant bands, then the spectroscopist can eliminate that hit as a possible choice. A computer may also be misled by sample or reference spectra with sloping baselines or with extraneous bands from contamination. A judgment call by an experienced spectroscopist is needed to determine the importance of spectral features and deviations. The quality of the spectra in the library, as well as the spectrum of the unknown sample, are critical to obtaining credible results. Spectral history—such as analysis method, run parameters, and use of correction routines—should be known for both the unknown spectrum and those in the computer library.

Characterization of the sample type (polymer, mineral, solvent, etc.) helps in the selection of the computer libraries to be searched.

Since numerous libraries are available, search time may be minimized by limiting the number of libraries. When a library does not contain reference materials in the same chemical class as the unknown sample, it is not likely that the computer will obtain a good or even a reasonable match. In particular, search programs are very ineffective at characterizing material mixtures, because most libraries are composed of pure materials.

The results of a computer search routine are totally dependent on the searched spectral library. The best library is often one that contains reference spectra generated on one's own instrument. These spectra tend to match better, because the reference materials were prepared and analyzed in a manner consistent with the unknown sample. They are also more likely to contain materials related to the sample. It is time-consuming, however, to generate the large number of spectra required to compile a comprehensive library. Thus, an active exchange of IR data and spectral libraries between labs that are conducting similar experiments is the second-best choice for developing useful spectral libraries.

For the most part, IR spectra do not vary significantly from instrument to instrument. This is particularly true for FT-IRs, since they have the advantage of internal wavelength calibration. IR data exchange has been facilitated by the development of a universal data format by the Joint Committee on Atomic and Molecular Physical Data Exchange; the format is known by the committee's acronym JCAMP.DX (McDonald and Wilks 1988). Most IR manufacturers supply programs that use this standard form for external data exchange. Additionally, some stand-alone computer programs, such as Lab-Calc (see Suppliers, Galactic Industries Corp.), are available for conversion of spectra to and from the JCAMP.DX format. One commercial lab offers spectral searches on digitized spectra (see Suppliers, Photometrics Ltd.).

Spectral region examination

Since the functional groups for a sample are usually not known, interpretation is best started by dividing the spectrum into several frequency regions. The presence and absence of absorption bands in each region are then used to characterize the sample (Smith 1979). Thus, starting with the frequency regions at the high-wavenumber end, an unknown spectrum may be divided as follows:

OH-NH region (4000–2600 cm^{-1})

Hydroxyl groups generally produce a broad-envelope-type band centered around 3400 cm^{-1}. This band can be misleading, because absorbed water on the sample or on the alkali salt substrate will also produce an O-H band. Hydrogen bonding can change the band position and shape. The O-H band for carboxylic acids is very broad and centered about 3400 cm^{-1}, while the water of hydration for clay produces very sharp bands at 3700–3500 cm^{-1}. N-H stretching bands also occur in this region, with their center near 3350 cm^{-1}. N-H bands are usually sharper than O-H bands. For a natural product sample, the absence of bands in this region shows that the sample does not contain any carbohydrates or proteins.

C-H stretching region (3200–2800 cm^{-1})

C-H stretches from aromatic and vinyl hydrocarbons occur at 3100–3000 cm^{-1}. C-H stretches for methylene groups occur near 2925 (asymmetric) and 2850 (symmetric) cm^{-1}, with the corresponding C-H stretches for methyl groups near 2962 and 2872 cm^{-1}. Thus, with 3000 cm^{-1} as an imaginary dividing line, the hydrocarbon stretches for unsaturated carbon groups occur at higher wavenumbers, and the C-H stretches for saturated carbon groups occur at lower wavenumbers. Waxes, oils, and natural resins, for example, have strong hydrocarbon stretches.

Window region (2800–1800 cm^{-1})

This is usually a baseline region where few absorption bands occur. The most common band is for atmospheric carbon dioxide (doublet at 2340 cm^{-1}). Other bands that occur in this region are a carbon triple bond near 2120 cm^{-1}, nitrile stretch near 2240 cm^{-1}, isocyanate stretch near 2265 cm^{-1}, and thiocyanate near 2160 cm^{-1}. Prussian blue, ferric ferrocyanide, has a strong absorption band in this region.

Carbon double bond region (1800–1500 cm^{-1})

The highly polar carbonyl bond produces a strong absorption band at 1850–1650 cm^{-1} that varies with carbonyl type—for example, 1740 cm^{-1} for ester, 1710 cm^{-1} for ketone. The carbonyl band for the amide I group is shifted down to about 1650 cm^{-1}, with corresponding amide II (1550 cm^{-1}) and amide III (1450 cm^{-1}) bands occurring in stair-step-type intensities. In carboxylic acid salts, the asymmetric stretch for the carbon-oxygen bonds occurs near 1650–1540 cm^{-1}. A carbon double bond stretch occurs about 1640 cm^{-1}. Aromatic carbon-carbon stretching vibrations occur at 1600–1585 and 1500–1400 cm^{-1}.

Fingerprint region (1500–500 cm^{-1})

Most functional groups also have bands in the fingerprint region. The fingerprint bands, when taken in combination with the group frequency bands, can confirm an assignment. The absorption pattern in the fingerprint region is frequently complex, with interacting vibrations and overlapping bands. It is, however, unique for each material and thus most valuable for specific identifications made by direct comparison to reference spectra.

Negative interpretation

Just as the presence of functional group bands can be used to select potential matches, the absence of a functional group band is used to eliminate potential groups of materials. Entire chemical classes of materials can be ruled out just on the absence of one or more basic group frequencies, such as hydroxyl, hydrocarbon, carbonyl, and amide. This key factor should not be underestimated. The positive identification of a material can never be based on the presence of just a single band, because multiple materials can produce similar bands. However, the absence of an absorption band is unambiguous: if one absorption band is not present, then that material is not present. For example, if the spectrum for a paint binder does not have a carbonyl band between 1750 and 1650 cm^{-1}, then it does not have any measurable amounts of oil, egg yolk, glue, natural resin, acrylic, alkyd, or any number of other synthetic resins that contain a carbonyl group. This result, in itself, may be sufficient to answer the analytical question about the sample, thus freeing important analysis time for other samples.

Spectra-structure correlations

An essential part of the spectral identification process is the examination of absorption band frequency to determine corresponding functional groups. Table 5.2 provides a list of molecular functional groups and their corresponding absorption band frequencies.

Hydrocarbons: Aliphatic

Commonly found in spectra of most organic materials, the C-H stretching vibrations occur in the region of 3000–2800 cm^{-1}. As seen in Figure 5.6, a spectrum of paraffin, the methyl group (CH_3) vibration produces two small bands at 2962 and 2872 cm^{-1}, because of the asymmetric and symmetric stretching modes. The methylene groups (CH_2) produce sharp asymmetric and symmetric stretching vibrations at 2926 and 2850 cm^{-1}.

The bending vibrations for the CH_3 groups are found at 1450 and 1380 cm^{-1} for asymmetric and symmetric, respectively. The in-plane bending or scissoring band for CH_2 is found at 1465 cm^{-1}. In

Table 5.2

Some important group frequencies.

Functional group	Formula	Associated absorption band frequencies
Hydroxyl	-OH	Hydroxyl groups generally produce a broad-envelope-type band centered at 3400 cm^{-1}. Hydrogen bonding can change position and shape.
Unsaturated CH	-H-ϕC -H-C=C-	CH stretches from aromatic hydrocarbons occur at 3100–3000 cm^{-1}. CH stretches from a carbon double bond occur about 3030 cm^{-1}.
Aliphatic CH-methylene	$-CH_2-$	CH stretches for methylene groups occur near 2925 (asymmetric) and 2850 (symmetric) cm^{-1}. The vibrations for bending occur near 1465 cm^{-1}, for rocking, ca. 730 cm^{-1} (only for CH_2 sequences greater than 4 and may be split in solids).
Aliphatic CH-methyl	$-CH_3$	CH stretches for methyl groups occur near 2962 (asymmetric) and 2872 (symmetric) cm^{-1}. The vibrations for CH bending occur near 1450 and 1380 cm^{-1}.
Carbon-carbon multiple bonds	-C=C- -C≡C-	A carbon double bond stretch occurs ca. 1640 cm^{-1} and a carbon triple bond near 2120 cm^{-1}. Aromatic carbon stretching vibrations occur near 1600 and 1500 cm^{-1}. Out-of-plane bending mode vibrations may produce strong sharp bands at 1000–650 cm^{-1}, depending on the pattern of substitution.
Carbon-nitrogen	-C≡N -N=C=O -S-C≡N	A nitrile stretch occurs near 2240 cm^{-1}, an isocyanate stretch near 2265 cm^{-1}, and a thiocyanate near 2160 cm^{-1}.
Carbonyl	-C=O	The highly polar carbonyl bond produces a strong absorption at 1850–1650 cm^{-1} that varies with carbonyl type—e.g., 1740 for ester, 1710 for ketone.
Amide	-CONH-	N-H stretch about 3350 cm^{-1}. Amide I occurs near 1650 cm^{-1}, amide II near 1550 cm^{-1}, and amide III near 1450 cm^{-1} in stair-step-type intensities.
Acid salts	$-CO_2^-$	Asymmetric stretch for the carbon-oxygen bonds occurs near 1650–1540 cm^{-1}, depending on structure.
Carbonate	$-CO_3^=$	Broad stretching band near 1450 cm^{-1} with sharp bands at 900–700 cm^{-1}, depending on cation.
Carbon-oxygen	-C-O-	C-O stretch occurs at 1200–1000 cm^{-1}; varies with hydrogen bonding and molecular structure.

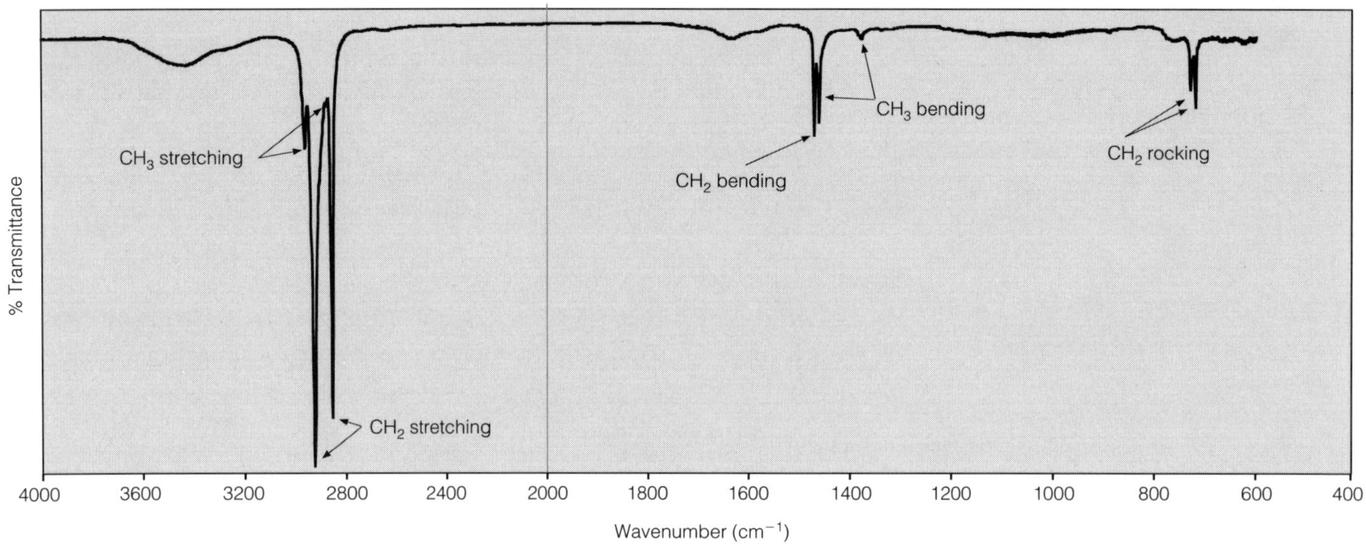

Figure 5.6

An IR transmittance spectrum of paraffin.

long-chain hydrocarbons, with more than four methylene groups, there is also an in-plane rocking band that is found near 730 cm^{-1}. In solid semi-crystalline materials, such as paraffin and beeswax, the absorptions appear sharper, with characteristic peak splitting forming doublets at the methylene absorption positions. Other weaker methylene bands are observed in the 1350–1150 cm^{-1} region.

Hydrocarbons: Aromatic

Aromatic C-H stretching bands occur between 3100 and 3000 cm^{-1}. The strongest, and sometimes most informative, bands in the spectra of aromatic compounds occur in the low-wavenumber range between 1000 and 655 cm^{-1}. These strong absorption bands result from the out-of-plane bending vibrations of the ring C-H bonds. The absence of any major bands in this region generally shows that the material is nonaromatic. Skeletal vibrations from the C-C interactions in the ring absorb in the 1600–1585 and 1500–1400 cm^{-1} regions. A spectrum of polystyrene, seen in Figure 5.7, illustrates these absorptions.

Alcohols and hydroxyl absorptions

The characteristic bands observed in the spectra of alcohols result from O-H and C-O stretching vibrations. These functionalities are susceptible to hydrogen bonding, which produces broad IR absorption bands. The O-H stretch generally occurs from 3600 to 3200 cm^{-1}. The C-O absorption in alcohols occurs from 1260 to 1000 cm^{-1}. Other weaker, O-H bending vibrations may be seen at 1420 to 1330 cm^{-1}. Figure 5.8 shows a spectrum of honey, a complex sugar containing multiple C-O and O-H stretches.

Carbonyls: Aldehydes, ketones, and esters

Ketones, aldehydes, acids, esters, anhydrides, and amides show a strong carbonyl (C=O) stretching absorption in the region of 1870–1640 cm^{-1}. The carbonyl band has a relatively constant position and high intensity,

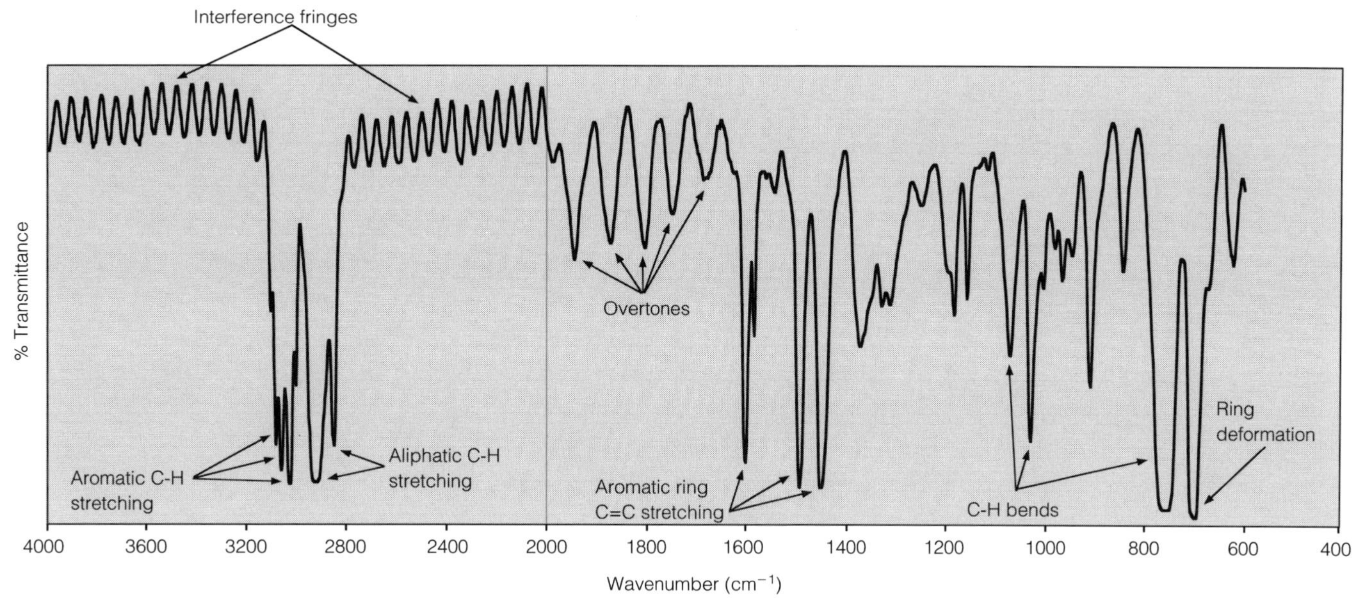

Figure 5.7
An IR transmittance spectrum of polystyrene.

which makes it one of the easiest bands to recognize in the IR spectrum. The exact position of the C=O stretch is determined by its environment in the molecule. A carbonyl in a ketone usually occurs at 1720–1690 cm^{-1}, while a carbonyl band in an ester usually occurs at 1750–1730 cm^{-1}. The carbonyl groups in aldehydes absorb near 1740–1720 cm^{-1}. The adjacent C-O stretching vibrations found in esters and aldehydes occur between 1400 and 1000 cm^{-1}. Ketones also have weak backbone stretching and bending vibrations in this region. Figure 5.9 shows a spectrum of polyester.

Carbonyls: Amides

The spectra for primary and secondary amides contain a strong carbonyl absorption band in the region of 1650 cm^{-1}, called the amide I band. Secondary amides display an additional band near 1550 cm^{-1}, called the amide II band, that is a combination of C-N and N-H vibrations. A C-H

Figure 5.8
An IR transmittance spectrum of honey.

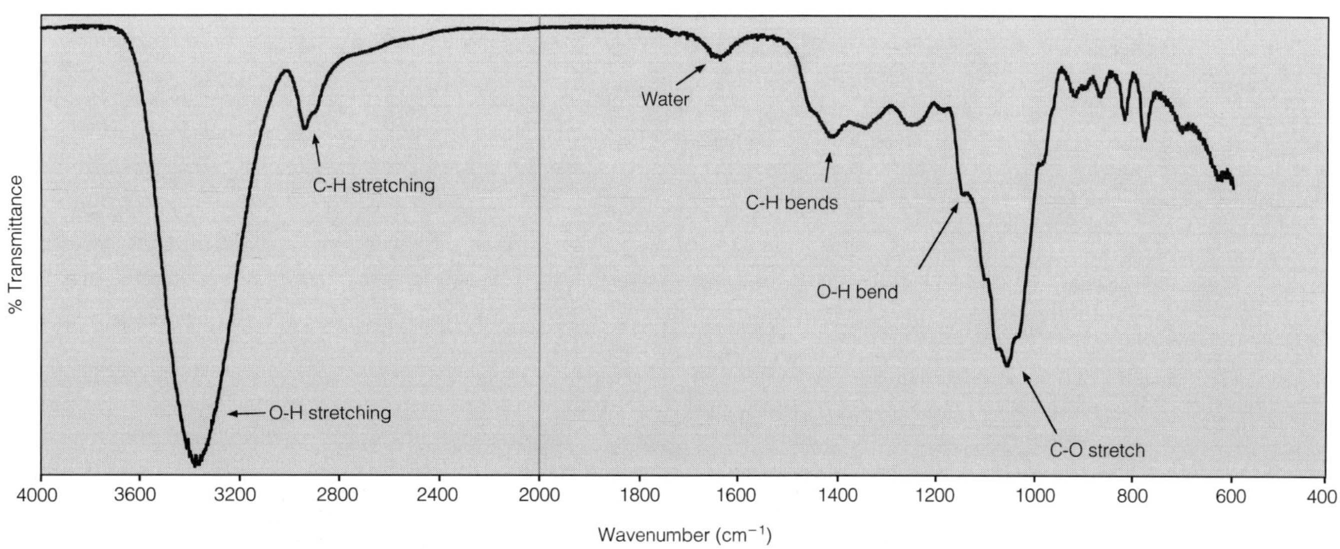

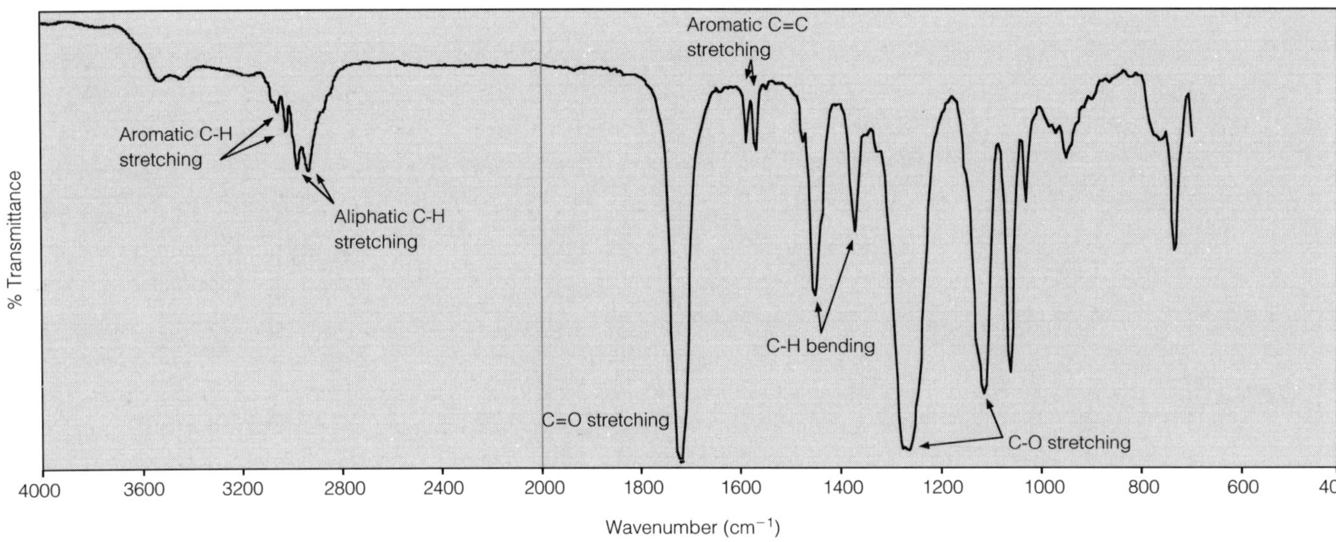

Figure 5.9
An IR transmittance spectrum of polyester.

bending vibration occurring near 1450 cm^{-1} has sometimes been called the amide III band. The relative intensities of the amide I, II, and III bands in polyamides (protein, nylon, etc.) occur in a stair-step pattern. The asymmetrical and symmetrical N-H stretching vibrations occur near 3350 and 3180 cm^{-1}, respectively. Hydrogen bonding may broaden the bands, giving the appearance of one band, although they are usually sharper than O-H bands. Often a stronger O-H band overlaps this region, and the N-H stretches appear as shoulders or peaks on the broader O-H band. Figure 5.10 shows a spectrum of gelatin, which exhibits the typical primary and secondary amide patterns seen in proteins.

Figure 5.10
An IR transmittance spectrum of gelatin.

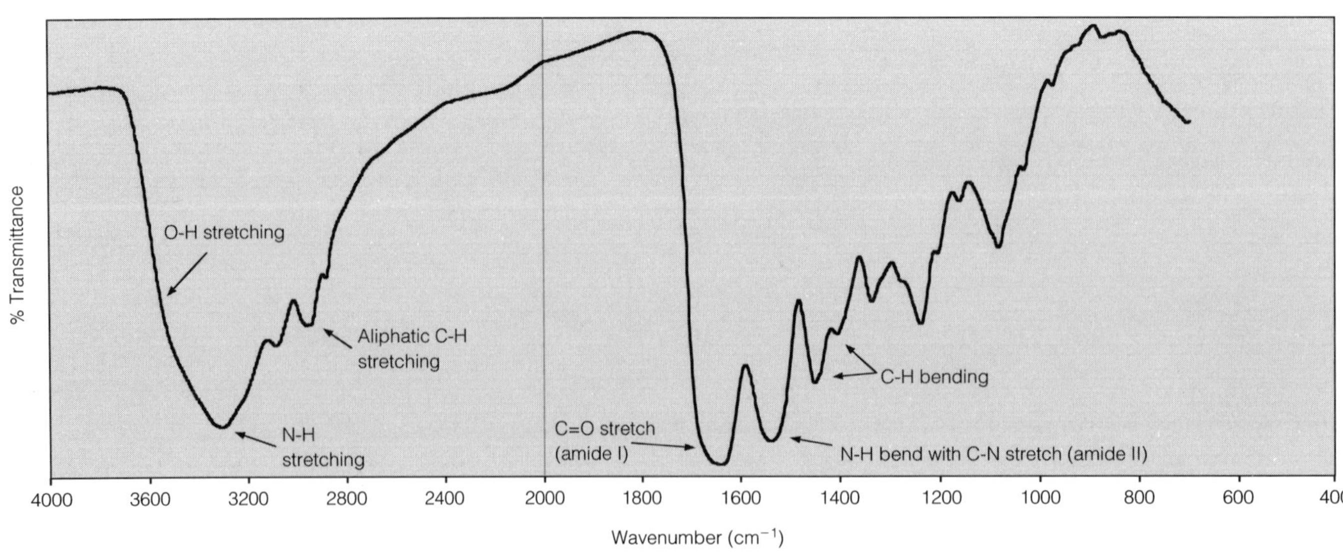

Correlation charts

Over the years, many researchers have done extensive studies to determine the absorption band ranges of specific functional groups. This information has been compiled into correlation charts, several of which may be found in the literature. A simplified version of a correlation chart is shown in Figure 5.11. The lines for each functional group indicate the most probable range for the absorption band. The precise wavenumber at which a specific group absorbs is dependent on its molecular environment and its physical state.

One limiting factor to correlation charts is that they represent only the strongest, most characteristic bands. Other important information used to identify an unknown spectrum, such as band shape and intensity, is not incorporated in the chart. The correlation chart is best used to classify the functional groups in a compound structure for only the major absorption bands, since minor bands may be due to vibrations not shown on the chart.

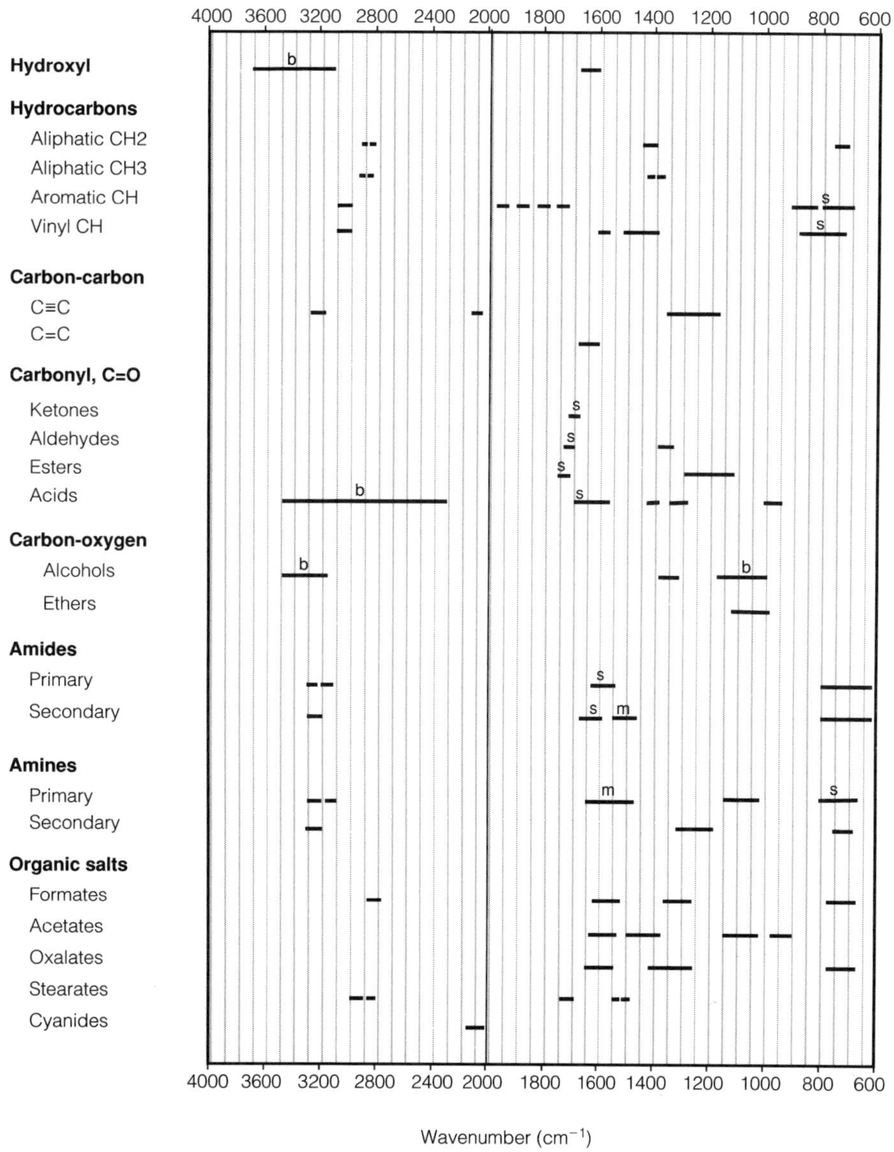

Figure 5.11
The general range of major absorption bands for specific functional groups. This chart can help in the selection of groups of materials for further examination by direct spectral comparison. Since the figure does not depict exact band position or intensity and does not include minor bands, it cannot be used as a sole means of identification (s = strong; m = medium; b = broad).

Identification of Materials Used in Art and Art Conservation

A variety of natural and synthetic materials have been used in creating works of art. Though ranging from acrylics to polysaccharides and from carbonates to oxides, the classification, if not the specific identification, of these materials in their pure form, as well as in simple mixtures, is well within the realm of IR spectroscopy. Appendix II provides reference IR spectra for many materials used in art and art conservation.

Natural organic materials

Natural products, derived from plants and animals, are rarely pure materials; rather, they are mixtures of many components (major, minor, and impurities), because of the number of reactions that occur simultaneously in biological systems. The components may range in complexity from simple molecules to multicomponent mixtures of organic and inorganic compounds. A mixture containing slight variations of many components produces an IR spectrum with some broad, ill-defined, and overlapping bands, especially in the fingerprint region. The presence of ill-defined bands indicates that a material is a mixture and possibly a natural product.

Flowchart for natural organic materials

For identification purposes, natural products encountered by analytical chemists working in the art conservation field may be grouped into the following classes that contain similar IR active functional groups: resins (tree), resins (insect, shellac), oils, gums, waxes, and proteins. Figure 5.12 gives examples of spectra for these categories. More information on vibrational group assignments for these natural materials may be found elsewhere (Bellamy 1980; Omecinsky and Carriveau 1982).

Classification of spectra for these general classes of natural organic materials can be simplified based on the position, intensity, and shape of the carbonyl and hydrocarbon absorption bands. Figure 5.13 is a flowchart that separates each material based on the presence or absence of characteristic bands. The schematic trail is followed until one of the final blocks is reached.

Positive identification of a material should always be made by final, direct visual comparison of the unknown spectrum to reference spectra. However, there are only a few good reference spectral collections, books, and digitized spectra available for natural products. Appendix I lists some of these (see Infrared Users' Group 1995; Hummel and Scholl 1981; and Snodgrass and Price 1993).

Waxes

Waxes are long-chain hydrocarbon materials that can be produced by either plants or animals. Waxes are very stable, not changing significantly as they age (Mills and White 1994). Kühn provided a basis for the identification of waxes in works of art using IR spectroscopy (Kühn 1960). More recently, Besaninou presented a thorough history of waxes and their methods of analysis (Besaninou 1984). Parra and Serrano used

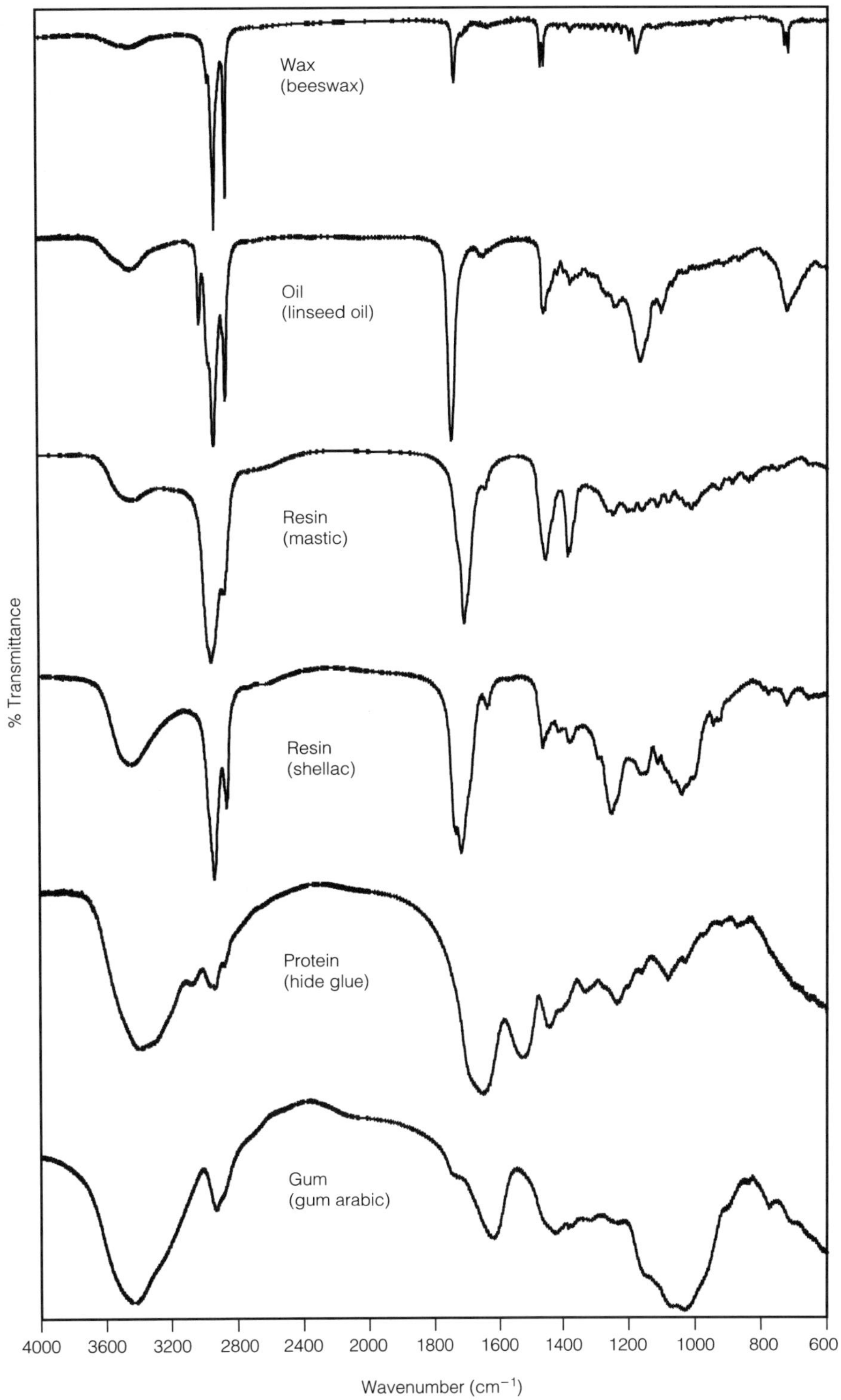

Figure 5.12
IR transmittance spectra for several types of unaged natural organic materials.

IR spectroscopy and other techniques to examine wax seals and their attached textile remnants (Parra and Serrano 1990). Other authors have used IR spectroscopy to identify beeswax in Egyptian paints (Birstein and Tul'Chinskii 1979), on a column of a German church (Bleck and Ziessler 1967), and in samples excavated in Sudan (Delbourgo and Gay 1968).

In a wax spectrum, the many CH_2 groups (methylenes) in the chain produce the characteristic and predominant stretches at 2926 and

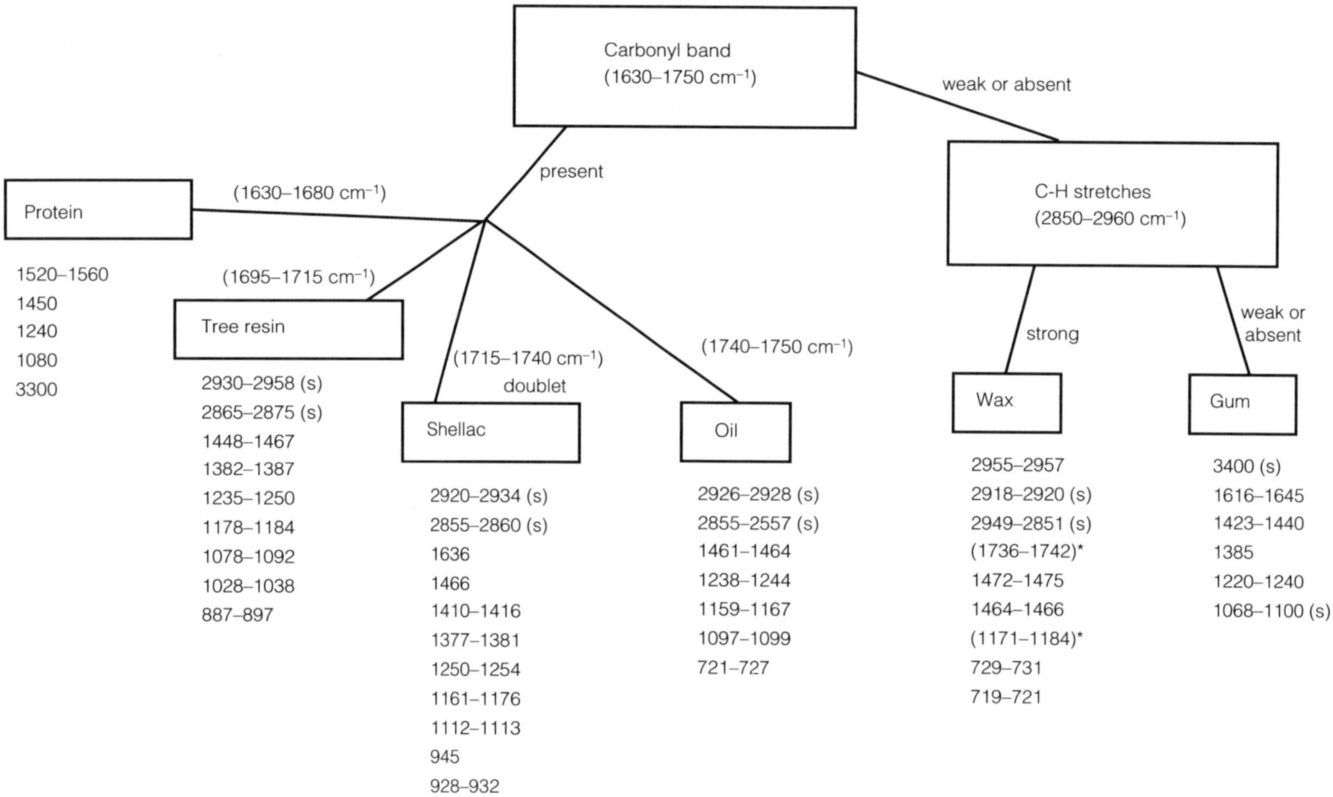

Figure 5.13
A flowchart to aid in characterization of several classes of natural organic materials based on their IR absorption band positions and intensities (s = strong; * = present only in ester-containing waxes).

2850 cm^{-1} (see Figs. 5.6 and 5.12 for paraffin and beeswax, respectively). Confirmatory bands for waxes are small, sharp doublets at 1466/1462 cm^{-1} and 730/720 cm^{-1}. The splitting of the bands near 730 cm^{-1} into a doublet indicates that there are at least four methylene groups in the chain. The appearance of the doublets indicates the semicrystalline structure of the wax (Ludwig 1965). Any pure, long-chain hydrocarbon will have CH_2 chains with CH_3 methyl end groups and present a spectrum similar to paraffin wax. Most natural waxes also contain esters of higher fatty acids with fatty alcohols. Beeswax is composed of about 70% higher aliphatic esters, with 13% free wax acids and only about 12% hydrocarbons (Fig. 5.12). The ester groups account for the weak C=O stretching band at 1740 cm^{-1} and for the C-O bands in the 1175 cm^{-1} region.

Oils

Oils, vegetable and animal, consist of glycerol esters of higher fatty acids with even carbon numbers; their diversity lies in the type and composition of the fatty acids (Mills and White 1994). Oils and fats are ubiquitous and have been used in many contexts; in art objects they are commonly found as binding media in paintings and as residues in archaeological samples. While chromatographic procedures are required for the specific differentiation of the fatty acid components in oils, IR can readily identify this class of materials (Shreve et al. 1950; Barclay 1989; Kosek and Green 1992). In addition, IR spectroscopy has been successfully used to study the drying of linseed oil (Baer and Indictor 1976) and its changes with UV radiation (Low and Baer 1977), as well as its aging characteristics when mixed with pigments (Meilunas, Bentsen, and Steinberg 1990). Hedley and coworkers used IR spectroscopy to evaluate changes in the dried oil film after cleaning treatments (Hedley et al. 1990).

Shown in Figure 5.12 is an example of a typical oil spectrum illustrating the significant CH_2 stretches. However, due to the polar molecular environment, the methylene bands, at 2926 and 2855 cm^{-1}, are at slightly higher wavenumbers than those of the waxes. In addition, there is a weak-to-medium olefinic C=C-H stretching band that occurs at 3020 cm^{-1}. The intensity of this band depends on the state of dryness of the oil, and in well-dried oils, this band will be very small. Oil spectra contain a strong, sharp carbonyl band at 1750–1740 cm^{-1}, because of the ester group. It is the only natural organic material of these five classes that has an intense carbonyl band in this region. This is a clear characteristic of oil. However, in mixtures with some pigments, the carbonyl band may be shifted to slightly lower wavenumbers. Other bands characteristic of the oils are aliphatic C-H bands at 1464, 1379, and 725 cm^{-1} and the C-O bands at 1240, 1165, and 1103 cm^{-1}. The three C-O bands occur in a characteristic maple leaf pattern, for which the band at 1165 cm^{-1} is the strongest.

Resins (tree)

Natural tree resins are primarily composed of aliphatic three-ring structures called resin acids. The softer resins, mastic and balsam, have been used as adhesives and varnishes, and the harder resins, such as copal and amber, have been used for decorative beads and sculpture. Tree resins fall into three main categories: (1) aromatic (e.g., benzoin, styrax), (2) diterpenoid (e.g., balsam, copal), and (3) triterpenoid (e.g., dammar, mastic) (Mills and White 1977). These resins vary in their degree of stability, and some are susceptible to oxidation (Mills and White 1994).

Gianno and coworkers used IR to analyze over one hundred southeast Asian natural resins and gums; those references served as a base to identify coatings and adhesives found on ethnographic objects (Gianno et al. 1987). Mastic and dammar resins, commonly used as painting varnishes, have been spectrally characterized by Feller (1954, 1959). IR spectroscopy was one method used by Burnstock and Learner (1992) to monitor changes in mastic varnishes after they were cleaned with alkaline reagents. IR spectroscopy was also used to study historic varnishes (Korte and Staat 1989; McCormick-Goodhart 1989), coatings on African ceramics (Hexter and Hopwood 1992), and archaeological residues (Shearer 1987). Derrick (1989) provided a schematic to differentiate five unaged resins found in furniture finishes using IR absorption band positions.

Tar and pitches have been used since early times as putties, paints, and waterproofing agents. Hadzi and Cvek (1978) used IR to identify "grave resins" found on urns in excavated graves as birch bark pitch. First analyzed with IR, then confirmed with chromatographic methods, pine pitches were found on an Etruscan shipwreck (Robinson et al. 1987) and in Mediterranean transport amphoras (Beck, Smart, and Ossenkop

1989). Asphalts and bituminous resins used as brown colorants in paint media have been characterized with IR by Wolbers (1984).

Early IR studies showed that various sources of amber, a fossilized resin, could be differentiated by their IR spectra (Schwochau, Haevernick, and Ankner 1963; Beck, Wilbur, and Meret 1964; Beck et al. 1965; Langenheim and Beck 1965). Reference sets of amber spectra were collected (Langenheim and Beck 1968) and later used to identify amber sources in Greek artifacts (Beck et al. 1971), in imported archaeological artifacts (Beck 1972), in Etruscan jewelry (Follette 1985), in samples from Japanese tombs (Fujinaga, Takenaka, and Muroga 1976), and in objects from other archaeological sites (Todd et al. 1976; Beck et al. 1978). Williams, Waddington, and Fenn (1990) used IR spectroscopy to examine the changes in amber after exposure to air pollutants. Grimaldi noted that the routine use of IR spectroscopy for the characterization of amber has significantly helped curators with the classification of artifacts (Grimaldi 1993).

Oriental lacquer, or urushi, is another significant plant resin used in art objects that has been well studied by IR spectroscopy (Masschelein-Kleiner and Heylen 1968; Fujinaga, Takenaka, and Muroga 1976; Kenjo 1978; Carriveau 1984). Kenjo used IR to study the effects of pH on the hardening of lacquer films (Kenjo 1976). More recently, Derrick and coworkers showed that IR spectroscopy can be used to distinguish between lacquer on Japanese furniture and the natural resin–based imitation coatings developed in Europe in the eighteenth century (Derrick, Druzik, and Preusser 1988).

The various types of natural resins exhibit several important spectral features. The cyclic ring structure of tree resin (mastic) gives a spectrum with strong C-H stretching vibrations at even higher wavenumbers than those in the oils (Fig. 5.12). These are generally found at 2958–2930 and 2875–2865 cm^{-1}. Because there is a variety of molecular environments for the methylene groups within the compounds in any given resin and because there are several methyl (CH_3) end groups, the C-H stretches are not as sharp and as well separated as those previously seen in the waxes and oils. Resins may be distinguished from the other groups by two bands. The first band, which is usually weak and broad, occurs at 2700–2500 cm^{-1} and is due to the O-H vibrations of a dimerized carboxyl group. The second distinguishing band that all tree resins contain is a strong carbonyl (C=O) stretch at 1715–1695 cm^{-1}. This band broadens with resin degradation and oxidation, but the band maximum remains within this wavenumber region (see, for example, spectra of light-aged mastic in Fig. 5.14).

Bands in the fingerprint region are characteristic for each particular tree resin and may be used to distinguish among them. Figure 5.15 shows exemplary IR absorbance spectra for five commonly used, unaged natural resins in furniture finishes. Figure 5.16 is an IR absorption band identification key for these resins; included are family names and major components (Derrick 1989). Once a natural material has been classified as a natural resin by use of the information in the natural products identification flowchart (Fig. 5.13), then the type of resin

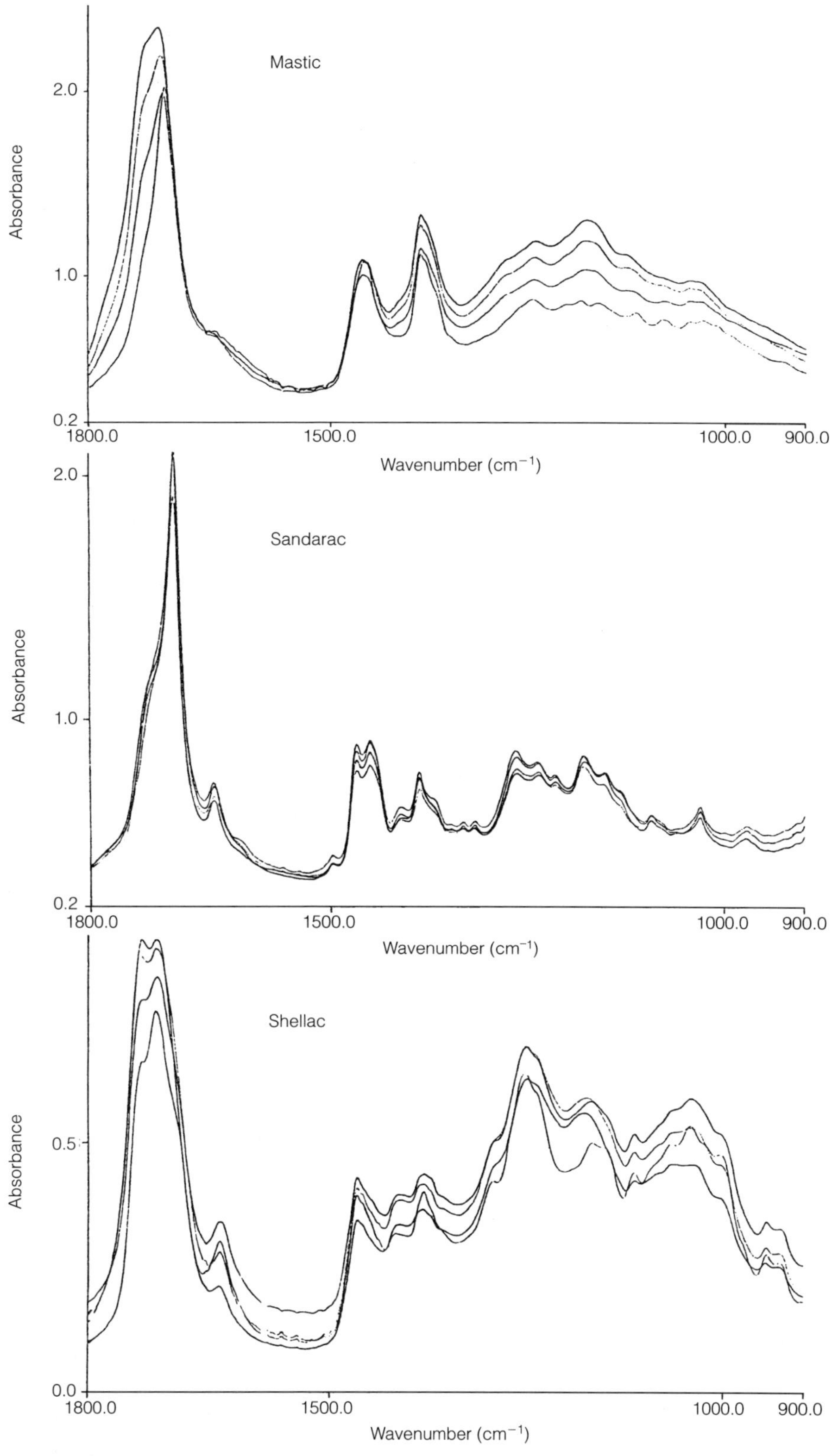

Figure 5.14

IR absorbance spectra (1800–900 cm^{-1}) for three sets of fresh and deteriorated natural resins (mastic, sandarac, and shellac). Each plot is a spectral overlay of samples that were exposed to 0, 37, 75, and 151 kilojoules of energy from a xenon arc lamp in a temperature- and humidity-controlled environment. At the longest exposure time, all samples exhibited discoloration and cracking. Note that the absorption band position maxima remain stable, while the band shapes and relative intensities change. The most significant changes are seen in the shapes and relative intensities of the carbonyl bands for mastic and shellac; for each of these carbonyl bands, the shortest band corresponds to the fresh resin sample, and the tallest band corresponds to the most deteriorated resin sample.

may be distinguished with the aid of the additional information provided in Figure 5.16.

This schematic provides a simplified structure for the easy identification of the presence of a resin in a spectrum by a list of its

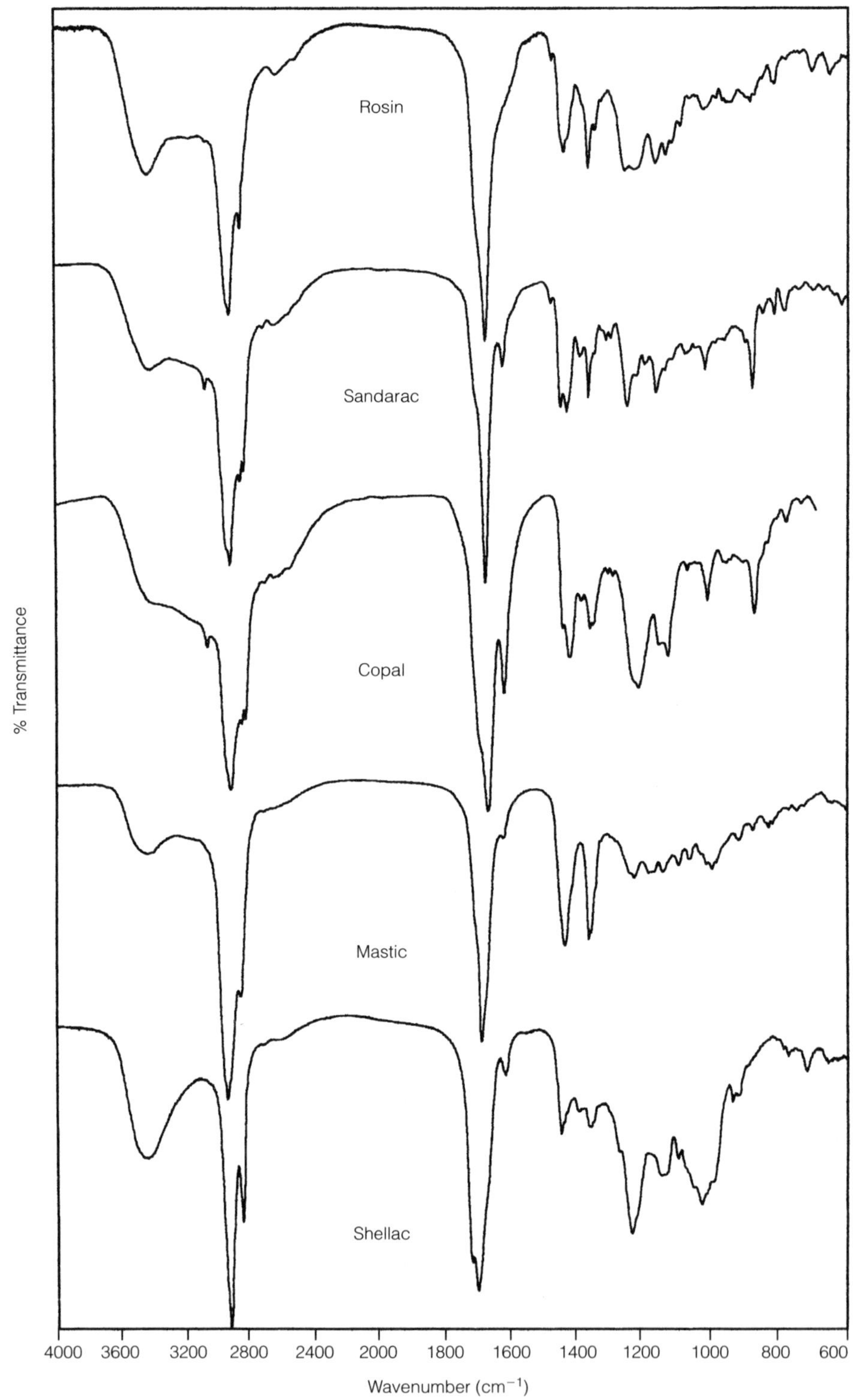

Figure 5.15
Exemplary IR transmittance spectra for five types of unaged natural resins.

absorption band positions. The strongest bands in the resin spectra are generally the carbonyl and hydrocarbon stretching frequencies. These bands, listed at the top of the key, can provide a distinction for four of the resins. Sandarac and copal are chemically and spectrally very similar and may be distinguished only by some smaller bands in the fingerprint region (2000–500 cm^{-1}). The presence of at least half of the listed bands

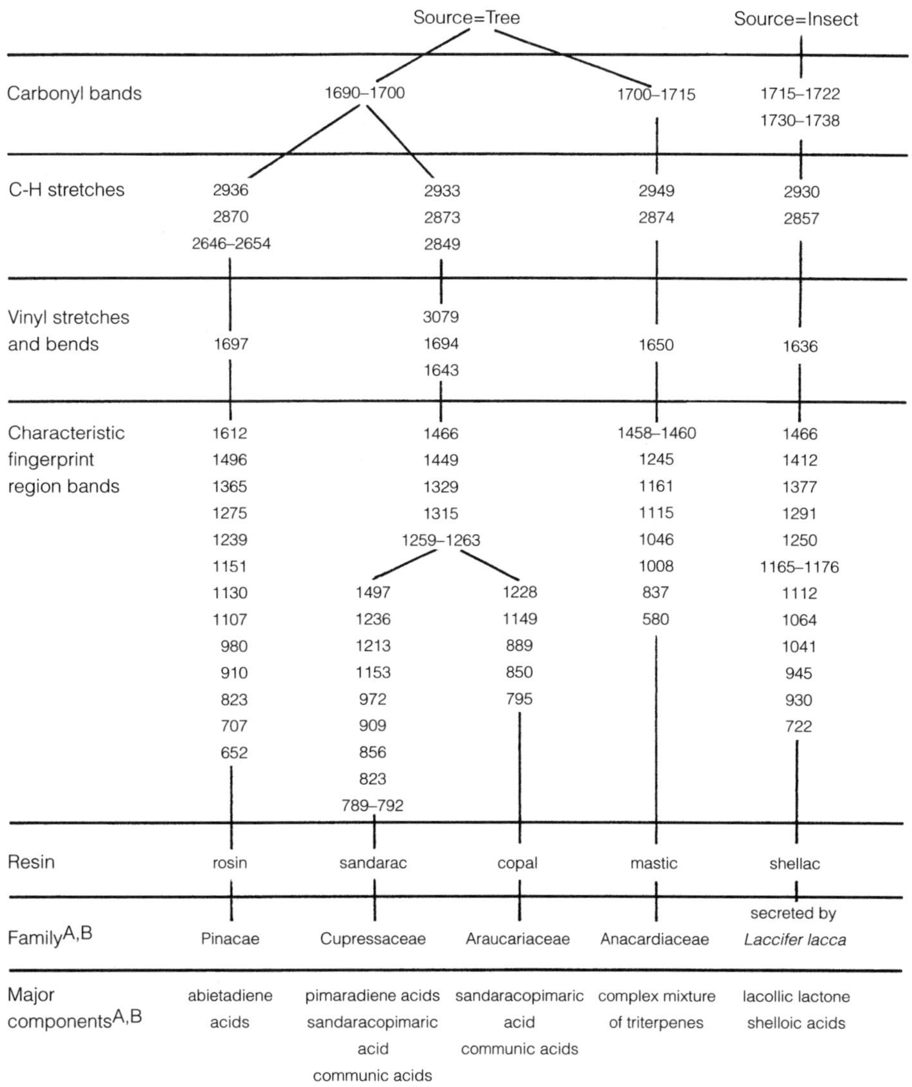

Figure 5.16

A flowchart to aid in characterization of five types of natural resins based on their IR absorption band positions (absorption bands are represented as wavenumbers [±2.4 cm^{-1}]; A = Mills 1977; B = Mattiello 1941). (For more information, see Derrick 1989.)

in an unknown sample spectrum is a strong indication that a particular resin is present. Final confirmation comes through visual comparison of the sample spectrum with a reference spectrum of a known material.

Resins (insect)

Shellac is a resin excretion from the lac beetle. Chemically, shellac resin is a complicated mixture of lactones, esters, and ethers of aliphatic and aromatic polyhydroxy acids. The C-H stretching bands (Figs. 5.12, 5.15) fall in positions similar to those of oils, at 2934–2920 and at 2857 cm^{-1}. The carbonyl band in fresh shellac is a characteristic doublet from the ester (1735 cm^{-1}) and from the acid (1715 cm^{-1}). A small olefinic band occurs at 1635 cm^{-1}. Several C-O bands are present, most noticeably at 1240, 1163, and 1040 cm^{-1}; they are due to the ester, acid, and alcohol groups. A double band at 730/720 cm^{-1} (from partially crystalline long-chain hydrocarbons) occurs in many shellacs and is probably due to accompanying shellac wax. Shellac is included in the resin identification schematic presented in Figure 5.16.

Proteins

Proteins are polymeric substances composed of amino acid monomeric units. The proteins present in living organisms consist of various combinations and proportions of twenty naturally occurring amino acids. Proteinaceous materials used in art objects may be from animal tissues (parchment, leather, hair, ivory) or from their by-products (glue, egg, casein, albumin, blood). All types of proteins, as a general material class, are readily identifiable by their IR spectra (Perron 1989). Close examination of absorption band position and intensities can show the denaturation of collagen to gelatin (Birstein and Tul'Chinskii 1981; Payne and Veis 1988; Derrick 1991). FT-IR analysis has also been used to differentiate between modern elephant ivory and ivory from other sources, such as walrus, hippopotamus, or ancient mastodon (Lee 1991).

Protein spectra form a consistent recognizable pattern of absorption peaks (Fig. 5.12). Proteins are characterized by the presence of amide I and amide II bands near 1650 and 1550 cm^{-1}, respectively. These two bands along with another, occasionally referred to as an amide III, found near 1450 cm^{-1}, form a consistent stair-step pattern. Additionally, the presence of an amide may be confirmed by the N-H stretching band near 3350 cm^{-1}. While IR is very useful for identifying the presence of a protein, few spectral differences are seen between various protein types, including materials as different in amino acid composition as fish glue and albumin. Thus, a secondary method, such as liquid or gas chromatography, must be used to determine the exact amino acid composition.

Gums

Carbohydrates are natural polysaccharides composed of various proportions of several monosaccharide units. They are typically water soluble. Some common examples are sugar, starch, cellulose, and plant gums. An excellent review of plant gums and their use as art materials was given by Twilley (1984). Birstein used IR spectroscopy to identify natural gums as binders in central Asian wall paintings (1975) and as protective coatings in late antique and early-middle-age Egyptian tombs (Birstein and Tul'Chinskii 1977). Masschelein-Kleiner and Tricot-Marckx detail methodology for the IR analysis of gums and illustrate spectra obtained neat and after acid hydrolysis (Masschelein-Kleiner and Tricot-Marckx 1965).

Gums and cellulose are long-chain polymers of sugars (polysaccharides). Sugars have a high proportion of O-H groups bound to the carbons. This structure produces a very characteristic IR pattern for the polysaccharides (Fig. 5.12), which have two strong, broad bands: one at about 1080 cm^{-1} due to C-O and the other at about 3300 cm^{-1} due to the O-H groups. These bands are typically about equal in intensity. The C-H stretches tend to be weak and poorly resolved. All polysaccharides also contain a moderately strong band at 1620 cm^{-1} that is partially associated with intramolecularly bound water and partially due to the presence of a carboxyl group. Some gums, such as gum tragacanth, also contain a weak-to-moderate band at 1735 cm^{-1}, which is associated with an ester-containing component.

Synthetic resins (polymers)

Over forty years ago, many industrial labs purchased IR spectrometers for the analysis of rubbers. As applications for other polymers grew, so did the breadth of samples analyzed by IR. Unlike natural materials, which are complex mixtures of many components, synthetic resins tend to be pure, specific molecular structures that provide sharp, well-defined IR absorptions. These very specific, recognizable IR patterns make IR spectroscopy the method of choice for the identification of polymers.

IR analysis has been used successfully to analyze polymers in objects of art and in art conservation practice (Freeman 1979; Martin 1988; Pratt 1991; Shearer and Doyal 1991). The capabilities and limitations of IR were examined for analyzing acrylic (Stringari and Pratt 1993) and alkyd (Hodson and Lander 1987) paint media. In poly(vinyl acetate) films prepared with different solvents, IR spectroscopy showed conformational variations in the polymer (Hansen et al. 1991).

Figure 5.17 shows spectra of several important polymers selected as synthetic equivalents to the natural products shown in Figure 5.12. Polyethylene is a long-chain hydrocarbon, similar to natural waxes such as paraffin. Many types of polyester resins, such as alkyds and acrylics, are used as modern artist materials and as conservation materials. Some polycyclohexanones, such as MS2A and Laropol K-80, have been proposed as substitutes for natural resin varnishes. Nylon, a synthetic polyamide, is a chemically resistant material used in fabrics and liners. The spectrum of a water-soluble, modified cellulose, Klucel F, is also shown. More information on vibrational group assignments for many synthetic resins may be found in these Appendix I references: Bellamy 1975; Kagarise and Weinberger 1954; Stimler and Kagarise 1966; Zeller and Pattacini 1973; Chicago Society for Coating Technology 1980.

Flowchart for synthetic resins

An identification flowchart for classes of synthetic materials is presented in Figure 5.18. It is similar in design and use to the chart presented for natural materials. This schematic method aids in identifying the type or class of material present. The presence of plasticizers and fillers can complicate the identification of the base polymer. However, since the spectrum is the sum of its components, it is usually possible to find sufficient unobscured bands to characterize the main polymeric component.

Positive identification of a material should always be made by final, visual comparison of the unknown spectrum to those in reference spectral collections. Several good reference spectral collections, books, and digitized spectra—such as Sadtler Research Laboratories, Hummel and Scholl (1981), and Chicago Society for Coating Technology (1980)—are available for polymers and additives (see Appendix I). The Gettens Collection contains reference spectra for several early polymers from the 1930s and 1940s (Appendix I, Snodgrass and Price 1993).

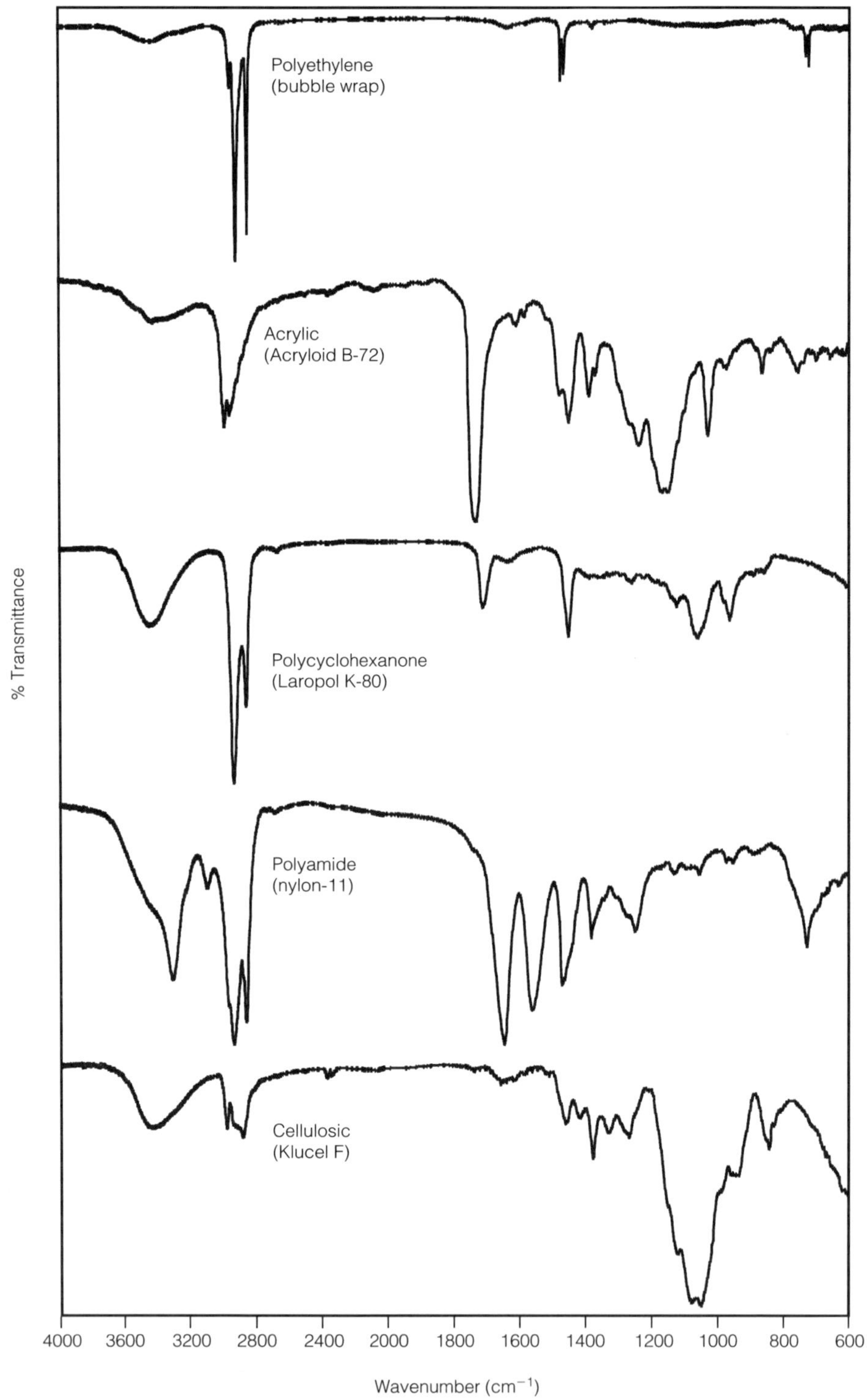

Figure 5.17

Exemplary IR transmittance spectra for five classes of unaged synthetic materials.

Characterization process for polymers

Polymers are usually composed of long chains of two or three primary functional groups. The following discussion uses the wavelength regions, mentioned earlier in this chapter, to classify several types of synthetic polymers based on the presence and absence of major functional groups in their IR spectra. All regions of the spectra should be examined and cross-checked. The goal of functional group analysis is to focus on one or more classes of materials for in-depth visual comparison.

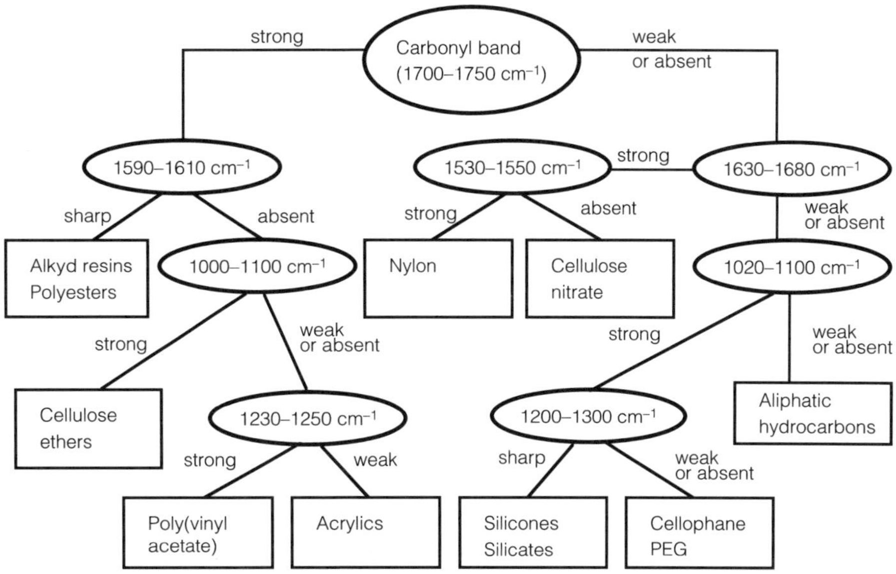

Figure 5.18
A flowchart for the characterization of several classes of synthetic polymers based on their IR absorption band positions and intensities.

OH-NH region (4000–2600 cm^{-1}). The absence of broad envelope O-H and N-H bands in this spectral region indicates that the sample does not contain amines, amides, alcohols, or organic acids. This rules out materials such as polyamide (e.g., nylon), phenolic resins (e.g., Bakelite), polyurethanes, polyethylene glycols (e.g., PEG-100), and polyvinyl alcohols (e.g., Elvanol), as well as some cellulose esters, alkyds, and epoxies. Note, however, that if the sample was prepared using a hygroscopic alkali salt, such as KBr, the presence of an O-H band in this region may derive from absorbed water rather than from the sample structure.

Hydrocarbon stretching region (3200–2800 cm^{-1}). Absorption bands between 3100 and 3000 cm^{-1} correspond to an aromatic or vinyl C-H group and may indicate compounds such as polystyrene, phenolic resins, some polyurethanes, epoxies, or phthalate plasticizers. A sharp band at 2980 cm^{-1} may indicate silicone resin or oil.

Strong sharp, methylene absorption bands (2925 and 2850 cm^{-1}) with no carbonyl band (see next section) indicate the presence of a long-chain hydrocarbon polymer, such as polyethylene, polypropylene, butadiene, or natural rubber. The spectra for some polyethylenes are only slightly different from the spectra for paraffin waxes.

Numerous aliphatic esters, such as polyesters, have moderate to strong methyl and methylene stretching bands. However, if the spectrum has weak or absent C-H vibrations, then compounds such as fluorocarbons (e.g., Teflon), polyimides (e.g., Kapton), poly(vinyl acetates) (e.g., AYAA), epoxies (e.g., Epon), or regenerated celluloses (e.g., cellophane, rayon) should be examined.

Window region (2800–1800 cm^{-1}). This region is associated with adjacent double and triple bonds. In polymers, compounds containing isocyanate or nitrile groups will absorb in this region about 2100 cm^{-1}. Examples of nitrile-containing polymers are acrylonitrile-butadiene-styrene

(ABS rubber), acrylonitriles (e.g., Orlon, Saran F-120), and some polyurethanes (e.g., Adiprene L-100).

Overtone and combination vibrations for a monosubstituted aromatic will occur as small, regularly spaced bands from 1950 to 1650 cm^{-1}. These bands, when seen in the spectrum for a polymer, may indicate that polystyrene or phenolic resin is a sample component.

Carbon double bond region (1800–1500 cm^{-1}). Because the carbonyl band (1850–1650 cm^{-1}) is one of the most important and usually one of the strongest, this region is sometimes the first examined. While it can be difficult to assign an exact type of carbonyl (i.e., ester, aldehyde, ketone, etc.) in every case, the presence or absence of any carbonyl band is a key feature of a spectrum.

Polymers with a strong carbonyl band in the region of 1750–1700 cm^{-1} are polyesters (e.g., Mylar), acrylics (e.g., Acryloid), alkyds (e.g., Glyptal), poly(vinyl acetates) (e.g., AYAA), plasticized polyvinyl chlorides (e.g., vinyl storage sleeves), polyurethanes (e.g., Adiprene L-100), and cellulose esters (e.g., cellulose acetate). A carbonyl band shifted down to about 1650 cm^{-1} may indicate that polyamines (e.g., Melmac), polyamides (e.g., nylon), or cellulose nitrates (e.g., Duco, collodion) are present.

Examples of polymer classes that have weak or absent carbonyl bands in the region of 1750–1650 cm^{-1} are polyolefins (e.g., polyethylene, polypropylene, etc.), polystyrenes (e.g., Styrofoam, Fom-cor), fluorocarbons (e.g., Teflon), phenolic resins (e.g., Bakelite), unplasticized polyvinyl chlorides (e.g., Geon), polyvinyl alcohols (e.g., Elvanol), acrylonitriles (e.g., Orlon, Saran F-120), regenerated celluloses (e.g., cellophane, rayon), cellulose ethers (e.g., methyl cellulose, Ethylcel), silicone resins, and some epoxies (e.g., Araldite).

Epoxies that are a mixture of epichlorohydrin and Bisphenol A consistently produce spectra with a strong absorption band near 1510 cm^{-1} and a weaker but sharp band at 1610 cm^{-1}. Esterified epoxies and plasticized epoxies will show a carbonyl band near 1720 cm^{-1}.

Phthalates are plasticizers that produce a strong carbonyl band at 1735 cm^{-1} and small but distinct doublet bands at 1600–1585 cm^{-1}. They are often combined with cellulose nitrates, polyvinyl chlorides, alkyd resins, or polyesters. Organic acids and acid salts, such as stearates, are also common additives. They usually have one or two sharp absorption bands near 1585–1545 cm^{-1}.

Absorbed water, in the sample or in an accompanying alkali halide salt plate, will produce a weak, broad absorption band near 1650 cm^{-1}.

Fingerprint region (1500–500 cm^{-1}). The fingerprint region is most useful when the sample spectrum is directly compared to reference spectra. When two spectra are overlaid, any missing or absent bands in either spectrum show that the compositions of the two materials are different, even if the functional group bands are the same.

A strong, broad absorption in the range of 1100–1000 cm^{-1} may indicate regenerated celluloses (e.g., cellophane, rayon), cellulose ethers (e.g., methyl cellulose, Ethylcel), silicone resins, or inorganic additives. Silicone resins have very characteristic sharp bands at 1260 and 800 cm^{-1}, one on each side of the broad band.

Aromatic compounds have sharp, strong bands in the region of 1000–650 cm^{-1}. The presence of sharp bands in this region may indicate compounds such as polystyrene, epoxies, phenolic resins, some polyurethanes, or phthalate plasticizers.

Colorants

Minerals, clays, and other inorganic materials are commonly used as colorants and fillers in paints. Many of these materials may be readily identified by IR spectroscopy as well as by several analytical methods, such as scanning electron microscopy with energy dispersive spectroscopy, X-ray diffraction, X-ray fluorescence, and polarized light microscopy. These other methods, which generally provide elemental analysis or crystal structure, are complementary to the molecular species information obtained from IR. An advantage for IR is that it supplies information on both the organic and inorganic components in a paint.

An excellent source for reference spectra on artists' colorants is the *Artists' Pigments* handbook series, volumes 1, 2, and 3, which provide in-depth chapters on the characterization of numerous specific pigments, each of which provides corresponding IR spectra (Feller 1986; Roy 1993; West-FitzHugh 1997). Additionally, a review article by Newman on IR spectroscopy for the analysis of painting materials provides interesting applications and comparative spectra, as well as an extensive bibliography (Newman 1980). He cites several articles on the IR spectral identification of pigments, including Riederer 1969; Kühn 1970; Gettens, West-FitzHugh, and Feller 1974; Gettens and West-FitzHugh 1974; Riederer 1974; and Siesmayer et al. 1975. More-recent articles use IR as one method to identify pigments in paints (Lear 1981; Guineau 1983), in grounds (Schulz and Kropp 1993), and in inks on medieval manuscripts (Orna et al. 1989). IR spectroscopy was also used in the following technological studies of pigment and paint sets: Winslow Homer's watercolor box was analyzed by Newman, Weston, and Farrell (1980); the Hafkenscheid collection of 139 pigments and painting materials dating from the early nineteenth century was analyzed by Pey (1987); and a collection of pigments used in a 1929 diorama at the Missiemuseum at Steyl-Tegelen was analyzed by de Keijzer and Karreman (1989).

IR spectroscopy is also useful in the identification of organic colorants and dyes, particularly when they occur as pure materials. For example, phthalocyanine blues and greens produce very distinctive sharp bands from 1700 to 700 cm^{-1} (Newman 1980). Other studies also used IR to identify synthetic organic pigments (Venkataraman 1977; Strauss 1984; Gillard et al. 1994). When organic colorants or dyes have been mordanted on a fiber or particle, IR analysis of the combined material produces a strong spectrum for the fiber or particle overlaid on the colorant absorption bands. This hinders the colorant's identification by IR spectroscopy, particularly if it is present in low concentrations (Kirby and White 1996). McGovern and Michel (1990) used IR, as one of several techniques, for the confirmation of royal purple dye in an archaeological vessel. Low and Baer (1978) studied alizarin complexes prepared on an alumina lake. Colorants in fragile historic and prehistoric colored fabrics were characterized by IR microspectroscopy (Martoglio et al. 1990; Jakes, Katon, and Martoglio 1990). Martoglio and coworkers also showed that IR characterization of colorants can be assisted by the use of UV/Vis microspectroscopy on in situ material (Martoglio et al. 1990). Another analytical method for the identification of colorants is first to extract the colorant from the substrate with acid or base and then to use IR, UV/Vis, or thin-layer chromatography for characterization (Schweppe 1979, Schweppe 1989; Saltzman, Keay, and Christensen 1963; Saltzman 1992; Hansen, Wallert, and Derrick 1995). Typically, chromatographic analysis methods are preferred for the analysis of dyes. McClure, Thomson, and Tannahill (1968) provide ninety-six reference IR spectra for organic colorants. Digitized spectral collections of organic dyes and colorants are distributed by Bio-Rad Sadtler and Aldrich-Nicolet (Appendix I). See the above references for more information on the identification of dyes, as further details are not given in this text.

Characterization process for pigments

Six exemplary IR spectra for inorganic materials used as pigments or fillers are shown in Figure 5.19. In general, absorption bands for inorganic materials are broader, are fewer in number, and occur at lower wavenumbers than absorption bands for organic materials. This is due to their external ion structure (i.e., solid, sometimes crystalline matrix), as well as to their internal ion composition (i.e., functional groups). This section will examine the basic characterization of absorption bands for inorganic materials. Additional information on vibrational group assignments of minerals and inorganic materials may be found in the following references (see Appendix I): Farmer 1974; Bellamy 1975; Gadsen 1975; Zeller and Juszli 1975.

External ion structure. Pigments, minerals, and clays used in works of art are solids at room temperature. With the exception of amorphous silica, the repeating set of molecular units (–A-B–A-B–A-B–) exists as a semirigid matrix in a layered, or three-dimensional, crystalline structure called a lattice. The lattice structure restricts many molecular transitions: translational, rotational, and even some vibrational motions. In some cases, each functional group (A, for example) may not have an environment identical, or equivalent, to other A functional groups. Depending on their position in the overall pattern of molecules, some A functional groups may freely vibrate, while other A functional groups may have vibrational motion limited by space or direction. This situation could result in a split, or degenerate, vibrational band that appears as two or more adjacent bands. Alternatively, the restricted motion can broaden absorption bands.

Vibrational transitions of the lattice structure, or lattice vibrations, usually occur at low wavenumbers in the far-IR region. Thus, this region is particularly useful for detailed studies of a mineral's structure.

Internal ion composition. A simple inorganic compound, such as table salt (sodium chloride, NaCl), has one cation, sodium (Na^+), and

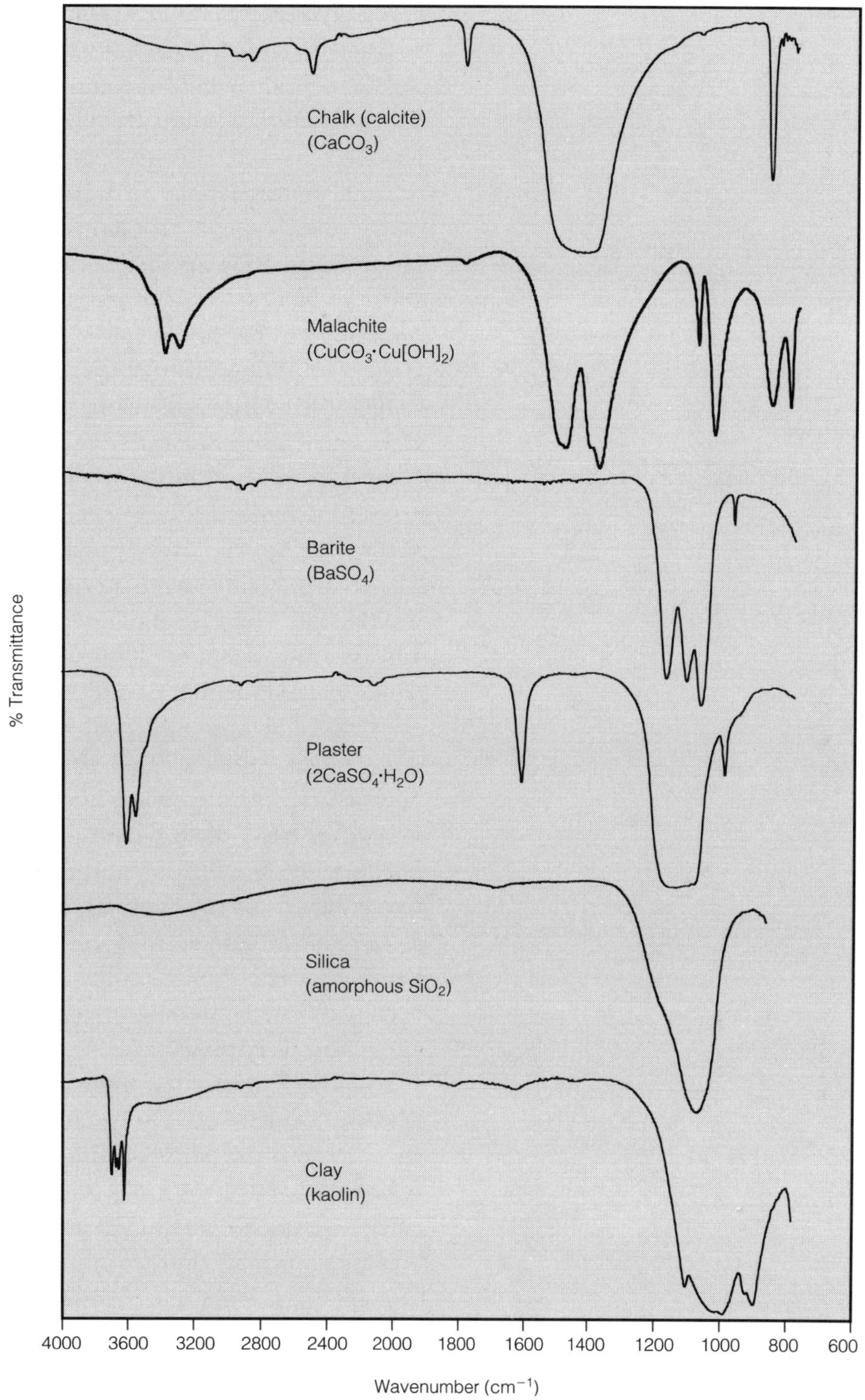

Figure 5.19

Exemplary IR transmittance spectra for six inorganic pigments and fillers.

one anion, chloride (Cl^-), and thus, it can be classified as a simple anionic compound. Most simple anionic compounds will not produce any vibrations in the mid-IR range, and their lattice vibrations occur in the far-IR region. Because of this lack of absorption bands in the mid-IR region, many materials with simple anions (e.g., NaCl, KBr, zinc selenide [ZnSe], silver chloride [AgCl], etc.) are used as IR transparent windows.

A more complex inorganic compound, such as chalk (calcium carbonate, calcite, $CaCO_3$), has calcium (Ca^{2+}) as the cation and carbonate ($CO_3^=$) as the anion. Carbonate can be classified as a complex anion, because the anion is itself a functional group. The covalent bonds in the carbonate tightly hold the anion together. Molecular vibrations within the anion functional group are termed internal vibrations. Complex anions produce characteristic absorption bands that are very useful for characterizing inorganics (Table 5.3). The attached cation, whether it is calcium, magnesium, or lead, will have only a slight effect on the position of the absorption bands for the complex anion. In general, heavier cations will shift the band to a lower frequency (Nyquist 1968). The effect is most prominent in the lower-wavenumber bending vibrations.

Water of hydration. Water adsorbed on the surface of a sample or incorporated in an amorphous structure will produce broad O-H stretches (3400 and 1650 cm^{-1}), such as were noted in organic compounds. Water molecules incorporated in the lattice structure of a crystalline molecule, however, will produce specific, sharp, characteristic absorption bands in the regions of 3800–3200 and 1700–1600 cm^{-1}. The environmental symmetry seen by each hydroxyl group determines whether the band is single or split. The unique patterns of the hydroxyl absorption bands near 3800–3200 cm^{-1} are very important for characterizing the composition of hydrated inorganics.

Hydrated layered silicates (e.g., kaolinite, talc) resemble a double-deck sandwich with the water molecules positioned in sheets between the silicate anions and interlayer cations. This lattice conformation restricts the direction in which the O-H groups can vibrate. Fortunately, fine distinct absorption bands are produced that are a sensitive indicator of the material. For example, O-H stretches for talc occur as one sharp band at 3676 cm^{-1}, with a smaller band at 3660 cm^{-1}. The hydroxyl stretches for kaolinite show as two sharp bands at 3700 and 3620 cm^{-1} with two slightly smaller bands between 3670 and 3652 cm^{-1} (Appendix I, Farmer 1974).

Carbonates. Carbonates, such as calcite ($CaCO_3$), cerussite ($PbCO_3$), azurite ($2CuCO_3 \cdot Cu(OH_2)$), or malachite ($CuCO_3 \cdot Cu(OH_2)$), show at least one strong absorption band from C-O stretching in the region of 1550–1350 cm^{-1} (Fig. 5.19). In anhydrous compounds, such as calcite, the band is smooth, symmetrical, and broad. In hydrated carbonates, such as malachite, this absorption band is split.

Table 5.3

Vibrational frequencies for selected complex anionic groups.

Anion	Stretch (cm^{-1})	Bend (cm^{-1})
Water of hydration ($\cdot H_2O$)	3800–3200	1700–1600
Carbonate ($CO_3^=$)	1550–1350	900–650
Nitrate (NO_3^-)	1500–1250	850–700
Sulfate ($SO_4^=$)	1200–1050	680–600
Phosphate ($PO_4^{\equiv}$)	1300–900	600–550
Silicate ($SiO_3^=$)	1200–800	800–400
Formates, acetates, and oxalates	1700–1300	1050–700
Cyanates, cyanides, and thiocyanates	2200–2000	—

Carbonate bending vibrations produce sharp bands in the region of 900–650 cm^{-1}. These bands show measurable frequency deviations corresponding to the attached cation. For example, the out-of-plane bending vibration for calcium carbonate (calcite) occurs at 872 cm^{-1}, while the same vibration occurs at 841 cm^{-1} for cerussite (lead carbonate) and at 820 cm^{-1} for malachite (basic copper carbonate) (Newman 1980). Because few organic compounds have strong absorptions in this region, these sharp bands are very useful for confirmation and identification of carbonates in a spectrum.

Hydrated carbonates—hydrocerussite ($PbCO_3{\cdot}Pb(OH_2)$), azurite ($2CuCO_3{\cdot}Cu(OH_2)$), or malachite ($CuCO_3{\cdot}Cu(OH_2)$)—have absorption bands due to hydration at 3535, 3425, and 3400/3320 (split) cm^{-1}, respectively (Appendix I, Farmer 1974). Additionally, small, sharp O-H bending vibrations occur at 1100–1000 cm^{-1} for the hydrated carbonates.

Sulfates. Sulfates, such as gypsum ($CaSO_4{\cdot}2H_2O$), anhydrite ($CaSO_4$), and barite ($BaSO_4$), show a strong S-O stretching vibration band in the region of 1200–1050 cm^{-1} (Fig. 5.19). This band is split if each of the S-O bonds in the tetrahedral sulfate anion sees a different environment, as in the barite. Another small band occurs near 1000 cm^{-1}, because of an S-O bending vibration, along with sharp, slightly stronger bands at 700–600 cm^{-1}.

Changes in the state of hydration make noticeable differences in spectra. For example, plaster ($2CaSO_4{\cdot}H_2O$) and gypsum ($CaSO_4{\cdot}2H_2O$) can be readily distinguished by the hydroxyl absorption bands near 3500 and 1600 cm^{-1}. Specifically, plaster has hydroxyl bands at 3615, 3465, and 1630 cm^{-1}, while gypsum has bands at 3555 and 1690 cm^{-1}. Souza and Derrick used IR spectra to quantitatively determine the proportions of plaster and gypsum in *gesso sottile* and *gesso grosso* layers (Souza and Derrick 1995).

Silica and silicates. Amorphous silica produces a strong Si-O stretching band near 1050 cm^{-1} that has a recognizable asymmetric shape from a shoulder near 1200 cm^{-1} (Fig 5.19). Nearly all types of glass (including smalt) produce an absorption band that is similar in appearance, even though glass is composed of a wide range and mixture of materials. The similar IR spectra occur because the basic glass structure contains an Si-O backbone. Since IR spectroscopy cannot differentiate glass types, it is not the method of choice for characterizing glass constituents. While the crystalline silica mineral quartz has the same primary absorption band near 1100 cm^{-1}, it also has a unique, but small, doublet band near 790 cm^{-1} that is very characteristic.

Silicates have a fully ordered crystalline lattice structure. This produces a well-defined Si-O absorption band at 1200–800 cm^{-1}. In layered silicates, such as kaolin, the band is split into two or more peaks, since some of the Si-O bonds are held perpendicular to the layers, while other vibrate in-plane with the layers. In three-dimensional silicates, such as ultramarine (lazurite), the Si-O absorption band is smoother, and splitting, when it occurs, is less defined. Bending vibrations for Si-O usually occur below 600 cm^{-1}.

The hydration absorption bands for layered silicates, mentioned earlier, are uniquely distinctive. Because of restricted molecular motion of the water molecules, these absorptions occur as sharp, well-defined bands in the region of 3700 cm^{-1}. Excellent reference spectra for these silicate hydration absorption bands are provided in Farmer (1974; see Appendix I). These bands are useful for identifying silicate species, even in instances where X-ray diffraction has failed.

In the production of ceramics, clay undergoes thermal transformations in its lattice structure. At 400–600 °C, clay dehydrates, losing all water molecules. An IR spectrum collected on a clay heated to this temperature will not have any hydroxyl absorptions. When clay is heated to higher temperatures, 970 °C for kaolinite, the lattice structure collapses, producing an amorphous, glasslike material. Accordingly, its IR spectrum will resemble glass, with a single broad absorption band at 1200–800 cm^{-1}.

Often the IR spectra for ochres or earth pigments, found in both ancient and modern paintings, correspond to spectra for silica and silicates (Newman 1996). As the natural earth pigments were simply dug out of the ground, they contain a mixture of minerals, including clay and quartz. The clay and quartz are readily identified by their absorption bands, while the compounds responsible for the pigment color rarely have any absorption bands in the mid-IR range. The color of red ochre is due to anhydrous iron oxide (hematite); that of yellow ochre is due to hydrated iron oxides (most often goethite); and brown earth pigments, such as umbers, contain manganese oxides. Of these colored compounds, only the absorption bands near 850 cm^{-1} for goethite appear in the mid-IR region.

Infrared spectral characterization of pigments

With a typical mid-IR FT-IR instrument, most inorganic compounds that contain complex anions (carbonates, sulfates, silicates, etc.) can be identified. Inorganic compounds that contain simple anions (oxides, sulfides, etc.) often cannot be identified. Since many inorganic pigments and fillers contain complex ions, they are amenable to identification by mid-IR region analysis. Table 5.4 lists selected pigments with their formulas and mid-IR absorption band regions.

Identification of a spectrum should always be made by final, direct comparison of the unknown spectrum to those in reference spectral collections. However, with pigments and minerals, it is important to evaluate information known about the sample and reference, especially when commercial pigments are used as references. Some commercial sources may label pigments on the basis of their color rather than their composition. Additionally, many pigments sold in the nineteenth and twentieth centuries are actually mixtures of different compounds, such as a colored compound with one or more fillers (Newman 1996). Since some components in a pigment mixture may have absorption bands in the mid-IR region while others do not, the IR spectrum of the mixture may not represent the total composition of the material. For example, titanium dioxide does not have any absorption bands in the mid-IR region, but a

Table 5.4 (opposite page)
Pigments and fillers correlated with IR absorption band ranges. Most pigments with complex anions absorb in the mid-IR region. All pigments have absorption bands in the far-IR region (below 700 cm^{-1}). The frequency range is marked for pigments that are active in the mid-IR region. More than one band may occur within each region (s = strong; m = medium; w = weak).

Pigment	Formula*	Mid-IR bands?	3800–3200 cm^{-1}	2200–2000 cm^{-1}	1700–1200 cm^{-1}	1200–1000 cm^{-1}	1000–700 cm^{-1}
White (with extenders)							
Aluminum hydrate	$Al(OH)_3$	yes	s		m	s	
Anhydrite	$CaSO_4$	yes				s	
Barium sulfate	$BaSO_4$	yes				s	
Chalk (calcite)	$CaCO_3$	yes			s		m
Clay (kaolinite)	$Al_2O_3 \cdot 2SiO_2 \cdot 2H_2O$	yes	w			s	m
Gypsum	$CaSO_4 \cdot 2H_2O$	yes	m		m	s	w
Lithopone	$BaSO_4$, ZnS	yes				s	
Quartz	SiO_2	yes				s	w
Silica (amorphous)	SiO_2	yes				s	w
Talc	$3MgO \cdot 4SiO_2 \cdot H_2O$	yes	w			s	
Titanium dioxide	TiO_2	no					
White lead (basic carbonate)	$PbCO_3 \cdot Pb(OH)_2$	yes	w		s		w
White lead (basic sulfate)	$PbSO_4 \cdot PbO$	yes				s	
Zinc oxide	ZnO	no					
Red and orange							
Cadmium red	CdS, CdSe	no					
Cadmium red (lithopone)	CdS, CdSe, $BaSO_4$	yes				s	w
Chrome orange	$PbO \cdot PbCrO_4$	yes				s	m
Realgar	As_2S_2	no					
Red lead	Pb_3O_4	no					
Vermilion	HgS	no					
Yellow							
Barium yellow	$BaCrO_4$	yes					s
Cadmium yellow	CdS	no					
Chrome yellow	$PbCrO_4$	yes					s
Massicot (litharge)	PbO	no					
Lead tin yellow	Pb_2SnO_4	no					
Strontium yellow	$SrCrO_4$	yes					s
Green							
Chromium oxide	Cr_2O_3	no					
Chrysocolla	$CuSiO_3 \cdot nH_2O$	yes	m				
Cobalt green	$CoO \cdot nZnO$	no					
Green earth	K,Al,Fe,Mg,hydrosilicate	yes	w			s	w
Malachite	$CuCO_3 \cdot Cu(OH_2)$	yes	m		s	w	m
Verdigris	$Cu(CH_3COO)_2 \cdot 2Cu(OH)_2$	yes	m		s	w	
Viridian	$Cr_2O_3 \cdot 2H_2O$	yes	w			w	
Blue							
Azurite	$2CuCO_3 \cdot Cu(OH_2)$	yes	m		s	w	m
Cerulean blue	$CoO \cdot nSnO_2$	no					
Cobalt blue	$CoO \cdot Al_2O_3$	no					
Egyptian blue	$CaO \cdot CuO \cdot 4SiO_2$	yes				s	w
Phthalocyanine blue	$C_{32}H_{16}N_8Cu$	yes			m	s	s
Prussian blue	$Fe_4(Fe(CN)_6)_3$	yes		s	m		
Smalt	potash glass	yes				s	
Ultramarine (lazurite)	SiO_2, Al_2O_3, Na_2O, S	yes				s	w
Brown and black							
Asphaltum (bitumen)	organic	yes			s	m	
Black oxide	Fe_2O_3	no					
Bone black (apatite)	$Ca_3(PO_4)_2$, carbon	yes		m		s	w
Charcoal	carbon	no					
Iron oxide pigments							
Yellow (goethite)	$Fe_2O_3 \cdot 2H_2O$	yes	w				s
Red (hematite)	Fe_2O_3	no					

*From Gettens and Stout (1966) and Afremow and Vandeberg (1966).

commercial formulation containing fillers, such as clay or barium sulfate, would produce a characteristic spectrum. In this instance, the recorded spectrum would indicate the fillers and not the actual pigment.

Several reference spectral collections, books, and digitized spectra (listed in Appendix I) are available for minerals and pigments; these are Sadtler Research Laboratories, Sprouse Scientific Systems, Inc., Nyquist and Kagel (1971), Afremow and Vandeberg (1966), Boldyrev (1976), Nakamoto (1978), and Price and Carlson (forthcoming).

Mixtures

Most conservation-related samples submitted for IR analysis are mixtures. Their IR spectra contain the superimposed spectra of each component present. The intensities of each component's bands are directly proportional to the concentration of that component. The resulting spectrum may be complex, with overlapping bands that obscure those needed for individual material identification. Thus, the interpretation of a spectrum from an unsuspected mixture may lead to erroneous conclusions. However, it is often possible to identify at least one component directly, and then eliminate its absorptions from further consideration. The pattern of other components, if not initially recognized, may then become clearer.

In paints, pigments are dispersed in an organic binder. This usually does not hinder the determination of the pigment, but it may make it difficult to determine the binder if it is present in low concentrations. Figure 5.20 illustrates the IR spectrum obtained from a 50/50 w/w% mixture of calcium carbonate and rabbit-skin glue. Even though the weight proportion of the two components is equivalent, the relative intensities of the strongest absorption bands for each component are not the same, because the absorptivity of each functional group depends on its dipole moment, bond strength, and molecular environment.

For the analysis of paints, it is helpful if the pigment component is identified and spectrally subtracted, thus providing a better indication of any binder absorption bands that may have been obscured by the overlapping pigment absorptions. When the inorganic component cannot be identified or when the subtraction does not work well, solvent extractions may be useful to physically separate the binder from pigments before each fraction is reanalyzed (for technique, see chap. 3). Figure 5.21 illustrates the IR spectrum obtained for a neat sample of red paint. The sharp absorptions bands due to a synthetic organic pigment overlapped the absorption bands for the binder, making it difficult to identify. After a drop of acetone was placed on the sample and allowed to evaporate, the dried residue of the acetone-soluble portion of the paint was analyzed. The absorption bands created by the red colorant were no longer noticeable. This spectrum corresponds well with an ethyl acrylate reference spectrum.

The average detection limit for most materials in a mixture is 5%. It may range from below 1% to 30%, however, depending on the absorptivity of the material and on the number of components in the mixture. The sensitivity to minor components may be enhanced by the use of techniques of spectral subtraction, deconvolution, and derivatiza-

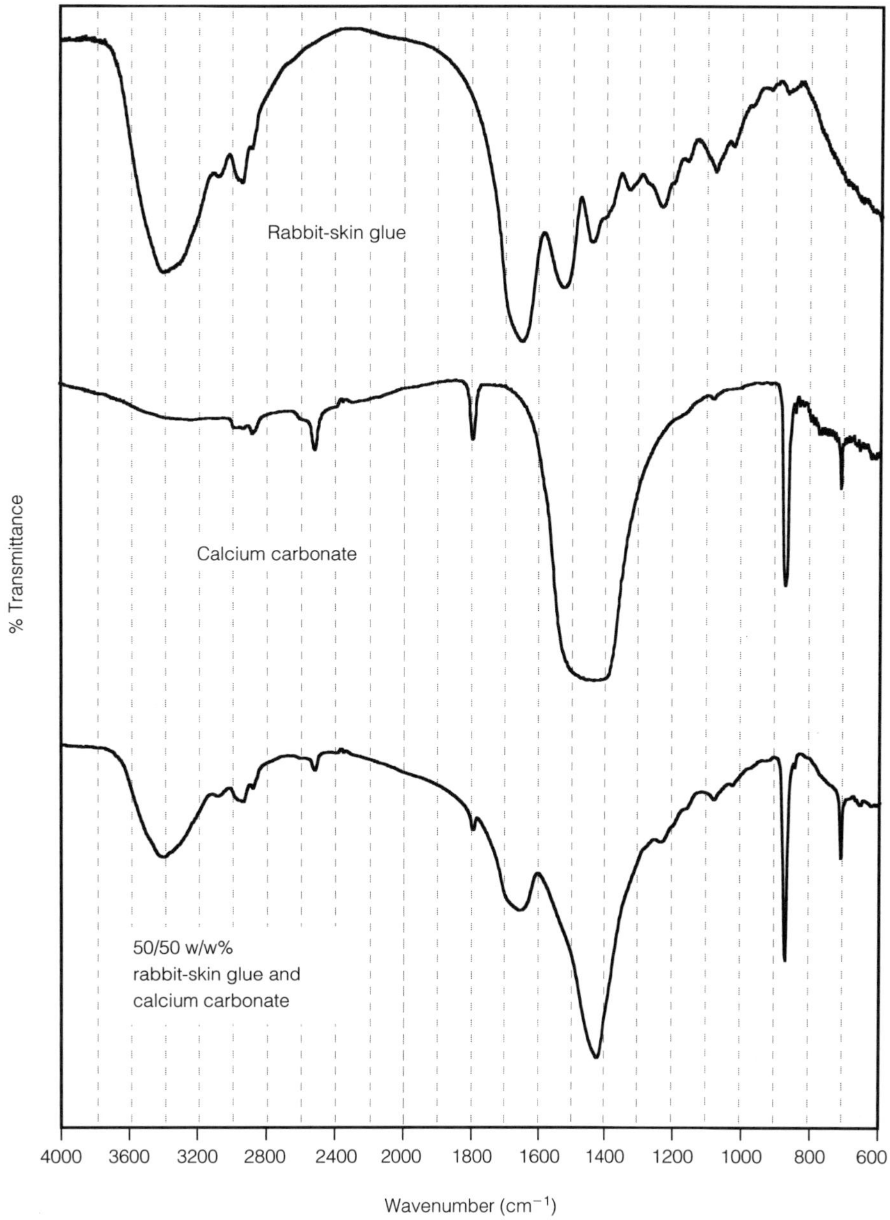

Figure 5.20
IR transmittance spectra of pure rabbit-skin glue, pure calcium carbonate (calcite), and a mixture containing 50% w/w% of each. The absorption bands for both individual compounds are present in the spectrum for the mixture, and their respective absorption band intensities correspond to the proportion of that material in the mixture. In material mixtures for which absorption bands overlap, it can be difficult to identify some components.

tion, as well as solvent extractions. More details about these spectral manipulation methods are given later in this chapter.

Quantitative Analysis

The intensities of spectral bands are used for quantitative (±1%) and semiquantitative (±10%) IR analysis. Because of potential spectral interferences, sample preparation variations, and sensitivity limitations, it is difficult to obtain highly accurate quantitative results with IR spectroscopy. Quantitative analysis of organic compounds is much more reliably achieved with chromatographic methods. Often, however, semiquantitative analysis is sufficient to answer the analysis question. In this case, IR has an advantage, because it can readily supply qualitative and

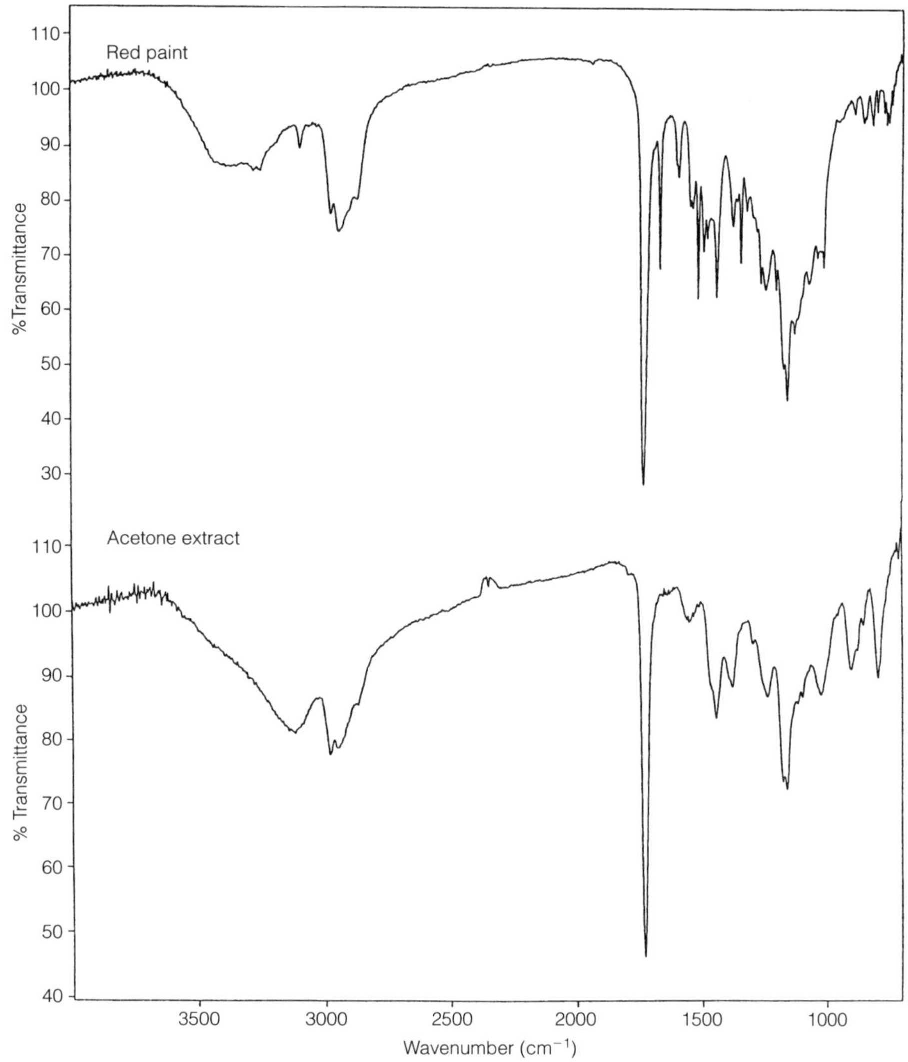

Figure 5.21
IR transmittance spectra for a neat red paint sample (top) and for the acetone-soluble portion of the sample (bottom). The absorption bands from the organic colorant hinder the identification of the binder. A solvent extraction separates some of the binder from the pigment, allowing for a "cleaner" spectrum to be collected. The spectrum for the acetone extract corresponds to an ethyl acrylate reference spectrum.

semiquantitative analysis results. In order for quantitative calculations to be performed, however, the material must first be identified.

The capacity of any component to absorb IR radiation is constant. This capacity is termed its molar absorptivity. Additionally, the intensity of any specific absorption band in relation to another is constant, because the intensity of an absorption band is directly proportional to the rate of change in the dipole moment of that particular vibration. A large change in the dipole moment of the atoms during a vibration will produce an intense band. Thus, very polar functional groups, such as those containing halogens, will exhibit intense absorption bands. An intense absorption band can also be produced by the presence of multiple functional groups within the molecule, such as CH_2 groups in paraffin wax, that each have the same vibrational energy, such that an additive effect is created.

There is a linear quantitative relationship between absorbance and concentration of absorbing molecules:

$$A = \in b\ c = -\log(1/T)$$

where: A = absorbance; $\in$ = molar absorptivity (a constant for the molecule); b = sample pathlength; c = concentration; and T = transmittance (%T/100).

This equation is called the Beer-Lambert law—or simply Beer's law—and it is used to determine a relationship between measured absorptivity, band intensity, and concentration of IR-active functional groups in the sample. The linear range for the relationship holds for spectra with a maximum absorption band between 20 and 65%T (0.7–0.2 absorbance). The equation shows that there is a one-to-one relationship between the height, or intensity (in absorbance units), of an absorption band and the concentration of that molecule. Note that the linear relationship holds for absorbance and not for transmittance, which has a logarithmic relationship. Thus, for quantitative work, spectra should be plotted in absorbance units.

All quantitative IR analyses are done by comparing the intensity of a specific absorption band, in absorbance units, of the unknown material with the absorbance, or band height, of the same material in a standard of known concentration. In a mixture of materials, the absorbances are additive; thus, the total absorbance at any given wavelength is the sum of the absorptions of the individual components. Therefore, for quantitative analysis of a material, it is advantageous to select an absorption band that not only is characteristic of that material but also is isolated from absorption bands due to other materials in the sample.

Beer's law shows that sample pathlength is also a factor in the measurement. For one quantitative method, *direct calculation* of concentration, the pathlength must be either known or fixed. Thus, direct measurement is normally limited to liquids or solutions that can be analyzed in a fixed-pathlength liquid cell. In this method, the unknown concentration of an identified single component can be calculated from a calibration curve. The calibration curve is prepared by analysis of the same component in solutions, or mixtures, of at least four different concentrations. An absorption band is selected that is characteristic of the component of interest and that is free from interferences. Then a plot is made of the absorbance value for that band versus the concentration of the component in each solution. The concentration of the sample is determined by comparing the intensity of that particular band in its spectrum with the calibration curve. The intensity of the band is measured as the absorbance difference from its maximum to its baseline. The baseline is drawn where the pen tracing would go if the band were not present (Smith 1979). Integrated, or total-area, measurements of the absorption band are rarely needed, since intensity measurements can be made more reproducibly and accurately (Smith 1979). Because the pathlength cannot be determined precisely, this method is not used for films and pellets.

The *absorbance ratio* method is used when the pathlength of the sample cannot be readily determined. The method works well for films, pellets, and diffuse or internal reflection measurements. For this method, at least two components (A and B) must be in the sample matrix, and each must have an absorbance band that exhibits minimum interference. Because the components are present in the same sample, the pathlength is the same and is no longer a variable. The calibration curve is generated from at least four spectra obtained from mixtures of the components in different proportions. The ratio of the intensities of the two bands of interest (I_A/I_B) is plotted versus the ratio of their concentrations (C_A/C_B). Once the curve is generated, the ratio of the concentration of the components in the unknown sample can be determined, since the sum of their concentrations equals unity, or 100%. Thus, the specific concentrations for each component can be easily calculated.

The absorbance ratio method was applied to the study of archaeological wood deterioration by Kirillov and Mikolajchuk (1990). Ferrus, Pages, and Diez (1981) used quantitative analysis for the examination of kaolin-casein coatings on papers. Other methods, such as the internal standard method used by Biscontin and Volpin for the analysis of calcium oxalate films (1989), are also available. For other methods of quantitative analysis, see the IR spectroscopy atlas published by the Chicago Society for Paint Technology (1980:53–58) and Smith (1979:219–68).

Mathematical Manipulations of Spectra

The linear proportionality of a material's concentration to its absorbance band intensity (Beer's law) is the basis for many of the IR data processing algorithms, such as spectral subtraction, spectral searching, and factor analysis. Other routines, such as Fourier self-deconvolution, can aid in the resolution of bands.

Subtraction techniques

Computer spectral subtraction methods are used to separate spectra of components in a mixture, to remove bands due to impurities, to confirm the identity of a sample, and to detect small changes in a sample. When one component is identified, its pure spectrum (in absorbance units) is then multiplied by a scaling factor and subtracted out, and the remaining spectral bands are examined for any differences. This is known as scaled absorbance subtraction.

Typically, an Autoscale subtraction computer routine estimates a scaling factor based on the integrated areas of several peaks. While this may be a good estimate, the best method for spectral subtraction is to select visually a single band of the component to be removed in both spectra, then null out that band (Griffiths and de Haseth 1986). The resultant difference spectrum should be evaluated closely to see which bands still exist and to determine if there are other potential components not initially considered.

When subtraction is used to confirm the identity of a material, its reference spectrum is subtracted from the sample spectrum; if the difference is zero, then it is clear that the two spectra are of the same compound. This is one method used by search routines to check variations between an unknown spectrum and the reference spectra.

Difference spectra, resulting from spectral subtraction, are useful for the evaluation of changes in a material due to chemical reactions or aging. The difference spectrum will show only the regions of changing absorption, while the molecular features that remain constant are removed by the subtraction operation.

For optimum subtraction results, the two initial spectra should be recorded from samples prepared and analyzed in the same manner. Their maximum absorption bands should be no more than 0.7 absorbance units (no less than 20%T). This guideline ensures that the components' concentrations are within the linear range covered by Beer's law. Slight changes in band positions or shapes can result in derivative-like bands or positive residuals occurring in the difference spectrum. These artifacts may be incorrectly attributed to the presence of minor components. For more information on the limitations of spectral subtraction and its applications, see Koenig (1992) and Bartick, Corbett, and McClure (1982).

Figure 5.22 illustrates the use of spectral subtraction for the identification of a minor component. The spectrum on top (A) is shellac plus a small amount of an unknown material, indicated by the bands at 1800 and 870 cm^{-1}. The middle spectrum (B) of pure shellac was spectrally subtracted from the top spectrum, and the resultant spectrum (C), which corresponds to calcium carbonate, is shown at the bottom.

Resolution enhancement methods

IR absorption bands are relatively broad and may overlap one another. This is particularly true for the spectra of mixtures. Resolution enhancement methods can artificially narrow the width of the bands and thus minimize overlap. Various mathematical methods, such as derivative spectroscopy and Fourier self-deconvolution, are iterative optimization processes performed on the interferogram. The operator selects band shape and bandwidth parameters; the maximum amount of bandwidth reduction is determined by the initial resolution at which the spectrum was measured.

In derivative spectroscopy, the measured absorbance spectrum is first transformed back into an interferogram. Then the computer multiplies it by a function to compute the *n*th derivative. While any order of derivative may be calculated, the even-order functions (2, 4, etc.) provide the better resultant spectra. The second derivative is the most commonly used. The resultant spectrum will have sharper bands than those of the original spectrum. Drawbacks are that strong bands will have side lobes and that spectrum noise is increased. Figure 5.23 illustrates the use of derivative spectroscopy to increase the resolution of overlapping hydrocarbon bands.

A deconvolution program may also help elucidate information in a spectrum. It operates on two or more overlapping bands to reduce the line width of the individual components and therefore to improve the resolution of each band up to three times (Kauppinen et al. 1981; Koenig 1992). Convolution is a broadening function that occurs in the production of a spectrum and changes the intrinsic line shape of the absorption

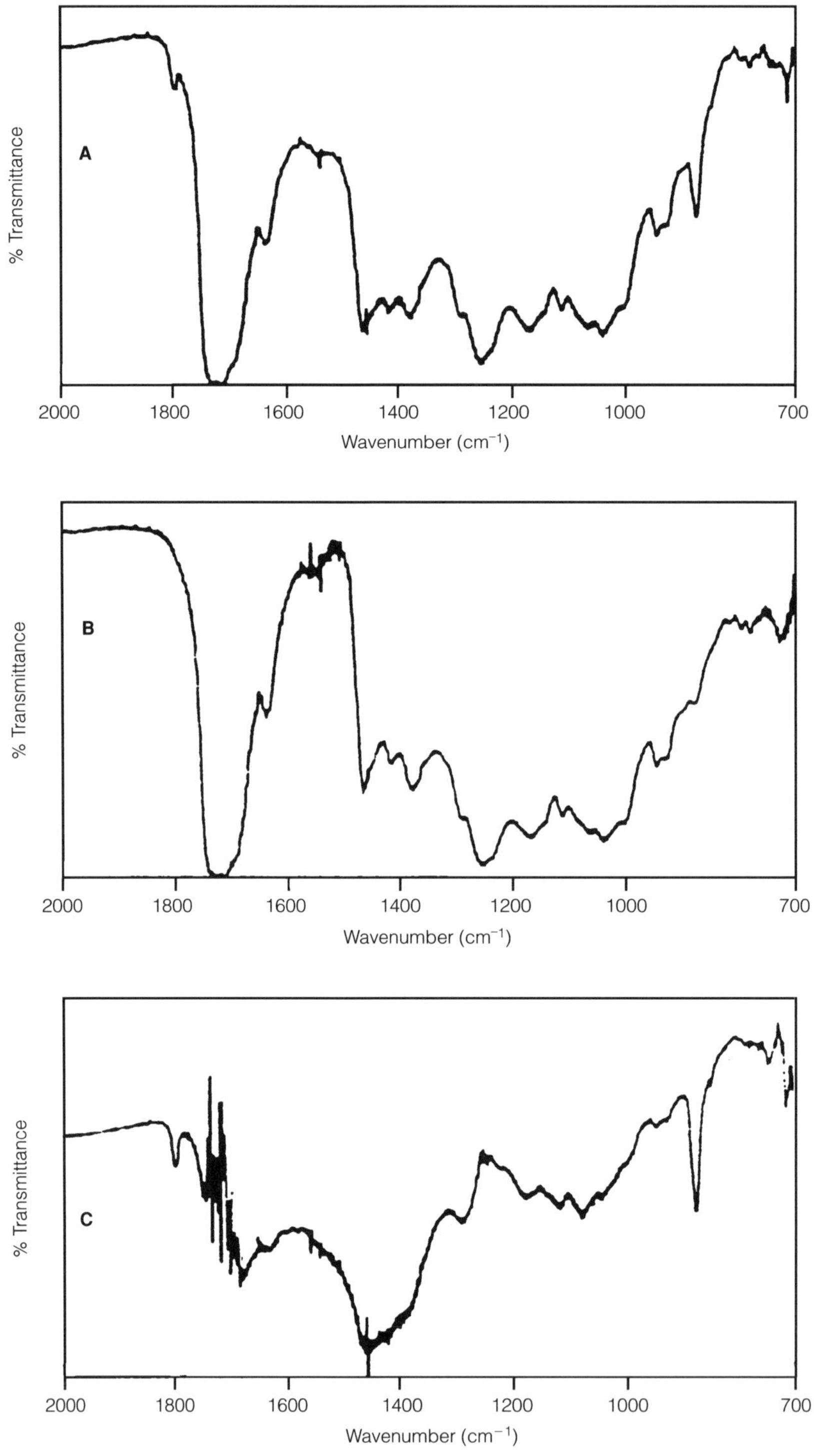

Figure 5.22
An example of spectral subtraction (A – B = C) for the identification of a minor component. Slight differences between the top spectrum (A) of a shellac and calcium carbonate mixture, and the middle spectrum (B) of shellac were noted. The middle spectrum (B) was subtracted from the top spectrum; the resultant spectrum (C) corresponds to reference spectra for calcareous materials.

into peaks (bands). Deconvolution is designed to correct for the broadening function and thus narrow the bands. The optimum use of deconvolution requires information about the true width of the absorption band, which is usually not known. If the spectrum is under-deconvoluted, little improvement in resolution will result. If the spectrum is over-deconvoluted, extraneous side lobes, as well as distortions in band intensities, will be produced. Since the different IR bands have different

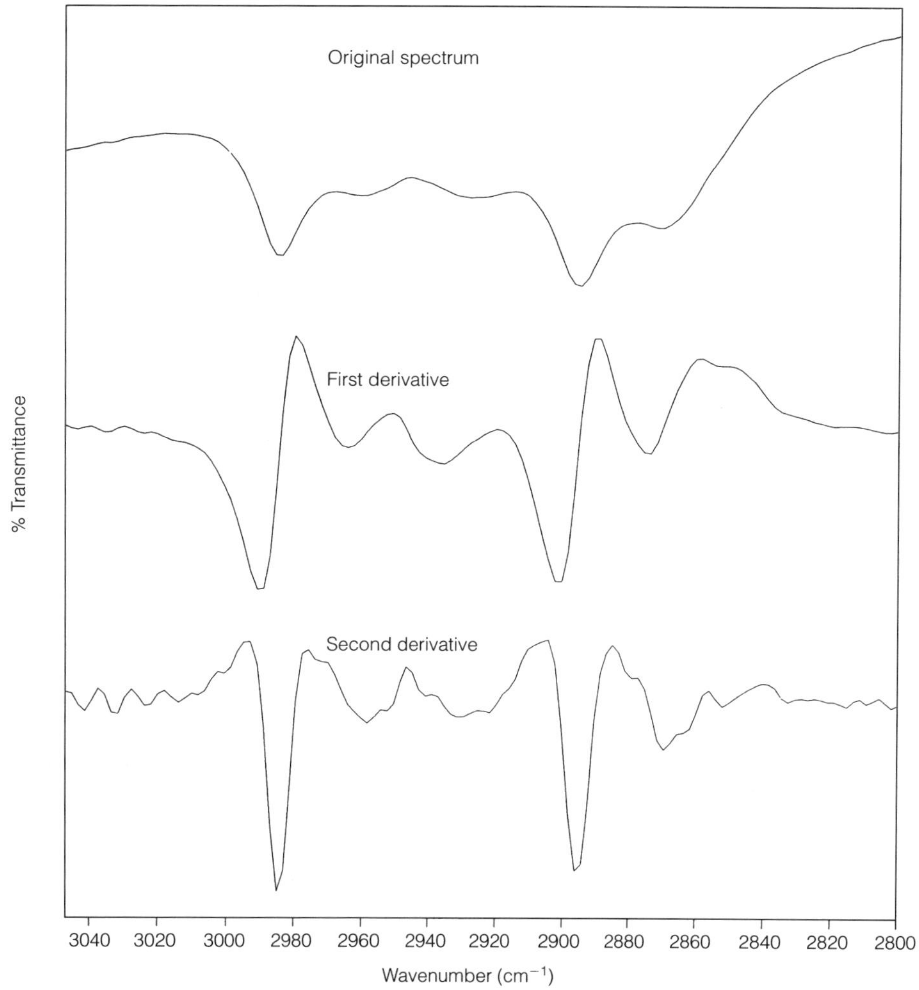

Figure 5.23
An example of derivatization, one method used to increase the effective resolution of a spectrum. The overlapping hydrocarbon bands for a silicone oil and hydrocarbon oil mixture are shown (3050–2800 cm^{-1}) for the original transmittance spectrum, the first derivative, and the second derivative. The second derivative is most often used, since the absorption band maxima are near their original positions. Derivatization changes the relative intensities of the bands, as well as increases the noise level of the spectrum.

inherent bandwidths, it is impossible to optimally deconvolute all bands. Thus, since the same deconvolution width is used for the entire spectrum, some bands may be overprocessed, while others are underprocessed. Negative bands may appear, and the signal-to-noise ratio is decreased.

An example of deconvolution is shown in Figure 5.24. A sample of a mastic-oil film from the Gettens and Stout Collection, dated 1934 (prepared on a glass plate as a dried film made from 5 cc oil in 100 cc ethanol, with 33 g mastic), was analyzed. Even though the components of the sample were known, because of the low amount of oil, the bands specific for oil were only slightly recognizable after deconvolution was performed.

Summary

Characterization of a material class for a pure sample is well within the realm of IR analysis. This chapter, as an introduction to skills and methods important in spectral interpretation, includes graphs, tables, and discussion aimed at facilitating spectral analysis. Initial examination of

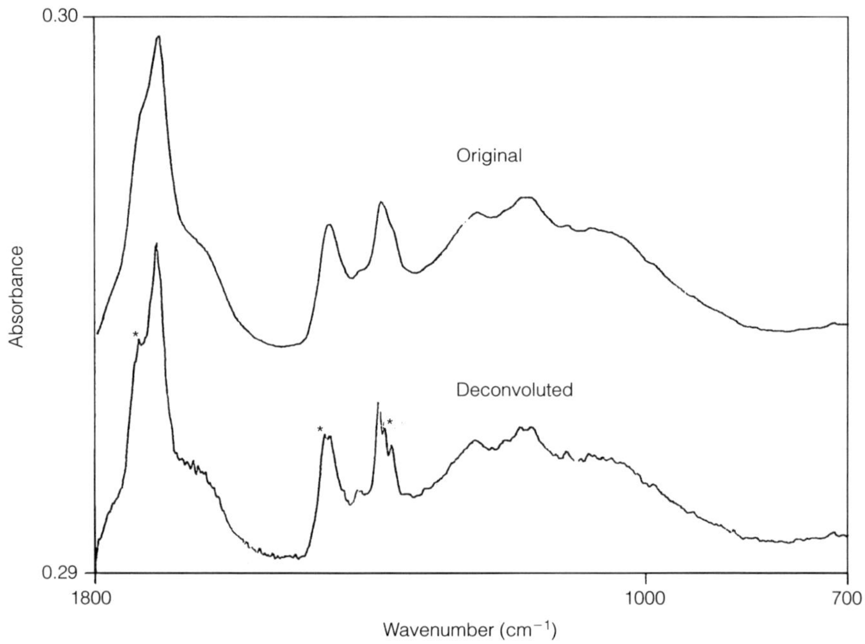

Figure 5.24
An example of deconvolution, another method used to increase the effective resolution of a spectrum. The original (top) and deconvoluted (bottom) absorbance spectra are shown for a sample containing a mixture of oil in mastic. Resolved bands due to the oil are marked with asterisks. However, because of the low concentration of oil in the sample, its absorption bands are poorly defined, even after deconvolution.

spectra quality is followed by a discussion of information gleaned from the spectral regions and spectra-structure correlations. In-depth information is then presented for the interpretation of IR spectra for natural products, synthetic resins, and colorants that are often found in samples from art materials and conservation treatments.

Samples from works of art, however, may be complex mixtures of components that are difficult to completely identify by IR spectroscopy alone. However, IR spectral analysis can characterize the material class(es) present within the sample and thus supply a basis for the selection of a secondary analysis method for further separation of sample components and their identification. For example, knowledge of a sample's component classes (oil, protein, carbohydrate, etc.) is fundamental to the selection of an appropriate chromatographic analysis protocol.

Positive identification by IR spectroscopy is often limited by the availability of relevant reference spectra. It is important to have a spectra collection that simulates the probable set of unknown materials. This is especially true for complex samples and mixtures. For the field of art conservation, it is also desirable to have a spectral collection of aged reference materials. Thus, an extensive set of relevant citations are presented.

While IR spectroscopy is a valuable method for the identification of an unknown, the spectra and the method of analysis must be examined critically before any conclusions are made. Computer spectral search routines can be very helpful, but the computer can never make the final decision. Instead, the analyst applies judgment to draw conclusions based on information derived from sampling, analysis, and interpretation.

Additional Reading

Allen, R. O., and P. Sanderson
1988. Characterization of epoxy glues with FTIR. *Applied Spectroscopy Reviews* 24:175–87.

American Society for Testing Materials (ASTM)
1991. *Standard Practice for General Techniques of Infrared Quantitative Analysis.* ASTM E 168-88. Annual Book of ASTM Standards, vol. 14-0. Philadelphia: ASTM.

Bellamy, L. J.
1975. *Advances in Infrared Group Frequencies.* London: Chapman and Hall.

1980. *The Infrared Spectra of Complex Molecules.* 2d ed. London: Chapman and Hall.

Chicago Society for Coating Technology
1980. *An Infrared Spectroscopy Atlas for the Coatings Industry.* Philadelphia: Federation of Societies for Paint Technology.

Colthup, N. B., L. H. Daly, and S. E. Wiberley
1990. *Introduction to Infrared and Raman Spectroscopy.* 3d ed. Boston: Academic Press.

Farmer, V. C., ed.
1974. *The Infrared Spectra of Minerals.* Mineralogical Society Monograph 4. Surrey, U.K.: Adlard and Son.

Silverstein, R. M., G. C. Bassler, and T. C. Morrill
1991. *Spectrometric Identification of Organic Compounds.* 5th ed. New York: John Wiley and Sons.

Chapter 6
Case Studies

Over thirty years ago, Jacqueline Olin stated that "infrared spectrophotometry is becoming an increasingly important method in the analysis, authentication, and preservation of paintings and ancient artifacts" (Olin 1966). Since that time, IR spectroscopy has been employed in several significant studies of museum materials, and it is now the leading choice for analysis of most organic, and of many inorganic, materials in art conservation research. IR spectroscopy enjoys such standing because it can be used for the analysis of small amounts of nearly any substance, thus encompassing the broad range of materials and restrictive sample sizes encountered in artifact analysis. The applications of IR are multifaceted, covering not only the identification of materials but also the evaluation of condition and the monitoring of chemical reactions.

Identification and Characterization of Materials

IR spectroscopy has long been a primary tool for the identification of art and archaeological materials. When Gettens promoted the technical examination of art in 1952, IR spectroscopy was included as a technique for the analysis of artifacts and paintings (Gettens 1952). In 1968 Masschelein-Kleiner and coworkers selected IR analysis as one method in a scheme for the identification of binding media, varnishes, and adhesives (Masschelein-Kleiner, Heylen, and Tricot-Marckx 1968). In 1969 Olin and coworkers reviewed favorably the emerging use of IR spectroscopy for the analysis of art objects (Olin, Salmon, and Olin 1969). Then, in 1977, Low and Baer showed the increased sensitivity of Fourier transform infrared (FT-IR) spectroscopy for the analysis of many natural organic materials (Low and Baer 1977). In 1989 Roelofs presented an analytical schematic for the identification of varnishes, media, and dyestuffs, using IR in a combination of modern techniques (Roelofs 1989).

One important application for IR spectroscopy is the determination of components in paint samples. Birstein used IR in several studies to determine the binders and pigments in Asian wall paintings and Egyptian tomb paintings, as well as to examine the aging characteristics of gelatin (Birstein 1975, 1978; Birstein and Tul'Chinskii 1977). Delbourgo used IR to analyze fresco painting cross sections (Delbourgo

1972). Both egg tempera and drying oils were found as media in Gothic panel paintings that were part of the main altar in the church of Tingelstad, Norway (Plahter, Skug, and Plahter 1974). Riederer used IR to examine the techniques and materials of early medieval mural paintings in Turkistan (Riederer 1977). Kobus describes the use of both IR and scanning electron microscopy with energy dispersive spectroscopy (SEM-EDS) for the microanalysis of small paint samples (Kobus 1987). A technical examination of Gérard Dou's *The Young Mother* used IR and gas chromatography (GC); results indicated the presence of bitumen (van der Loeff and Groen 1993).

Reference spectra of ethnographic materials are often generated as part of a corresponding technological study, since commercial libraries do not contain spectra of these materials. To examine the introduction of rice to the Pacific islands, Hill matched the extract of an archaeological specimen from Borneo to reference spectra of rice extracts (Hill 1983–84). As part of an investigation in the use of vegetable resins on Congo ceramics, a series of reference spectra for native resins were collected and compared to samples from decorative coatings on the ceramics (Hexter and Hopwood 1992). Pigments and media from 1,346 samples of historic Native objects were analyzed by the Canadian Conservation Institute to form a database characterizing material usage by tribes, regions, and time periods (Miller, Moffatt, and Sirois 1990).

Several technical studies have used IR spectroscopy in conjunction with other techniques to examine archaeological objects. Delbourgo and Gay analyzed organic remains found in archaeological excavations in Sudan (Delbourgo and Gay 1968). IR spectroscopy was used by Maniatis and coworkers to determine the hydration state of clays in low-fired terra-cotta statues from Greece (Maniatis, Katsanos, and Caskey 1982). Duraiswamy and coworkers later used this technique to determine the firing temperature of clays from excavations in India (Duraiswamy, Ramaswamy, and Venkatachalapathy 1992). X-ray fluorescence, atomic absorption, and IR studies were performed on tile and wall fragments from ancient Roman settlements (Blasius et al. 1983). IR was used for the initial examination of cosmetics from a Thracian tomb in Bulgaria; that examination was followed by chromatography for the final identification of individual components (Zolotovich and Popov 1969).

IR spectroscopy is ideal for the identification of synthetic resins used as coatings and binders. Jayme and Traser collected reference spectra for paper coatings and fillers and used them for the identification of coated papers (Jayme and Traser 1971). Cellulose nitrate has been identified as an emulsion used on early microphotographs (Newman and Stevens 1977). Stringari characterized acrylics used in artist paints (Stringari and Pratt 1993). Sharpless used IR to identify the plastics and their additives when he extensively characterized the effects of storage materials and cleaning products on the deterioration of coins in museum collections (Sharpless 1980).

IR spectroscopy can also help in the identification of some organic colorants and dyes (Masschelein-Kleiner and Maes 1978). Berberine, a traditional yellow Japanese dye, has been well characterized by fluorescent and IR studies (Matsuda 1986). Garrido used IR to differentiate between Maya blue (indigo on clay) and malachite for multiple blue painted Mayan objects (Garrido 1969). IR was also used as one method in the identification of royal purple dye, an indigo colorant used in antiquity found in the interior of jars in a thirteenth-century-B.C.E. Lebanese archaeological site (McGovern and Michel 1990). Roelofs and coworkers discussed IR in their comprehensive study of the advantages and disadvantages of several analytical methods for the identification of dyestuffs (Roelofs et al. 1987). Very small samples of archaeological fibers and dyes have been analyzed by IR microspectroscopy (Lang et al. 1986; Matsuda and Miyoshi 1989; Jakes, Katon, and Martoglio 1990; Martoglio et al. 1990; Jakes, Sibley, and Yerkes 1994).

Corrosion and patinas on metal surfaces are often a mixture of materials that can be characterized by IR. Tennent and Antonio used IR to characterize bronze disease and its products (Tennent and Antonio 1981). Giangrande presented in a paper a selection of reference spectra for bronze corrosion products (Giangrande 1988). IR was one method used by Jakes and Howard to characterize the mineralization of textiles found encrusted on bronze implements from the Chinese Shang dynasty (Jakes and Howard 1968). Schrenk found corrosion below protective coatings on Benin bronzes (Schrenk 1991). Koltai used IR to determine corrosion products on a Roman silver fibula (Koltai 1984). Another study used IR to examine the corrosion products of lead exposed both to controlled conditions and to an outdoor environment (Tranter 1976). Corrosion processes occurring on stained-glass windows have also been examined with IR spectroscopy (Bettembourg 1988; Fuchs, Romich, and Schmidt 1990). IR spectroscopy has been used to investigate the corrosion inhibition capabilities of metal coatings (Tobe et al. 1974; Ito and Takahashi 1985). Efflorescence on mollusk shells was characterized by Tennent and Baird, who used IR, X-ray diffraction (XRD), thermal analysis, and nuclear magnetic resonance (Tennent and Baird 1985).

IR spectral imaging is a special technique useful in the IR spectroscopic analysis of materials. Imaging technologies use multidimensional spectroscopic processing to create compositional maps of samples. Familiar examples are elemental imaging on an electron microscope and magnetic resonance imaging for medical diagnoses. Characterization of materials with IR mapping microspectroscopy (also known as functional group mapping) has been developed for several years (for history and more information, see Harthcock and Atkin 1988). Derrick and coworkers illustrated the application of IR spectral imaging to the characterization of paint cross sections (Derrick 1995). IR mapping combines the analytical capabilities of an IR microscope with a computer-controlled, motorized stage and appropriate software programs, in order to identify, as well as locate, various types of components, anomalies, and defects in materials.

Deterioration Studies

While it is easy to think of IR spectroscopy as a tool for the identification of materials, it is often used for other studies, such as the monitoring of chemical reactions, the determination of chemical changes and degradation, the ascertaining of damage from specific conditions, and the evaluation of material stability. Common exposure conditions (ultraviolet [UV] radiation, elevated temperatures, oxygen, high humidity, reactive vapor, and dirt) cause deterioration and damage to some materials. However, not all materials react to any one parameter in the same fashion. IR spectroscopy can be used to increase understanding of degradation processes, causes, and rates of change; this knowledge can, in turn, aid in the development of conservation strategies to extend the lifetime of an object.

Feller examined the Fade-Ometer–induced alterations occurring in the IR absorption spectra of natural resins and afterward concluded that it is necessary to have reference spectra of aged resins for comparison of samples from aged objects (Feller 1959). In another accelerated aging study, Kenjo, who used IR spectrometry to examine the effects of increased temperature and UV light on Japanese lacquer, found that the outer surface of lacquer-covered objects exhibited measurable oxidation (Kenjo 1976). The mechanism by which soluble nylon becomes insoluble has been studied: the change in solubility was found to be due in part to photooxidation and hydrolysis (Bockoff et al. 1984; Fromageot 1990; Fromageot and Lemaire 1991). Oosterbroek and coworkers used IR to relate chemical and mechanical changes (crack formation and stress development) in organic coatings (Oosterbroek et al. 1991). Both IR and UV/visible (UV/Vis) spectroscopies were used to study the effect of heat and light aging on three formulations of Paraloid B-72 adhesives (Butler 1988). In some wood degradation studies, the ratio of cellulose to lignin was measured by IR (Kosik, Luzakova, and Reiser 1972; Pecina and Kommert 1985; Mikolajchuk et al. 1987; Kirillov and Mikolajchuk 1990). Hon used IR spectroscopy to study the degradation of paper documents (Hon 1989). IR has also been used to study the structural stability of hair samples from Egyptian mummies (Lubec et al. 1987).

The weathering of stone due to the outdoor environment and to pollutants has been well studied with IR analysis. Reflection IR measurements were used to examine the effects of temperature, relative humidity, and pollutants on marble (Šrámek 1978; Faraone 1987; Rao 1982, 1984). The effects of sulfur dioxide pollutants and the resulting stone deterioration were further assessed (Eastes and Salisbury 1986; Zappia et al. 1992; Connor and Girardet 1992). Accretions, salt deposits, and weathered surfaces on stone sculptures and buildings were identified with IR spectroscopy (Frediani and Matteoli 1978; Bradley 1987; Abd El-Hady 1988; Domaslowski and Kesy-Lewandowska 1988; Shoeib, Roznerska, and Boryk-Jozefowicz 1990; Blanco et al. 1991). Calcium oxalate has been investigated as a weathering product formed by the action of lichens on stone (Gorgoni, Lazzarini, and Salvadori 1992). Other studies have quantitatively measured calcium oxalate

(Biscontin and Volpin 1989) and examined the stability of calcium oxalate hydration states (White and Ai 1992), as well as determined its potential for protecting underlying stone (Matteini, Moles, and Giovannoni 1994). In the study of the reversibility of fluoropolymer treatments of stone, Camaiti and coworkers used IR spectroscopy to determine the extent of polymer removal (Camaiti et al. 1991).

The Case Studies

Ten case studies are presented below. Note that for Case Studies 1–6, IR is used to characterize unknown samples. Case Studies 7–10 show how IR can be used to monitor changes and examine degradation in several types of materials.

Case Study 1: Ultramarine pigments

The most common method for the identification of an IR spectrum of an unknown sample is its direct comparison to a reference material spectrum. When the materials are the same, the intensity and position of all absorption bands will correspond. If the unknown spectrum is missing bands that the reference contains, then that reference can be unequivocally eliminated. However, if the two spectra correspond except for the addition of one or more bands in the sample spectrum, then the interpretation becomes complex. As shown in this case study, the analyst must determine whether these additional bands are due to a different material, to an added component, or to a contaminant.

Background

Ultramarine blues are composed of complex sodium aluminum sulfosilicates and are one of the oldest blue pigments (Moser 1973). Natural ultramarine is an expensive blue pigment produced from the semiprecious stone lapis lazuli. The colorant can be separated from extraneous minerals in the stone by a time-consuming process. As described by Cennino Cennini in the fifteenth century, the powdered lapis was mixed with a wax-oil-resin mixture and kneaded in a weak lye solution (Cennini 1960). The dough retained the extraneous particles, and the blue particles settled out. In the 1820s, a commercial process for synthetic ultramarine was developed that produced a very pure, deeply colored, fine-particle material. Optical microscopy is used to distinguish between the natural and synthetic ultramarine particles based on the size and shape of the particles. Other analysis methods (X-ray fluorescence, SEM-EDS, XRD) commonly used for the identification of pigments cannot readily differentiate between natural and synthetic ultramarine.

During routine IR microanalysis of pigments and binders from a sixteenth-century Italian painting (*Venus and Adonis* by Titian, J. Paul Getty Museum, Los Angeles), it was noted that the IR spectrum obtained from a natural ultramarine blue particle corresponded to a reference spectrum for natural ultramarine, with the exception of an unexpected additional absorption band at 2340 cm^{-1}. Further analyses showed that all the blue particles in that sample and in paint samples from two other paintings (*Presentation in the Temple* by Mantegna, Staatliches Museen, Berlin, and *Madonna with Child* by Mantegna, Accademia Carrara, Bergamo, Italy) contained this absorption band (Fig. 6.1). Since this region of the IR spectrum contains rarely found functional groups (e.g., cyano and carbon-carbon triple bonds), the first concern after the unusual absorption band was observed was that Prussian blue (ferric ferrocyanide, $Fe_4[Fe(CN)_6]_3$, developed in the eighteenth century) was in the sample.

Analysis question

What material is producing the absorption band at 2340 cm^{-1}, and is it normally found in spectra for natural ultramarine blue particles?

Method

IR microanalysis was used for this study. The size of the individual blue particles ranged from approximately 10 to 40 μm in diameter after flattening. Thus, each individual blue particle could be selected directly for analysis by use of the adjustable apertures on the microspectrophotometer.

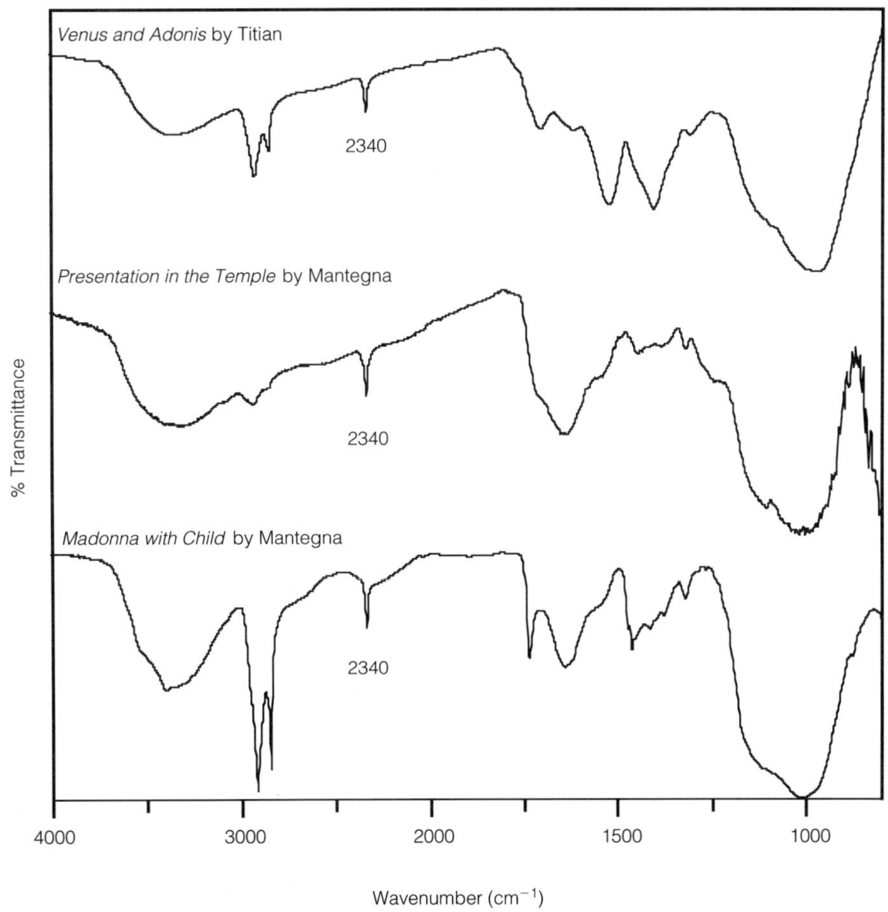

Figure 6.1

IR transmittance spectra for blue particles of natural ultramarine obtained from three fifteenth- and sixteenth-century Italian paintings (*Venus and Adonis* by Titian, J. Paul Getty Museum, Los Angeles; *Presentation in the Temple* by Mantegna, Staatliches Museen, Berlin; and *Madonna with Child* by Mantegna, Accademia Carrara, Bergamo, Italy). Each spectrum exhibits an absorption band at 2340 cm^{-1} that is not due to carbon dioxide.

Analysis

For this example (and all following case studies, except as noted), each particle was placed on a barium fluoride (BaF_2) window, flattened with a metal roller, then analyzed with transmitted radiation. A Spectra-Tech IRμS microprobe was used for the IR analysis (see Suppliers). It is equipped with a narrow-band, cryogenically cooled mercury-cadmium-telluride (MCT) detector. The spectra are the sum of 200 scans collected from 4000 to 800 cm^{-1} at a resolution of 4 cm^{-1}. The IRμS and the sample were continually purged with dry, carbon dioxide (CO_2)–free air. This factor is important, because the absorption band of interest occurs in the same region as the CO_2 doublet at 2340 cm^{-1}.

Results

Comparison of the sample spectra with the IR reference spectra for pigments containing cyano stretches (Prussian blue and bone black) showed that the small absorption band in the natural ultramarine at 2340 cm^{-1} was distinctly different from the cyano stretch that occurs near 2100 cm^{-1} (Fig. 6.2). A literature search showed that Orna and coworkers, in the analysis of samples from a fourteenth-century Italian manuscript, published a natural ultramarine spectrum that contained the small absorption band near 2340 cm^{-1} (Orna et al. 1989). Subsequent analyses at the Getty Conservation Institute (GCI) have identified this absorption

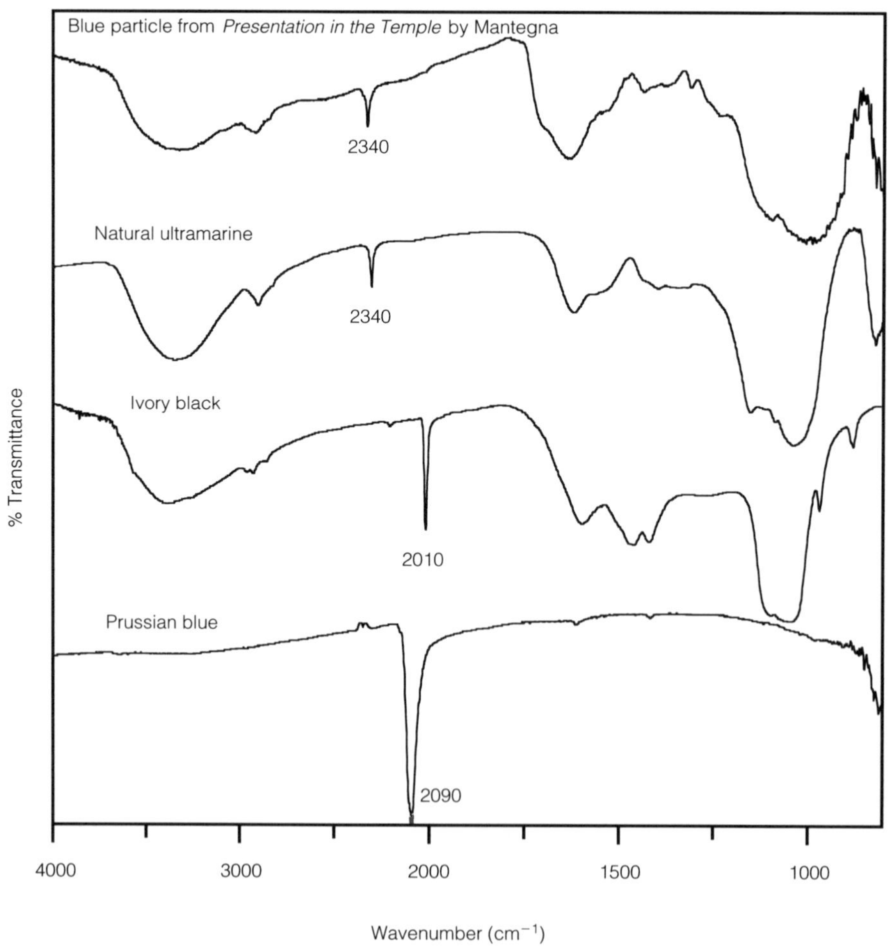

Figure 6.2

Comparison of IR transmittance spectrum obtained from a blue particle on a fifteenth-century Italian painting (*Presentation in the Temple* by Mantegna) with reference spectra for natural ultramarine, ivory black, and Prussian blue.

band in samples from four other Italian paintings from the fifteenth and sixteenth centuries.

Examination of over thirty samples of ultramarine showed an intriguing trend. The absorption band at 2340 cm^{-1} occurred only in lapis lazuli and natural ultramarine obtained from the Badakhshan mines in Afghanistan. The band did not occur in any synthetic ultramarine samples or in any samples of lapis lazuli and lazurite obtained from known sources in Siberia (former USSR) or Chile (Fig. 6.3). Thus, it appears that the presence of this particular absorption band in an ultramarine sample shows that it is a natural product whose source may be the lapis lazuli mines in Afghanistan.

The Badakhshan mines, the most famous source for lapis lazuli, have been worked for over six thousand years (Webster 1983). The blue color in the lapis is caused by two minerals, lazurite and hauyne. Geologically, the lapis from the Afghanistan mines contains predominantly hauyne (Banerjee and Hager 1992). Hauyne contains a

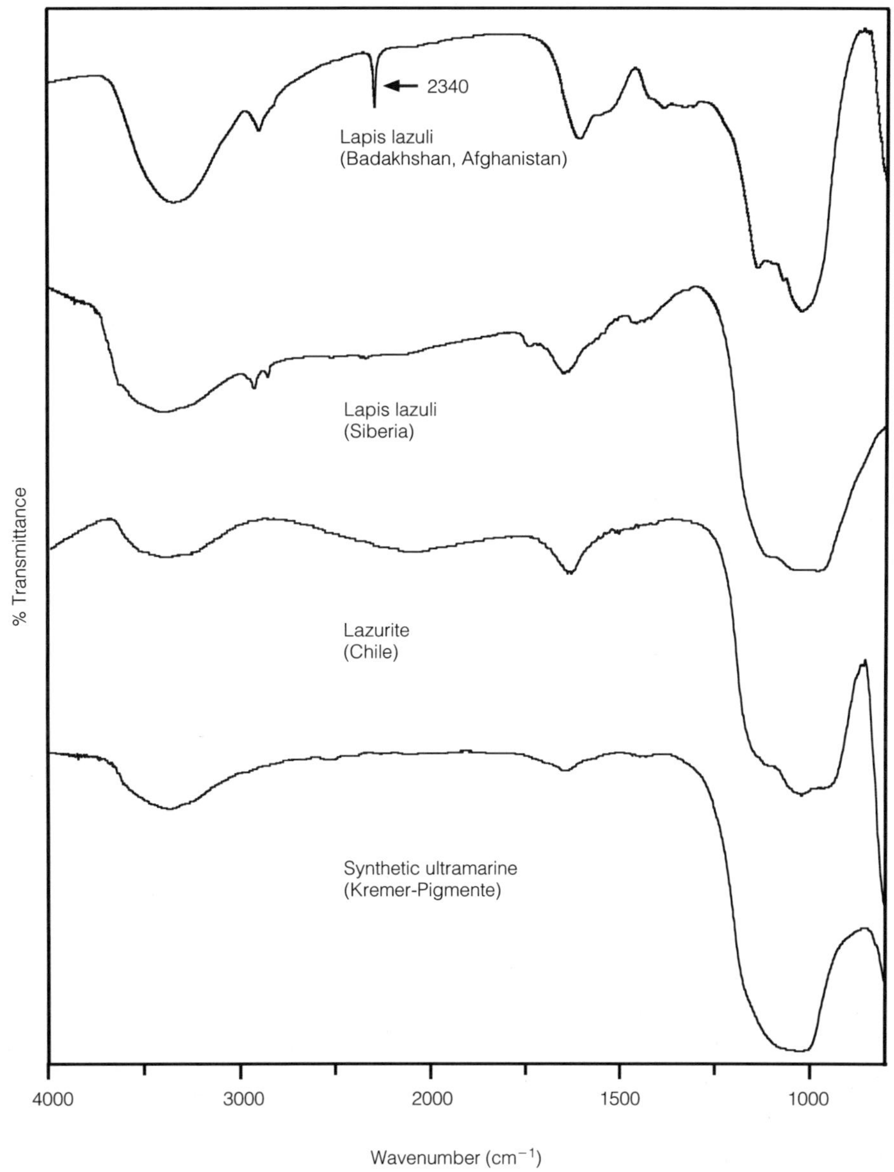

Figure 6.3

Comparison of IR transmittance spectra for lapis lazuli samples obtained from Badakhshan, Afghanistan; Siberia; and Chile, with a spectrum for synthetic ultramarine blue. Only samples from the mines of Badakhshan exhibit the absorption band at 2340 cm^{-1}.

higher concentration of calcium and sulfur than other types of sodalite minerals, such as lazurite. It is most likely the sulfur, S^{+6}, that produces the unique IR absorption band at 2340 cm^{-1}. While the literature states that the largest European source of lapis was Afghanistan, there are a few small mines of hauynite found in Italy (Taylor 1967). Samples from the Italian mines have not yet been analyzed.

Resolution

Based on the extensive number of IR spectra collected of natural ultramarine samples, it was concluded that the blue particles in the fifteenth-century paint samples were natural ultramarine. Furthermore, it is possible that the natural ultramarine in these samples came from the Badakhshan mines. This finding is consistent with known supplies and trade routes for that time period.

Case Study 2: Creosote lac resin

Prior to the analysis of an unknown material, a logical sequence is to obtain information both on the history of the object and on the suspected composition of the sample(s). The second step is to locate reference samples or spectra that correspond as closely as possible to that projected set of materials. The following example shows a sample that would have been difficult to identify without the contextual information about the object and without the correct reference materials (all provided at the time of the analysis request by Holcomb and Dean 1993).

Background

Native Americans in the southwestern United States used a variety of natural materials as adhesives and coatings on their pots, baskets, and other vessels. These include pitch or sap from several pines, junipers, and brittlebush; animal fats and glues made from horns, skin, or bone from deer or mountain sheep; and resin from creosote bush lac scale insect (Sutton 1990). Another source also states that "holes or cracks in pottery were repaired with creosote bush lac or pitch" (Felger and Moser 1985).

The creosote bush grows in the deserts of California, Arizona, New Mexico, Baja California, and northern Mexico. The resin is found on the outside of branches on infected plants and can be easily removed by twisting it off the branch. The resin is hard and brittle, but it is thermoplastic and becomes workable when heated; thus it must have been applied hot. It hardens on cooling to form a strong bond. Records show that its was used to adhere stone arrowheads in their shafts and to mend broken bowls (Coville 1892).

Found on the branches of the creosote bush, this insect resin is sometimes incorrectly labeled as creosote gum, when in actuality, it is a secretion produced from the female creosote lac scale insect. Lac scale insects compose a small family of species that exude a resinous substance known as lac. Shellac is the most common commercial product, but exudations from these insects are also used to produce medicines and dyes.

In the eastern Mojave Desert, several previously undiscovered Native American artifacts were recovered by cultural resource specialists of the U.S. Bureau of Land Management. A reddish-brown material was found on the exterior surface of one ceramic pot, near a small crack. The repair, on an otherwise intact pot, must have been done when the pot was in use. Thus, it was important to identify the adhesive in order to increase understanding of the culture, technology, materials usage, and perhaps even trade patterns of the native inhabitants in the area.

Analysis question
What is this adhesive material, and does it correspond to any materials indigenous to the region?

Method
IR microanalysis was used for this study because of convenience and availability. Sufficient sample sizes—almost a milligram of sample and a gram of reference material—were provided for analysis, so other IR analysis methods, such as potassium bromide (KBr) pellets or DRIFTS could have been used if needed. The preparation of small fragments of the samples on BaF_2 windows for microanalysis can be readily followed with selective solvent extraction on the same substrate, since BaF_2 is inert to most solvents.

Analysis
The instrument parameters and analysis conditions were the same as those specified in Case Study 1.

Results
IR spectra were collected on a small sample of the adhesive from the object. Figure 6.4 shows the spectrum obtained from the sample, as well as reference spectra for creosote lac, juniper pine resin, and piñon pine resin. After the bulk sample was analyzed, solvent extraction tests were done to check the solubility of the sample and to separate any impurities. A small drop of ethyl acetate was placed on the sample and an additional spectrum obtained from the soluble portion of the resin that collected in a ring around the sample (Fig. 6.5). A parallel reference spectrum collected for the ethyl acetate–soluble portion of creosote lac is also shown. The solubility test and IR spectra show that the extracted residue of the sample corresponds well to the extracted residue of creosote lac.

The extraction step is often necessary for the identification of aged, natural products. These materials may contain extraneous materials, such as dirt and insect and animal residues, as well as some insoluble fractions of the material that have chemically altered over time (i.e., oxidized). Separation of the soluble fractions aids in the interpretation of a mixture of materials, as well as provides solubility information.

Resolution
The IR spectra indicated that the adhesive used on the Native American ceramic pot did indeed correspond well to creosote lac. The subsequent analysis of the solvent-soluble portions of both samples confirmed the identification. Without the submission of the reference creosote resin for

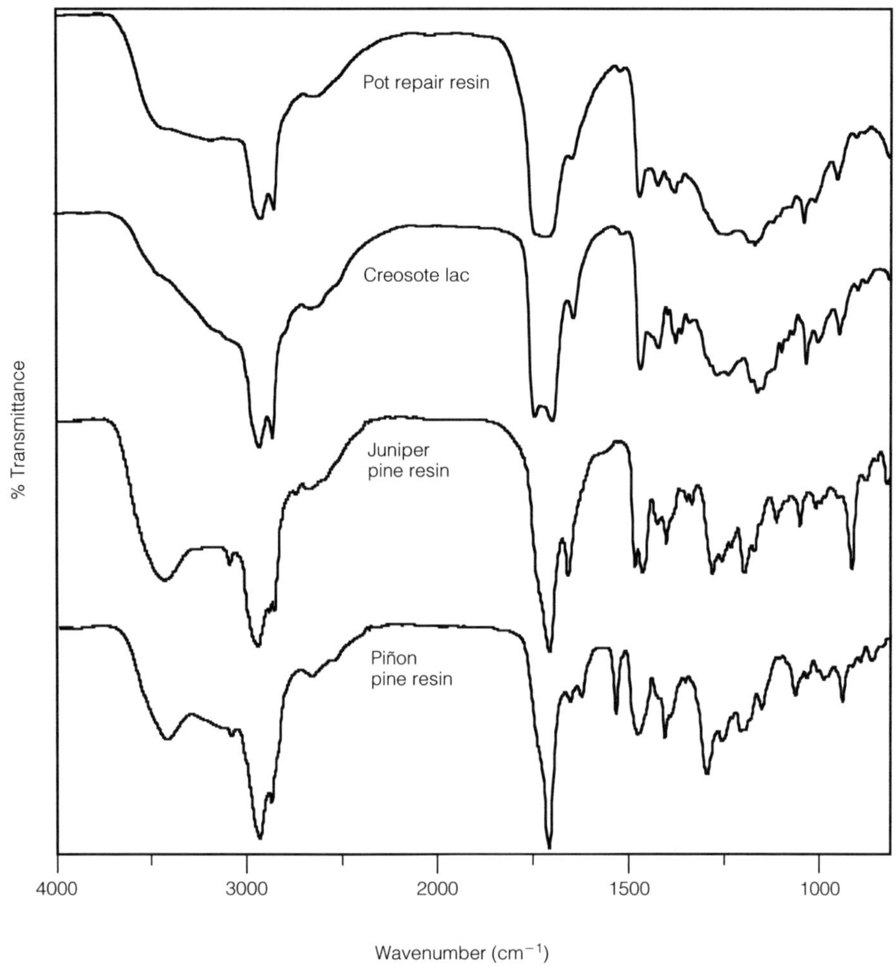

Figure 6.4

IR transmittance spectra for bulk adhesive samples obtained from a Native American ceramic pot, along with reference spectra for creosote lac, juniper pine resin, and piñon pine resin.

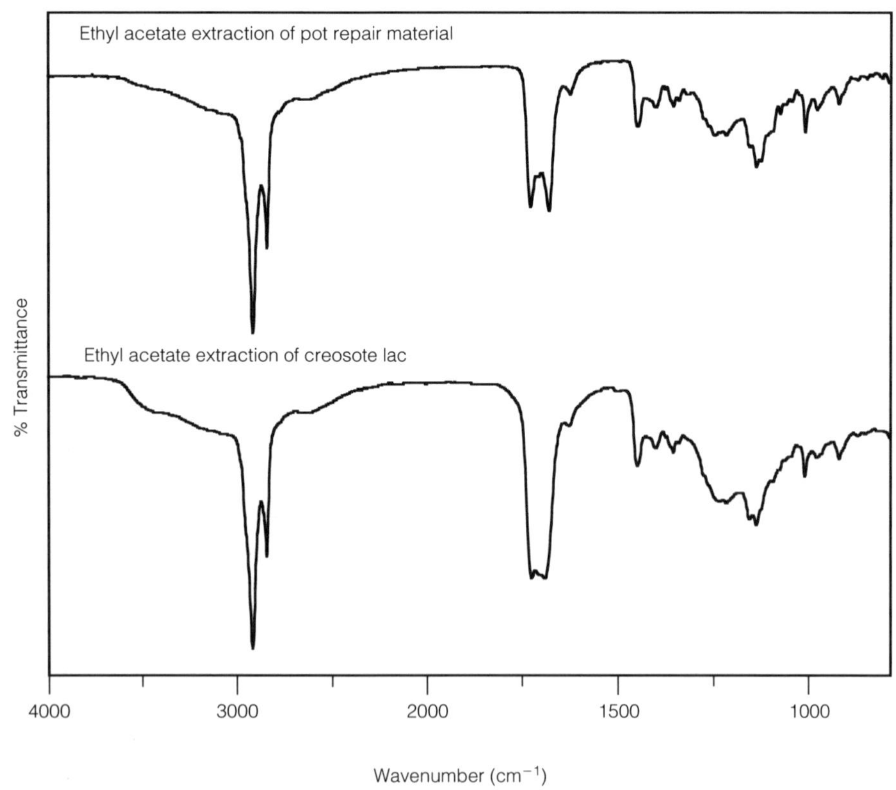

Figure 6.5

IR transmittance spectra of the ethyl acetate–soluble portions of the adhesive from the ceramic pot and the creosote lac resin. The two spectra correspond well.

direct comparison, this analysis would only have been able to classify the sample as a natural resin. IR reference spectra for natural products, such as creosote resin, are rarely contained in commercial IR libraries.

Case Study 3: Chumash Indian paints

IR spectroscopy functions well as a screening tool to characterize and compare many types of materials. This screening can be applied to divide large sample sets into similar compositional groups, as well as to supply a basis for the selection of secondary analysis methods. As shown in this example, IR is used to determine a sample's component classes (oil, protein, carbohydrate, etc.) prior to further specific identification.

Background

The south-central region of the California coast was inhabited by the Chumash Indians for over a thousand years. The Chumash were skillful artisans, and their rock art is well known. Most often found in the Santa Barbara and Ventura County mountains, some of these paintings reach sizes as large as 40 feet in length and have polychrome designs of six colors (Grant 1993). Some of the motifs are detailed, showing lizardlike creatures and other animals.

Ethnographic information suggests that the Chumash and other Indian groups of central and southern California may have used the paintings as part of a ritualistic procedure and that the paintings may have been related to astronomical observations. Early investigators were struck by the durability of the paintings and were curious as to the techniques used to make them. In 1883, Garrick Mallery was the first person to record and examine Chumash rock art sites; he looked at two sites near Santa Barbara.

The first set of samples analyzed was from a group of seven black pigment cakes collected from excavations of Chumash Indian sites in the nineteenth century. The pigment cakes submitted for analysis were obtained from the American Museum of Natural History, New York (Scott 1994). The pigments had been prepared by mixing them with a binder; the mixture was then formed into small cakes for convenience. Harrington describes the black paints as being prepared from soot mixed with deer marrow and made into a dough (Harrington 1942).

An additional sample came from the Painted Rock site near Santa Barbara, which is often described as one of the most important Chumash pictograph sites. The site is in the Carrizo Plain area near Soda Lake, a stopping point for migratory birds, and is adjacent to large forests of Digger pines, both sources of food for the Chumash. Historically, it was also home to considerable populations of deer and antelope. By comparing the images currently at the site to historic photographs, Grant showed that the site has suffered from extensive exfoliation and vandalism—including graffiti, gunshot damage, and modern paint (Grant 1993; Scott and Hyder 1993). The sample, possibly paint, was obtained from a black deposit in a painted region along a vertical crevice in the rock; the black material has previously been described as a tarry organic deposit, as bird droppings, or as a natural vein in the rock.

Analysis questions

Sample set 1. What are the binders in the seven samples cakes? Are they the same?

Sample 2. Does this sample contain any organic materials that may correspond to a paint binder?

Method

IR microanalysis was used for the first set of samples because of convenience, availability, and potential for use of selective solvent-extraction techniques. Additionally, while sufficient sample was available, it was important to retain as much sample as possible for future analytical studies. IR microanalysis was required for the second sample because of the minimal sample size—it contained only a few, barely visible black particles.

Analysis

The instrument parameters and analysis conditions were the same as those specified in Case Study 1.

Results—sample set 1

Of the seven paint cakes, the IR spectra showed that two clearly contained proteinaceous binders, while the other five appeared to contain oil-and-resin binders. As an example, the IR spectrum for paint cake sample T17482 is shown in Figure 6.6, along with a reference spectrum for dried blood of unspecified origin. Proteins are readily recognizable by a distinct pair of absorption bands for secondary amides that occur near 1650 and 1550 cm^{-1}. The presence of protein is confirmed by the sharp, triangular N-H stretching band near 3350 cm^{-1}. While the sample spectrum corresponds well to the reference spectrum of blood, IR analysis cannot differentiate between various proteinaceous media, such as blood, gelatin, or albumin. The similarity in the two spectra, with the exception of the hydrocarbon stretches at 2800–3000 cm^{-1}, indicates that the protein is the primary IR absorbing compound present in the sample. The hydrocarbon stretches imply that there may be a small amount of fat or oil present in the sample. However, within the detection limit of the method, no indications were found that the sample contains resinous or bituminous compounds. This finding was further confirmed by a negative result for the extraction of chloroform and ethyl acetate–soluble components. Additionally, no water-soluble components were found in the sample, indicating that the protein is probably not a gelatin (glue).

Resolution—sample set 1

Based on the IR analysis that classified sample T17482 as a protein, GC was used to identify the specific concentrations of amino acids. The GC results showed that the ratio of amino acids present was in close agreement with a reference standard for blood of unspecified origin (Scott et al. 1996). Now that the binder in the sample was identified as blood, there was great interest in the source animal species. Thus, Scott submitted the sample for immunological analysis. Using crossover electrophoresis, Newman found positive reactions for both pronghorn antelope antiserum

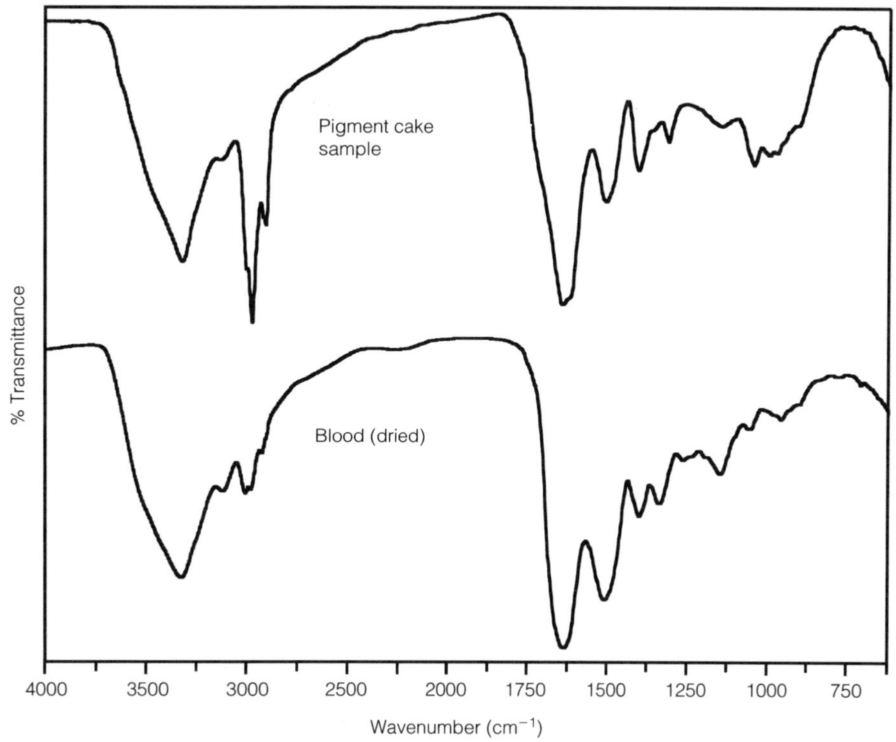

Figure 6.6
IR transmittance spectra for a Chumash Indian pigment cake sample (T17482) along with a reference spectrum of dried blood of unknown origin.

and for human antiserum (Newman 1994). Thus, this example shows the analysis sequence from a general determination to a very specific one.

Results—sample 2

In the second case, the bulk sample of the black deposit material from the Painted Rock site was analyzed first, followed by an ethyl acetate extraction of the sample. The IR spectra for the bulk sample and the extraction of the sample are shown in Figure 6.7. Solvent extraction was used selectively to remove components present in this mixture of materials. The spectrum for the bulk sample indicates that it contains some silicates and possibly carbohydrates. The ethyl acetate extraction of the sample revealed that one component of the paint sample is an esterified hydrocarbon (i.e., oils and fats). Thus, historical references were checked for the use of oil as a binder in rock art paintings.

The oil and pulp of the seed of the wild cucumber *(Marah macrocarpus)* has been cited as one type of binder in California rock art (Bishop 1994; Watchman 1993). This plant, commonly found in southern California, was also used ritualistically as a body paint in a Chumash curing procedure (Walker and Hudson 1993). The pulp of the cucumber seed contains a mixture of components, primarily oils, proteins, and carbohydrates. The oil pressed from the seeds produced a spectrum that corresponds well with other plant oils and is primarily composed of long-chain hydrocarbon fatty acid esters.

Other natural sources of oils or fats, such as deer fat, black walnut oil, or piñon oil, were also native to the region. Thus, for comparison, a sample of mule deer fat was obtained. Mule deer fat is often

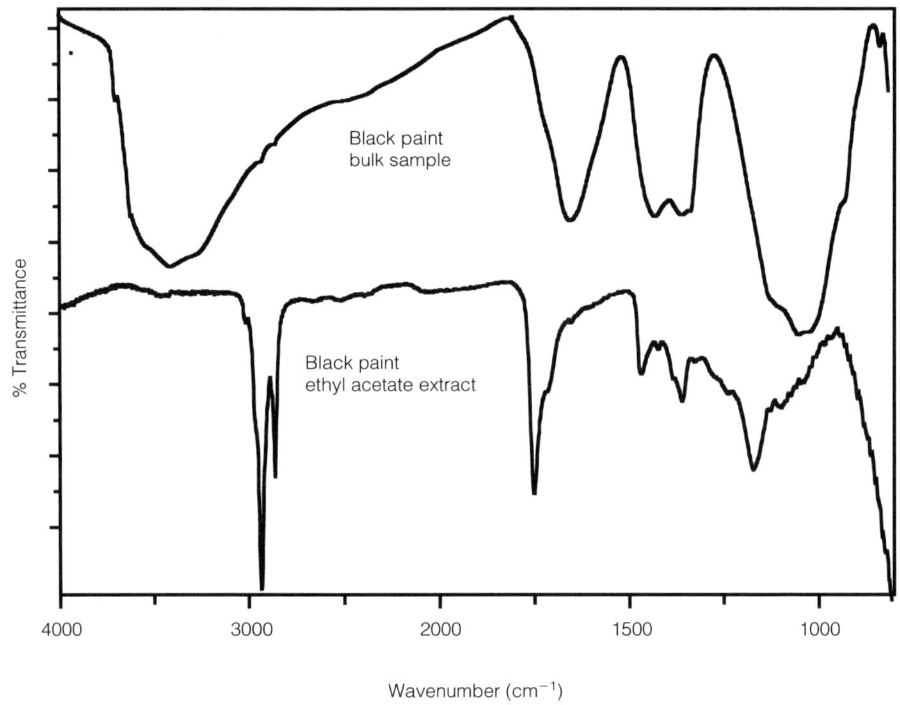

Figure 6.7
IR transmittance spectrum for a sample of black paint from Painted Rock along with the spectrum for the ethyl acetate–soluble portion of the sample.

cited as another paint binder, although it was probably used for body painting. Figure 6.8 shows the full IR spectra for the ethyl acetate extract of the rock art sample, along with unaged reference samples of cucumber oil and mule deer fat. In general, the spectra are very similar, as would be expected for samples from the same chemical family (i.e., long-chain hydrocarbon fatty acid esters).

Resolution—sample 2

IR spectra provided characterization of the major components in the sample described as a black deposit or paint found in a crevice. Based on its chemical functional groups, the sample was determined to contain some silicates, possibly some carbohydrates, and a long-chain hydrocarbon fatty acid ester. Ester compounds in this category are typically fats and oils that may potentially have been used as paint binders. However, before specific identification can be made of the oil in the paint sample, further analytical studies by GC must be done.

Case Study 4: Varnish on a desk

As in archaeology, the stratigraphy of paint and varnish layers can be a key to the interpretation of an object's treatment history. Thus, it is often important to examine the composition of individual layers. This case study illustrates how IR spectroscopy can be used to reveal additional information by cross section analysis.

Background

Identification of resins used for furniture finishes is important for art-historical analysis of an artifact, as well as for an initial material survey in restoration or conservation treatments. The materials and techniques

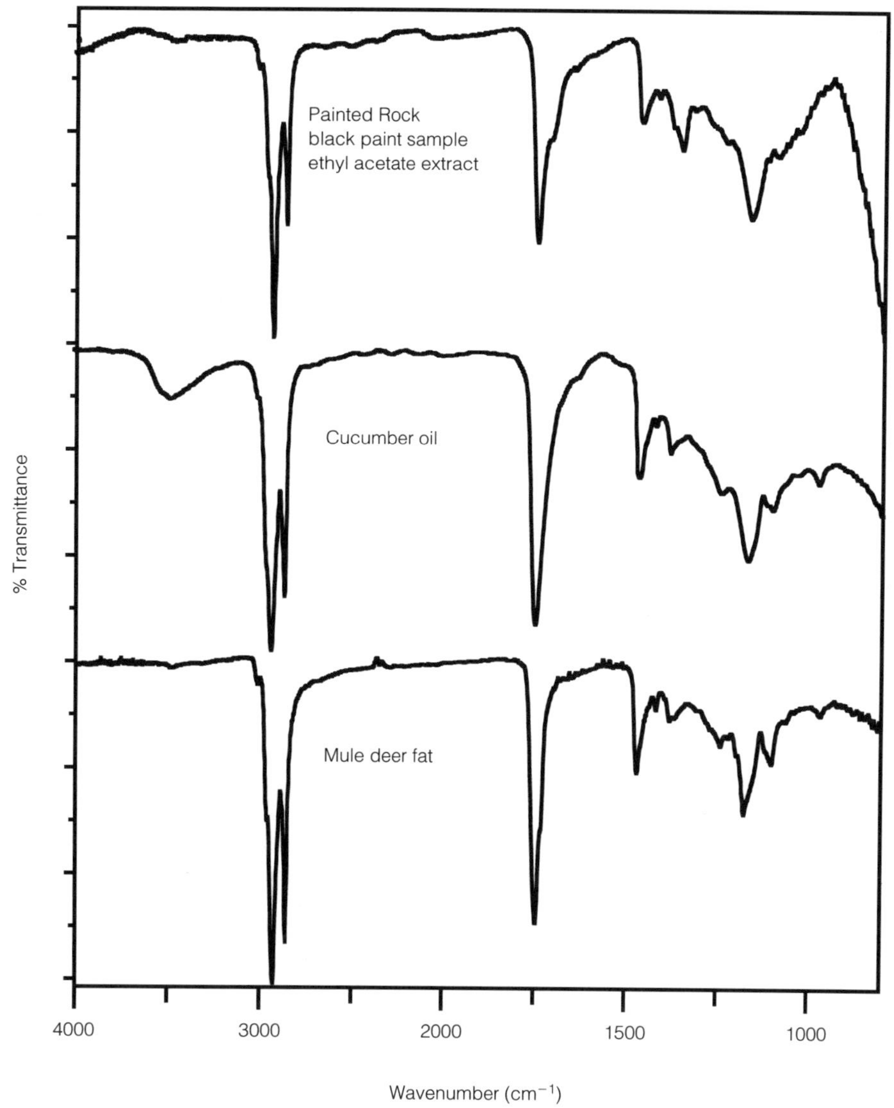

Figure 6.8
IR transmittance spectra for the ethyl acetate extraction of the black paint sample from Painted Rock, along with reference spectra for cucumber oil and mule deer fat.

for furniture surface treatment have, for the most part, been developed empirically throughout the ages. Collections of old recipes provide some insight into the chemistry of varnishes used for furniture surface treatment (Brachert 1978–79). Even though the natural resins used for surface finishes (shellac, copal, pine resins, etc.) may be difficult to analyze because of their complex composition, natural variability, and susceptibility to oxidation, IR microspectroscopy has proved to be a good method for the characterization of many pure natural resins used in furniture finish recipes (Derrick 1989).

An unusual mahogany rolltop desk with hidden drawers and elaborate mechanical devices is in the Decorative Arts collection of the J. Paul Getty Museum (72.DA.47). The desk, although not stamped with a maker's name, is attributed to David Roentgen and is thought to date from around 1785. It has a highly polished finish and beautiful filigree gilt-bronze mounts. While the finish on the desk is in good condition, conservators questioned whether the desk had been refinished, and if so, whether any of the former finish remained below the new layers.

Previous IR analysis of a sample from the top surface of the German desk had shown that it contained shellac, cellulose nitrate, and wax. However, since this sample had been removed by scraping a small amount of finish from the surface of the piece, in was unclear as to whether these components had been applied as layers or as a single mixture.

Analysis questions

Are there multiple layers in the finish? If so, what is the composition of each layer?

Method

A new sample was collected as a cross section from the top of the shelf near a small crack. The cross section was mounted in polyester embedding media, then microtomed with a glass knife to produce a thin section. The thin section was placed directly on a BaF_2 window for photodocumentation followed by IR analysis. IR microanalysis was selected in order to obtain a clear characterization of each layer in an intact thin section.

Analysis

The layers of the thin section were characterized by FT-IR microanalysis; the analysis conditions were the same as those specified in Case Study 1. Each layer was isolated for analysis by a rectangular aperture approximately 20×100 μm in size.

Results

A photomicrograph of this cross section after microtoming is shown in Figure 6.9. Visually the thin section contains a bottom clear yellow area beneath a darkened opaque layer, which then was covered with a clear layer. The top clear layer is very brittle, and small portions from this layer were lost during the microtoming step.

IR analysis showed that the bottom yellow layer (Fig. 6.10, spectrum C) was shellac. Further examination of the spectrum obtained from the middle, opaque layer (Fig. 6.10, spectrum B) indicated that it contained a few additional bands not present in the spectrum from the bottom, clearer region. Subtraction of the spectrum of the bottom, clear region from the spectrum of the middle, opaque layer resulted in a spectrum that corresponds to calcium carbonate (Fig. 6.10, spectrum D). It is possible that the chalk was used in the final polishing stages of the shellac finish and that it became embedded in the shellac. Analysis of the top clear layer (Fig. 6.10, spectrum A) showed that it consisted of a mixture of shellac and cellulose nitrate. This was a common commercial waterproof varnish sold in the 1930s and 1940s.

Resolution

For this sample, IR microanalysis of the thin cross section showed that the desk has been revarnished at least once with a cellulose nitrate and

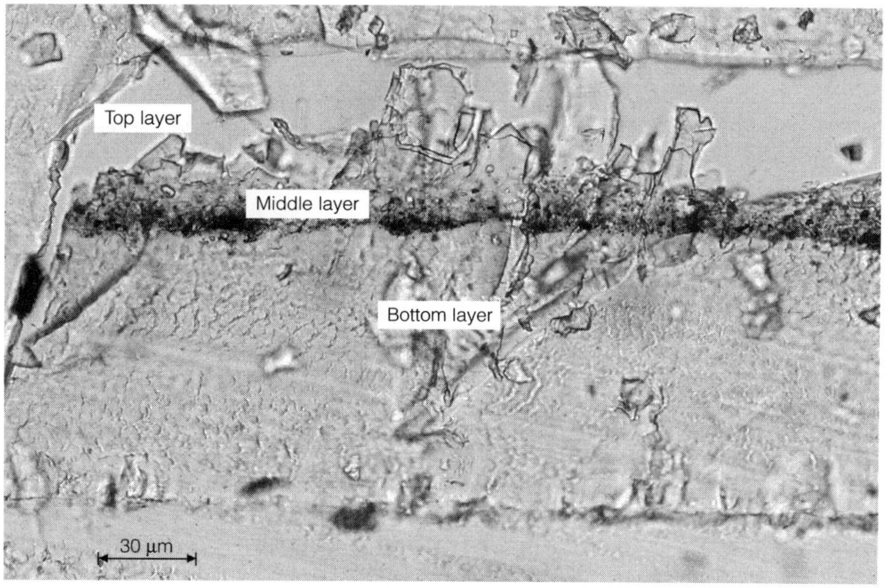

Figure 6.9
A photomicrograph of a thin-section sample from the Roentgen desk, embedded in polyester resin and microtomed to a thickness of 10 µm.

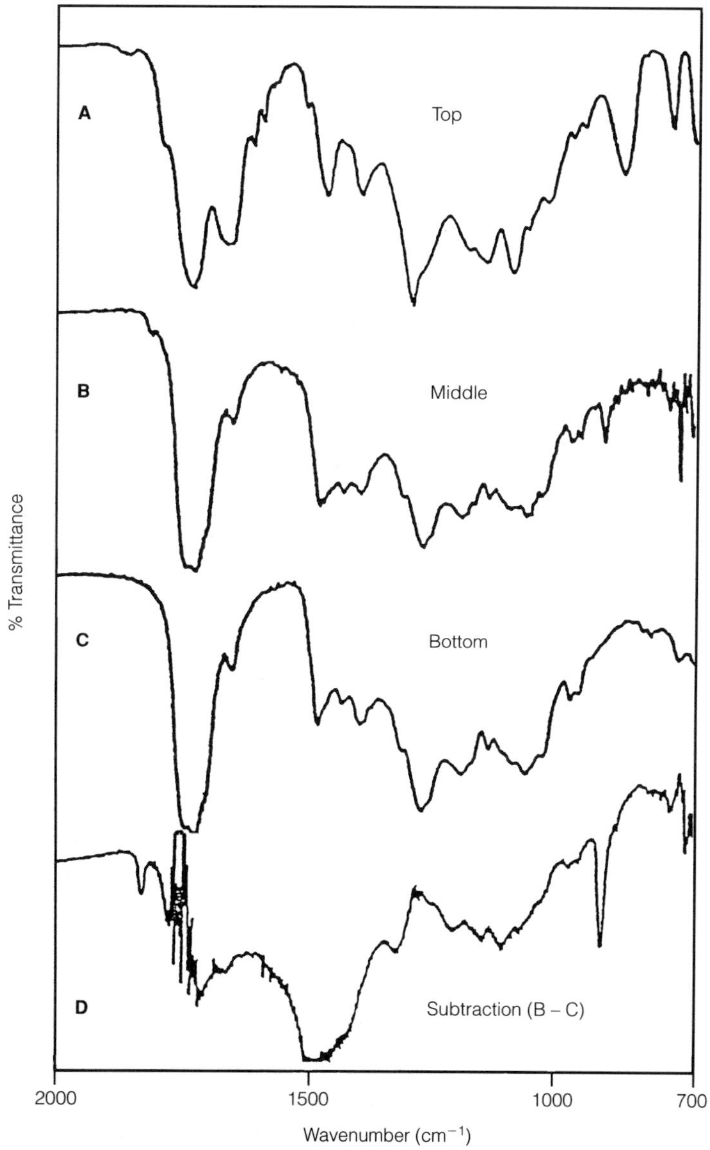

Figure 6.10
IR transmittance spectra (2000–700 cm^{-1}) for the sample shown in Figure 6.9: top layer (cellulose nitrate/shellac), middle layer (shellac/calcium carbonate), bottom layer (shellac). The subtraction (spectrum B – spectrum C) produces a spectrum (D) with absorption bands that correspond to calcium carbonate.

shellac varnish that was added on top of a thick shellac layer, and that the two layers were separated by a layer containing calcium carbonate.

Case Study 5: Reflection versus transmission

Cross sections are typically embedded for stratigraphic examinations. However, it is not always possible or easy to microtome thin sections from the embedded sample for IR transmission analysis. Instead, it would be beneficial to perform IR reflection analysis directly on the embedded sample. This case study compares the results obtained from both reflection and transmission IR analysis for an embedded cross section sample.

Background

Analysis was requested on a paint cross section sample from an eighteenth-century armchair in the Bayou Bend Collection of the Museum of Fine Arts, Houston (Shelton 1994). This highly ornate Neoclassical armchair, or "drawing room chair," is one of a set of eight extant armchairs attributed to an unknown maker working in Philadelphia in the late 1790s (Sands 1993; Brown and Shelton 1994). The distinctive decoration of the chair includes extensive use of applied composition ornaments and a high-contrast, water-gilded, and white painted surface.

Observations revealed that the water gilding on the composition ornaments and frame was broadly executed with a red clay bole bound in glue, and that a lead white pigmented paint was neatly cut in around the gilding (Shelton 1994). Previous microscopic characterization of the binders with reactive fluorescent dyes and elemental analysis of the pigments supported the possibility of a proteinaceous binder in the white paint layer. Visually, the cross section sample contained a thick gesso layer, a thin red-brown bole layer, and a layer of white cut-in paint directly over the bole layer. Above the white layer were remnants of a clear sealer or coating on the surface, covered with a thin layer of grime and two layers of restoration.

Analysis questions

What is the composition of each paint layer? Can comparable results be obtained by IR microreflection spectroscopy of the polished cross section and by IR transmission analysis of a microtomed thin section?

Method

The sample, as received, was embedded in a polyester medium. It was microtomed into a thin section for IR analysis in transmission mode. Additionally, IR reflection analysis was done on the portion of the sample that remained in the embedded block.

Analysis

The layers of the thin section were characterized by FT-IR transmission microanalysis by use of the conditions specified in Case Study 1. The microtomed surface of the remaining embedded cross section was smooth and flat. It was lightly polished, then analyzed by IR spectroscopy in reflection mode in a stepwise linear fashion, or line scan. The sample was placed on the motorized stage of the spectrophotometer, and an analysis window of 40 × 80 μm was selected. The stage was moved 100 times in approximately 5 μm increments in the x direction, in order to step the analysis window over the sample. One spectrum was collected at each step, such that 100 spectra were produced for the line scan (Fig. 6.12)

Results—transmission analysis

From the thin section, spectra were collected for the gesso, the bole, the layer above the bole, and the overpaint. These are shown in Figure 6.11. The IR analysis shows that the primary binder in the gesso and the bole layers is a protein (see amide I and amide II bands at 1650 and 1550 cm^{-1}). Carbonate shows up strongly in the gesso layer (to the point of saturation, 2550, 1800, 1450 and 870 cm^{-1}) and can also be seen to a lesser extent as a component in the bole. Since the analysis aperture was wider than the bole layer, spectra may show an overlap of components. Thus, the carbonate seen in the bole layer could be due to the gesso layer. The primary inorganic material in the bole layer is a clay (silicate: sharp bands near 3600 cm^{-1} and a broad, strong band at 1050 cm^{-1}). Oil appears to be a component in the layer above the bole (the white, cut-in paint) and in the overpaint, or restoration layers (see the carbonyl

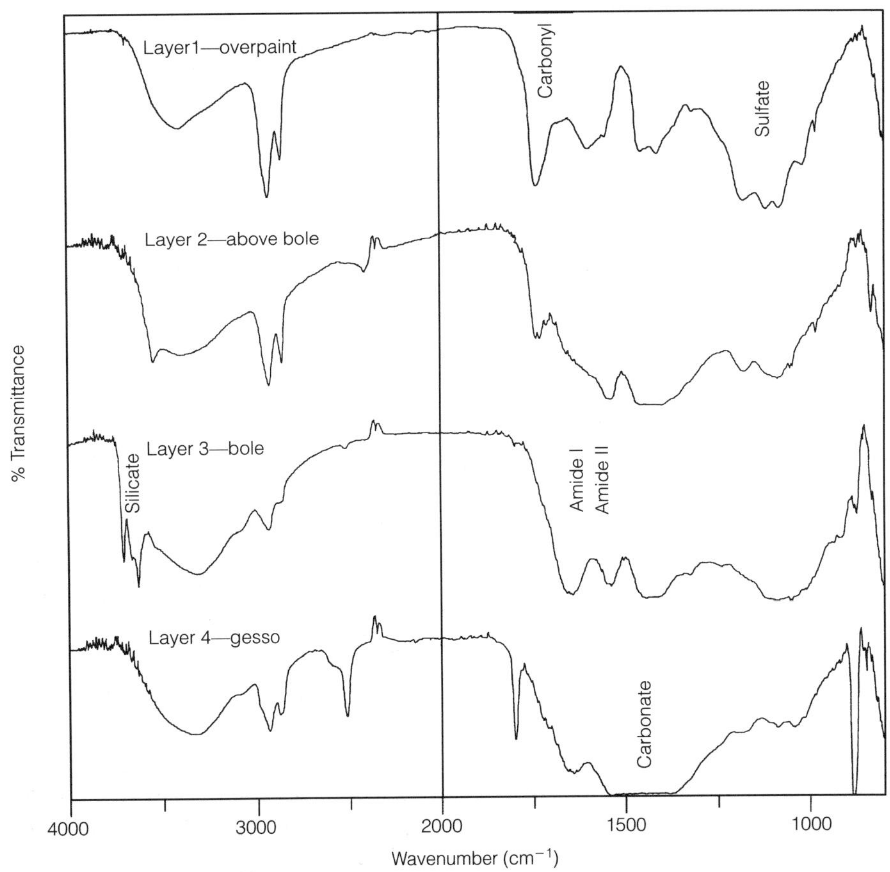

Figure 6.11

IR transmittance spectra collected for four visually distinct layers with IR transmission microspectroscopy on a microtomed thin section of paint from an eighteenth-century chair.

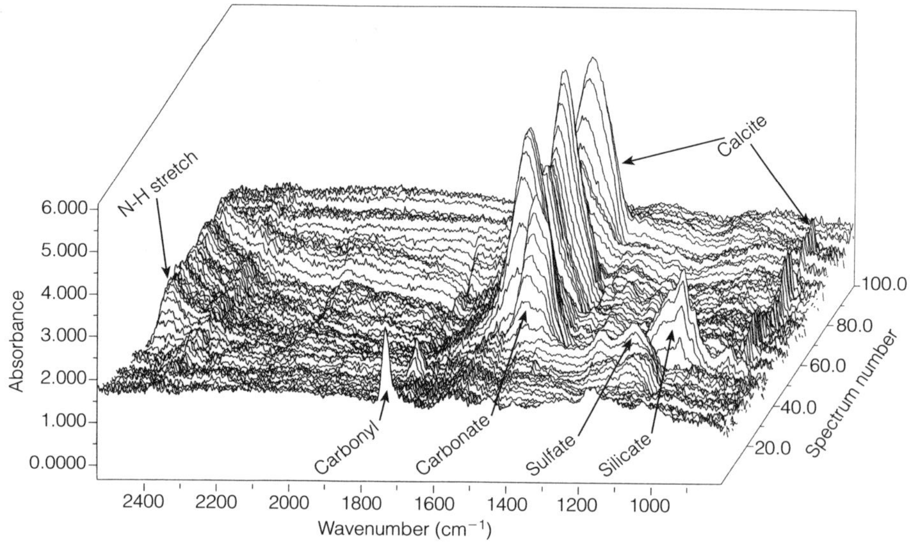

Figure 6.12
IR absorbance spectra collected from a linear step scan using IR reflection microspectroscopy on an embedded paint cross section from an eighteenth-century chair. The analysis window was 40 × 80 μm, and the step size was approximately 5 μm. As the composition of the sample changes, so do the corresponding IR spectra. These relative changes for selected bands can be plotted as a line scan, as shown in Figure 6.13.

band at 1740 cm^{-1}, along with strong hydrocarbon stretches at 2930 and 2860 cm^{-1}). However, caution must again be used, since the appearance of the oil bands in the layer above the bole may be due to an overlapping aperture with the coating or overpaint layers. The spectrum for the layer above the bole also appears to have absorption bands that correspond to protein and to basic lead carbonate (a broad band at 1400 cm^{-1}, with a band at 3600 cm^{-1}). Previous elemental analysis had indicated that lead-based pigments were found in this region. The overpaint layer contains primarily barium sulfate (a band near 1140 cm^{-1} with three tips and a sharp band at 3550 cm^{-1}).

Results—reflection analysis

From the spectra collected in reflection mode (Fig. 6.12), line plots were drawn that show the absorbance intensity of a particular absorption band versus its position in the sample (Fig. 6.13). The absorbance intensity of a band corresponds to its relative concentration. All spectra were collected on the same sample, so the relative height of a band from one position can be compared to its height at another position. Since the analysis aperture was fairly large, there are no well-defined edges for the beginning and end of each layer. However, the peak maxima for a material should correspond well to the center of the layer in which the material is found. Also, some voids in the surface of the sample may cause apparent decreases in the concentration that are not, in fact, real.

Examination of Figure 6.13 indicates that the band assigned to silicate (1040 cm^{-1}, line profile 1) exhibits the highest concentration in the region of the red bole. This finding corresponds to an identification of clay in this layer. The sulfate-assigned band (1143 cm^{-1}, line profile 2) also absorbs strongly in one portion of the overpaint region; this finding corresponds to the transmission results. Since the band selected for sulfate overlaps with the silicate absorption bands, the apparent sulfate in the bole region is likely due to the silicate band. The line profile assigned

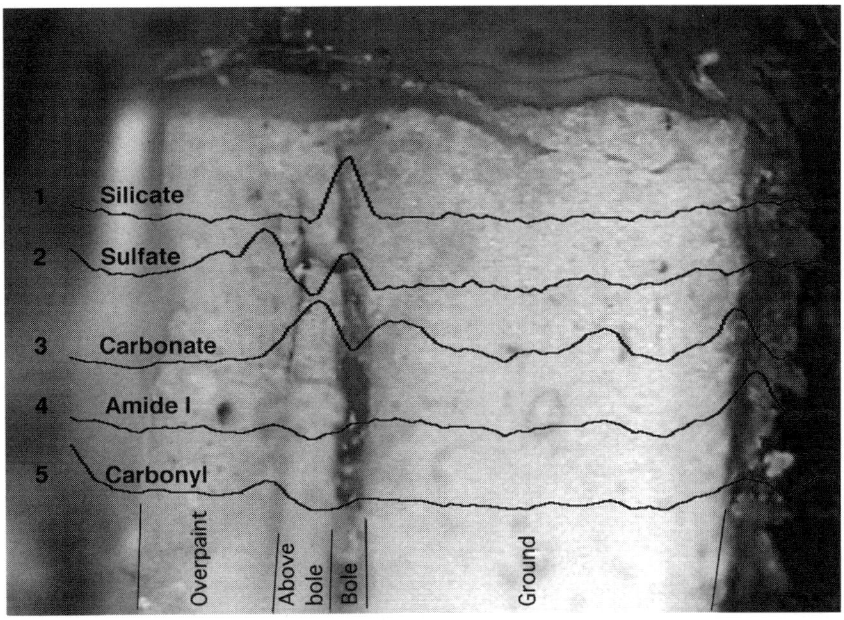

Figure 6.13
Line scans overlaid on a photomicrograph of the embedded paint cross section from an eighteenth-century chair. The line scans show the intensity of a selected absorption band versus its collection position on the sample.

to carbonate (line profile 3, 1450 cm^{-1}) is high in the layer directly above the bole. Also, the carbonate band intensity is variable from high to moderate in the gesso. It is possible that the gesso contains a fairly even distribution of carbonate (calcite) but that the intensity of the absorption varies due to the particles and voids encountered by the beam (i.e., that the variation is an artifact of the reflection analysis method).

Line profile 4 (1650 cm^{-1}) is assigned to the amide I band, which is likely due to proteins. The profile lines are Autoscaled to the maximum intensity; thus, the strong absorption for the protein at the bottom (right edge) of the gesso layer makes the rest of the sample appear as if it has little protein. Gesso typically has about 10–20% glue; this proportion should correspond to a small but recognizable absorption band, such as is seen in the transmission spectra. However, the line scans do not clearly indicate its presence.

The carbonyl band (line profile 5, 1730 cm^{-1}—indicative of oil and resin) is consistently low in the gesso and in the layer above the bole. The carbonyl band absorption increases in the region described as a coating on the layer above the bole. This increase could be due to a layer containing oil or a natural resin. A slight increase in the carbonyl intensity in the overpaint area, as opposed to the gesso region, may indicate the presence of oil in the overpaint. This finding would correspond with the transmission results. The increase in the carbonyl band intensity at both edges of the sample is probably due to the polyester embedding media.

Resolution

Obtained through direct reflection analysis performed on the mounted cross section, the line scans of the sample provide an effective means for examining and reporting the results. This method also provides some hints as to thin layers that may be present in the sample.

However, without the information obtained from the initial transmission spectra collected from the thin section, it would have been difficult to interpret the information in the line scan. Thus, both are important. Comparison of reflection versus transmission methods suggests that the organic components of the sample—such as the proteins and oils—do not reflect the IR beam as well as do the inorganic components. Therefore, the transmission spectral data are a better indicator for the binders in each of the layers.

Case Study 6: Painting cross sections

Reflection analysis showed some potential for the examination of embedded cross sections in Case Study 5. However, an area scan should provide a better visual image of the components than a linear scan. An area scan would give a total sample depiction identifying each component versus its location, such as is obtained with elemental mapping and, to some extent, with fluorescent staining. An IR reflection map of the entire sample area is illustrated in this case study.

Background

Andrea Mantegna was an Italian artist of the late fifteenth and early sixteenth centuries. Many of his paintings have the very characteristic matte appearance of glue (distemper) paint on a fine-quality linen support, and they were never meant to be varnished (Rothe 1992). The distemper technique, called *Tüchlein,* was popular in the Netherlands and Germany but had not been associated with any other Italian artist.

As part of a detailed investigation of Mantegna's techniques by Rothe, over thirty samples from ten paintings in museums in Europe and the United States were submitted for binding media analysis (Rothe 1991). Multiple methods—FT-IR microscopy, SEM-EDS, GC/mass spectroscopy (GC/MS), and high performance liquid chromatography (HPLC)—were used to characterize the components in these samples. The primary reason for analysis was to determine the presence of glue—as opposed to egg—in the binder; egg medium was typical for wood-panel paintings in that time period.

One sample, from Mantegna's *Adoration of the Magi* (J. Paul Getty Museum, Los Angeles), was selected for in-depth study by several analysis techniques. Initial IR transmission analysis on particles removed from each visually distinct layer indicated that protein was the primary binder. HPLC analysis of the amino acids in the protein showed that it contained significant amounts of hydroxyproline, thus indicating that animal glue is present in the sample. GC/MS analysis confirmed this finding and also noted the presence of a wax and the absence of cholesterol in the sample.

Analysis question

Can IR mapping microspectroscopy be used to identify and locate sample components? Of particular interest is the distribution of the wax that was identified by GC/MS.

Method

Initial IR transmission microanalysis was done on particles removed from each visually discrete layer as a control. Then, computer-controlled x–y mapping of a sample in the reflection mode was selected to characterize the components in a sample and to detect components or layers that may not be visually apparent in an optical microscope.

Analysis

Instrument parameters for both transmission and reflection methods were the same as conditions specified in Case Study 1.

For the area map, a portion of the sample, a four-layer cross section, was embedded and polished, then placed in the IR spectrophotometer, and an analysis grid of 10 × 15 points was selected. An array of spectra were collected by reflection of the IR radiation off the surface of the sample at each grid point. The effective resolution of the components in the sample was determined by the size of the analysis aperture and by the density of the grid. The size of the aperture for this analysis was 20 × 40 μm. The selection of size involved a trade-off between resolution and energy throughput. The step size was approximately 20 μm. The overlap of the windows in the x direction provided an effective increase in the resolution of the components in that direction. Each spectrum was the sum of 50 scans and took approximately 1 minute for collection and processing.

Results

An array of IR spectra were collected and contour maps produced that provide information on the concentration and location of compounds in the sample. This procedure was done by the selection of a wavelength of interest, such as a hydrocarbon band; then its intensity versus its position in the grid where it was collected was plotted. A total sample contour map was prepared by connecting lines for the band intensities of similar value. In these plots, variations in line thickness are used to represent changes in band intensity. The thickest lines correspond to the areas of strongest band intensity—that is, to the highest concentration of the material. The intensities are relative to one another, and the background intensity may not be zero, because of other absorptions in the region. In this particular sample, because previous extensive analyses had been done to determine its components, the selected IR absorption band and corresponding functional group could be related specifically to components in the sample. On other samples, it would be precarious to identify a material based on only one IR absorption band.

Figure 6.14 shows a photomicrograph of the Mantegna sample, and Figure 6.15 shows four IR reflectance maps. Map A is a plot for the carbonyl band at 1730 cm^{-1}. The highest intensities of the carbonyl band are due to the polyester embedding media surrounding the sample; they provide a general indication of the area analyzed. The absence of a carbonyl band in the region of the sample corresponds to previous analyses, which determined that the binder did not contain egg or oil. Map B is a plot of the intensities of the carbonate band at

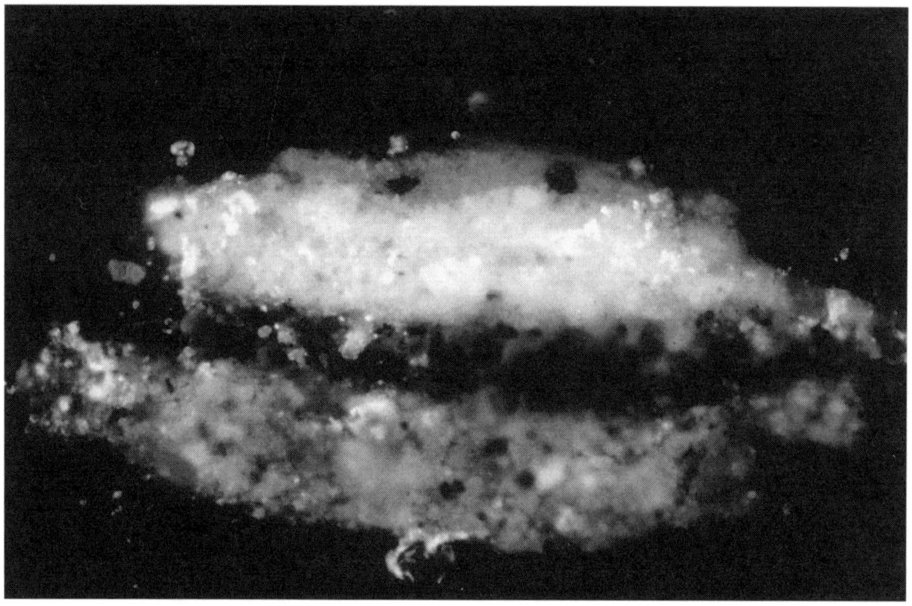

Figure 6.14
A photomicrograph of a painting cross section sample from the *Adoration of the Magi* by Mantegna.

$1416\ cm^{-1}$. This map shows that the entire sample contains carbonate, with the exception of the surrounding embedding media and a central point in the second layer where the sample contains a large particle. Map C plots the intensity of the band at $1092\ cm^{-1}$, which may correspond to sulfates or silicates, or to both. The highest concentration of this material(s) is in the third layer of the sample. Map D is a plot of the intensity of the hydrocarbon band at $2919\ cm^{-1}$. The highest concentrations occur in the ground (bottom) layer and correspond to wax.

Figure 6.15 (opposite page)
IR reflectance contour maps for painting cross section sample shown in Figure 6.14. The thickest lines correspond to the highest-intensity absorptions. Absorption bands mapped are (A) a carbonyl band at $1730\ cm^{-1}$; the high intensities represent the embedding media; (B) a band assigned to carbonate at $1416\ cm^{-1}$; this map shows a uniform distribution with the exception of lower levels in one region of the middle layer and the surrounding embedding media; (C) a band assigned to sulfate/silicate absorptions at $1092\ cm^{-1}$; this map shows a higher intensity in the middle layer; and (D) a hydrocarbon band at $2919\ cm^{-1}$; this band has the strongest intensities in the bottom layer.

Resolution

With the aid of the results from IR transmission studies, many of the components in each layer could be identified by the IR maps. Of most significance was that the wax, which was previously found in the sample by GC/MS bulk analysis, can now be designated according to its location in the sample. Two potential sources for the ground-layer wax are, first, that it could have been included in the original mixture of components in the ground layer or, second, it could have come from the wax lining. However, since the IR map shows a fairly even distribution of wax throughout the lower layer, it is more likely that wax was an original component in the ground layer. Additionally, there is no treatment history indicating that this painting has been relined.

Limitations to mapping

The method of IR reflection mapping shows potential for the determination of materials and their locations within a cross section sample. The technique is complementary to elemental mapping with an electron microscope and may be done on the same sample. At this point, the IR mapping method has two major limitations. The first limitation is the use of specular reflection as an analysis method; this method requires an

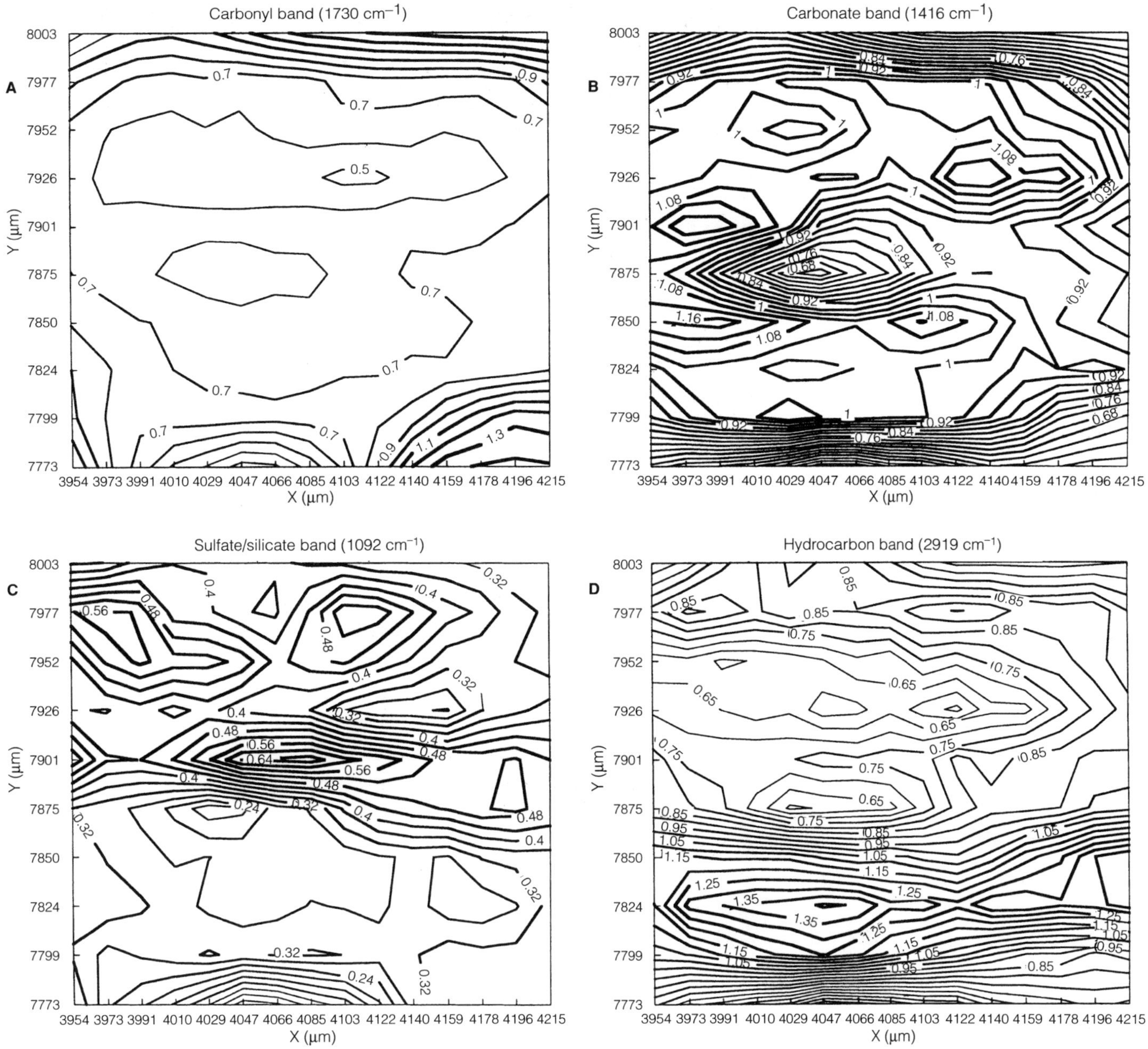

analysis window with a minimum size of 20 × 20 μm, because of energy restrictions. Specular reflection can also result in band distortions and shifts for which it is difficult to compensate. The second limitation is that a component cannot be reliably identified based on one IR absorption band. Thus, additional analyses are required to supplement the map and provide interpretation. Future computer programs should allow a map to be created based on the selection of multiple absorption bands that can help identify specific compounds.

Case Study 7: Vikane

Vapor-phase depletion measurements have been commonly used to study catalyst and pollutant reactions (Grosjean 1985). Normally these measurements are made by periodic sampling of the gas in a chamber and analysis of it by a thermal or ionic detector. In this study, IR spectroscopy was used as an in situ detector to determine the decrease in concentration of a gas-phase fumigant, Vikane (sulfuryl fluoride), in the presence of various materials.

Background

Vikane was developed by Dow Chemical Company in the 1950s, specifically for the control of drywood termites typically found in warm climates, such as the southern United States (Dow Chemical Co. 1982). It has since been widely used as a structural fumigant against a variety of destructive pests in homes, buildings, construction materials, furnishings, and vehicles. Studies conducted on Vikane show that it has several advantages; they include easy dispersal into a structure, rapid penetration into materials, formation of almost no residues, and ready dissipation after aeration (Meikle and Stewart 1962).

This study was a part of a joint investigation by the GCI, the Canadian Conservation Institute, and the Conservation Analytical Laboratory of the Smithsonian Institution, to examine the extent of interaction of Vikane with a wide variety of materials under controlled conditions. Final studies showed minimal measurable changes in most materials with Vikane (Baker et al. 1990).

Analysis question

Can IR analysis be used to determine which materials are susceptible to interaction with Vikane?

Method

Vapor-phase depletion measurements can be used as a fundamental method to determine the interaction of a solid material with a gas. The theory of this method is that when a gas is placed in a closed chamber with a test material, the only decrease in the gas concentration will occur by (1) leakage or penetration through seals or gaps in the chamber, (2) absorption or reaction with the chamber walls, and/or (3) absorption or reaction with the test substrate surface. Under controlled experimental conditions, the first two parameters should remain constant from run to run. Thus, the test substrate may be varied, and several materials may be compared for their specific absorption of and reactivity to the gas. Since the absorptions will be proportional to the surface area, deposition or uptake of the gas can be normalized for the surface area of each sample, so that any real differences in sample reactivity may be determined.

For these experiments, the concentration of Vikane was measured in situ with IR spectroscopy. This measurement was performed by placing a substrate in an IR gas cell, out of the path of the beam (Fig. 6.16), introducing Vikane, and then monitoring the concentration of the fumigant in the cell for several hours.

Analysis

IR spectra were obtained at 4 cm^{-1} resolution on a Digilab 15-E FT-IR spectrophotometer (see Suppliers) equipped with a Motorola 3200 computer and a dry nitrogen purge. A wide-range, cryogenically cooled MCT detector was used to examine the mid-IR region from 4000 to 500 cm^{-1}. Each spectrum represents an accumulation of 500 scans. The computer was programmed to run automatically, collecting a spectrum every 5 minutes for the first 5 spectra, then every 10 minutes for 25 spectra, then every 20 minutes for the remaining 20 spectra. For the 0% relative humidity (RH) runs, the test material was placed inside the cell, then the cell was flushed with pure, dry nitrogen and placed in the instrument.

A spectrum was taken of the purged cell after it was placed in the instrument to ensure that the cell was clean; this spectrum was then used as a background. Then, 2–3 ml of 100% Vikane was injected into the gas cell, and the automated IR collection program was initiated. The test materials had an exterior dimension of 2.5 × 5.0 cm (1 × 2 in.). After each run, the cell was repurged with nitrogen. At this point and 8 hours later, spectra were taken to determine whether the test materials were desorbing Vikane.

Results

The gas-phase spectrum of sulfuryl fluoride produces strong vibrational bands that are well resolved at 4 cm^{-1}. The intensity of IR absorbance

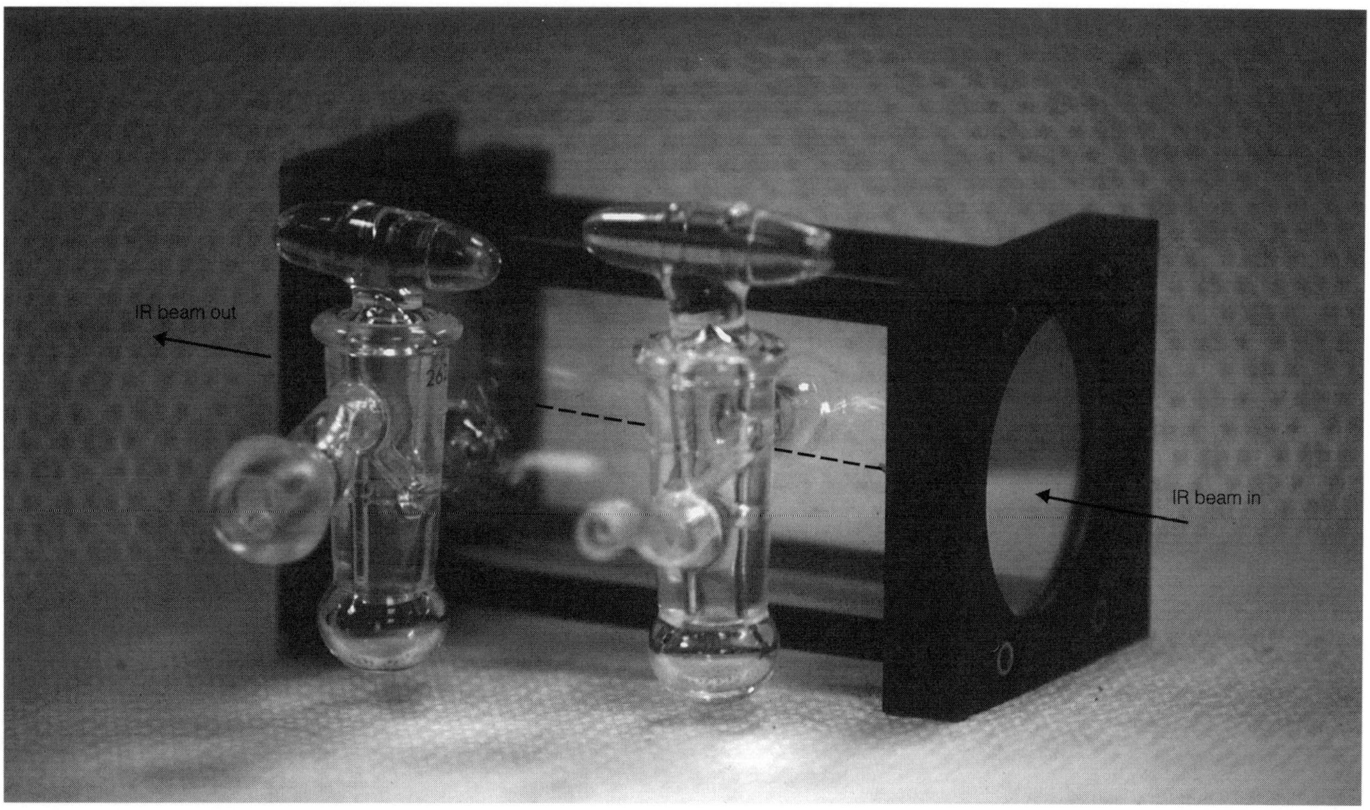

Figure 6.16
An IR gas cell for depletion studies. The sample is placed in the bottom of the airtight cell, and Vikane is injected; the concentration of Vikane over time is then monitored by the collection of IR spectra of the gas in the cell over a 10 hour period. The cell is 10 cm long and has a diameter of 5 cm. The windows in the cell are zinc selenide.

bands is proportional to the concentration of that species within the linear range of the detector. Comparison of several blank runs showed that the precision of the measurements for four runs with an empty cell (blank) taken over a period of four weeks was ±2%. Each of the curves is linear and horizontal and represents a relatively constant concentration for the 12 hours.

Figure 6.17 shows depletion curves for several materials tested at 0% RH. For the blank cell, the Vikane concentration decreases slightly in the first hour, then remains essentially constant for the next 12 hours. Similar results were seen for metals; initial slight decreases in concentration may be related to the additional deposition surface area in the cell. Vikane depletion is slightly greater (5%) for the cotton sample, but it still exhibits the same general pattern of an initial drop, then a leveling. In comparison to the other materials that were tested, the proteins showed drastic changes. Silk—showing a 37% decrease—is an exacerbated version of the pattern noticed with proteins.

In these experiments, samples analyzed at 0% RH were placed in a cell that was purged with dry nitrogen. However, since the lack of oxygen and water may inhibit some reactions with Vikane—particularly the corrosion of metals—additional studies were done with dry air humidified to 65% RH (with oxygen present).

At 65% RH, the curves for each metal sample exhibited a rapid initial drop, followed by a leveling. In each case, the curves were lower than on the comparable graph obtained in a dry nitrogen atmosphere (an example for brass is shown in Fig. 6.18). The depletion at 65% RH ranged from 15% (copper) to 6% (aluminum and silver), while at 0% RH, the depletions ranged from 1% to 3%. The presence of moisture or oxygen seems necessary for Vikane absorption or reaction with metals. Even with the depletion of the Vikane, no corrosion or tarnishing was visible on any of the metals after 24 hours of exposure. Interestingly, the organic materials, such as cotton and silk, showed much less Vikane

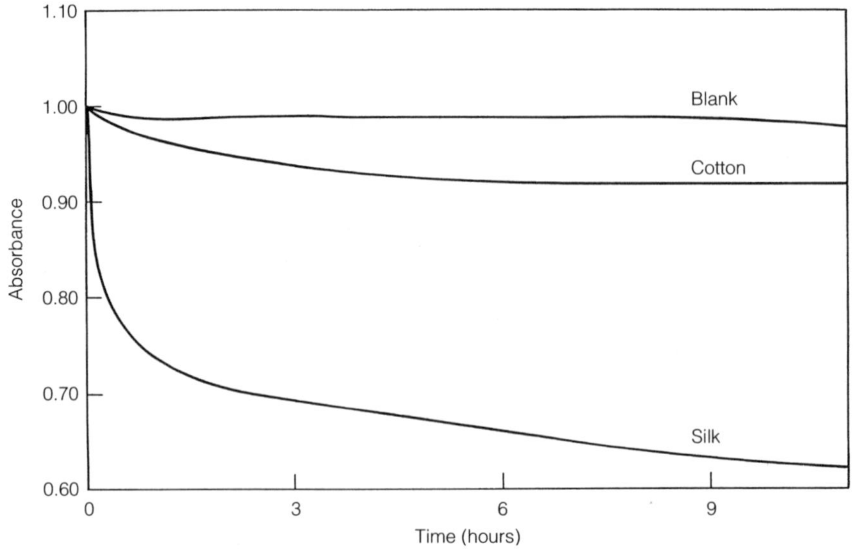

Figure 6.17
A plot of the IR absorption band of Vikane at 1492 cm^{-1} versus collection time. The three lines show experimental results for a blank cell, a piece of cotton, and a piece of silk. The cell containing a piece of cotton exhibited a 5% decrease, and the cell with silk showed a 37% decrease in Vikane concentration.

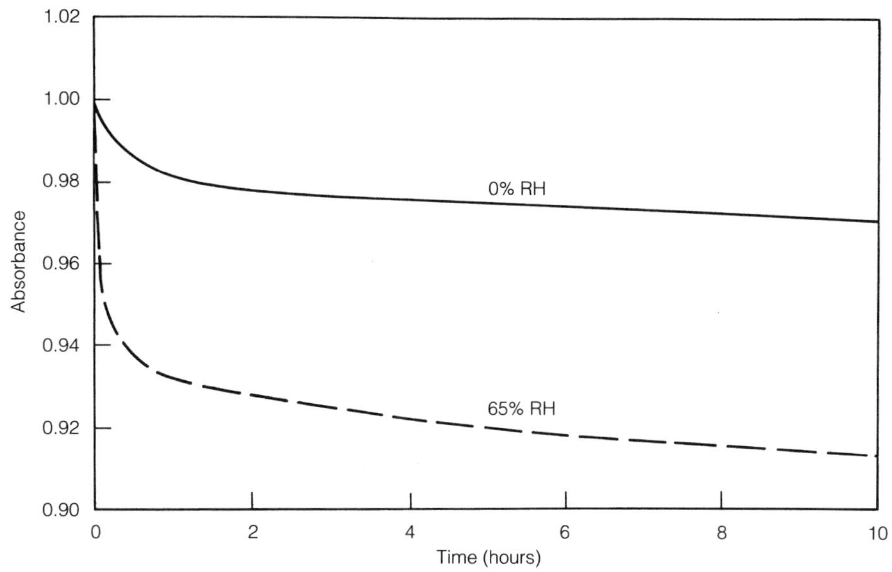

Figure 6.18
A plot of the IR absorption band of Vikane at 1492 cm^{-1} versus collection time. The first run involved a sample of brass placed in the cell with a dry nitrogen atmosphere; there was a 2% decrease in Vikane concentration. The second sample of brass was placed in the cell with 65% RH. That run showed a decrease of 8% in Vikane concentration.

depletion in the presence of the higher humidities and oxygen than had been found in the nitrogen atmosphere studies.

Resolution

In summary, depletion studies showed that IR spectroscopy is a useful tool for monitoring changes in the gaseous environment in an IR cell. While it is not possible to determine from these experiments whether the Vikane was reacting with the materials, the results did give useful information as to the types of conditions for which reactions are most likely. Results also showed that humidity does have an effect on Vikane's interactions with materials.

Case Study 8: Parylene

Artificial aging is used to hasten the degradation of materials. In the following study, IR spectroscopy is used to evaluate the chemical oxidation changes that occur as samples of Parylene are exposed to filtered UV/Vis radiation.

Background

Parylene, a generic name for polymers based on para-xylylene, is a conformal, vapor-deposited coating that was considered for use in the conservation field in the early 1980s (Humphrey 1984, 1986). The polymer has been used for the last-chance consolidation of friable, otherwise irreparable artifacts and other objects, such as ancient pinecones and feathers (Abbey Newsletter 1988). The polymer was found to have good thermal stability (Grattan and Bilz 1991).

Nowlin and coworkers have shown that an uptake of oxygen by three types of Parylene relates to a decrease in tensile strength during thermal oxidation studies at 120 °C to 200 °C (Nowlin, Smith, and Cieloszyk 1980). The oxidation kinetics were obtained by both neutron activation analysis and IR spectroscopy. Comparing these two methods,

Nowlin showed that the oxygen concentrations determined by neutron activation are directly proportional to the size of a carbonyl (C=O) absorption band at 1695 cm^{-1} in the IR spectra. Thus, for Parylene oxidation, the formation of the carbonyl band in the IR spectra coincides with a reduction in tensile strength, while also providing specific information on chemical alterations.

An initial experiment indicated that color changes occurred in free films of Parylene-C following exposure to a xenon light source filtered through nominal cutoff filters of 305, 345, 385, and 400 nm (Hansen and Ginell 1989). Yellowing occurred progressively, with both increasing time of exposure and wavelength minima for 500 hours exposure. However, since the color-change kinetics did not provide specific information on the chemical modifications or other physical properties (e.g., tensile strength), IR spectroscopy was used to provide information on the chemical degradation products in the Parylene.

Analysis questions

Can the chemical degradation products of Parylene be monitored with IR spectroscopy? What types of radiation exposure cause measurable degradative changes?

Method

Attenuated total reflectance (ATR) was chosen for this study, since it selectively analyzes the surface of the sample that is held in optical contact with the crystal. Since photooxidative degradation products are more abundant on the surface of a film, where the oxygen supply is plentiful, the sensitivity of the IR measurement to deterioration is increased by use of the surface technique.

Analysis

Parylene-C (dichloro-para-xylylene; see Suppliers, Union Carbide Corp.) free films (12 μm thick) were irradiated in a Hereaus Sun-Test chamber equipped with a xenon arc lamp filtered to yield a simulated solar spectrum. A water-cooled support plate in the chamber maintained the exposure temperature at 30 °C (±1 °C), which was determined by a thermocouple placed on a sample surface. Long-band-pass optical filters with nominal cutoffs of 305, 345, 385, 400, and 420 nm were inserted between the xenon lamp and the Parylene-C films to determine the wavelength threshold for photooxidation. Films were removed for IR and color measurements at intervals during the initial 500 hour exposure. Because the 420 nm minimum filter was excluded prior to completion of the 500 hour exposure, a second exposure experiment was conducted under similar conditions to 750 hours, for verification of the test results.

IR measurements were made on a Digilab 15-80 FT-IR spectrophotometer (see Suppliers) with a Harrick ×4 beam condenser in the ATR mode (see Suppliers). The instrument was continually purged with dry, CO_2-free air from a Balston clean-air unit (see Suppliers, Whatman, Balston Div.), to eliminate any atmospheric-induced interference bands. A cryogenically cooled MCT detector was used for maximum spectral sensitivity. Each spectrum represents the accumulation of 200 scans at a resolution of 4 cm^{-1}.

For the ATR measurements in this study, a KRS-5 (thallium bromoiodide) crystal was used at a 45° angle of reflection. This provides an approximate beam penetration of 5 μm. A sample (approximately 5 × 50 mm) was cut from the film, placed on one side of the crystal, and reproducibly pressed against the crystal surface with a torque wrench set at 2.8 gm-m (4 oz.-in.). Care was taken to ensure that the surface next to the crystal was the surface of the sample that faced the xenon lamp.

Results

Figure 6.19 shows a spectral series of Parylene-C exposed under several cutoff filters (345, 385, 400, 420 nm) for 750 hours, along with an unexposed control. The main changes in the IR absorption spectra detected during the photolysis of Parylene-C are (1) an increase in band intensities in the carbonyl region (1750–1670 cm^{-1}), due to the uptake of oxygen; (2) an increase in intensity of the hydroxyl absorptions (a broad band at

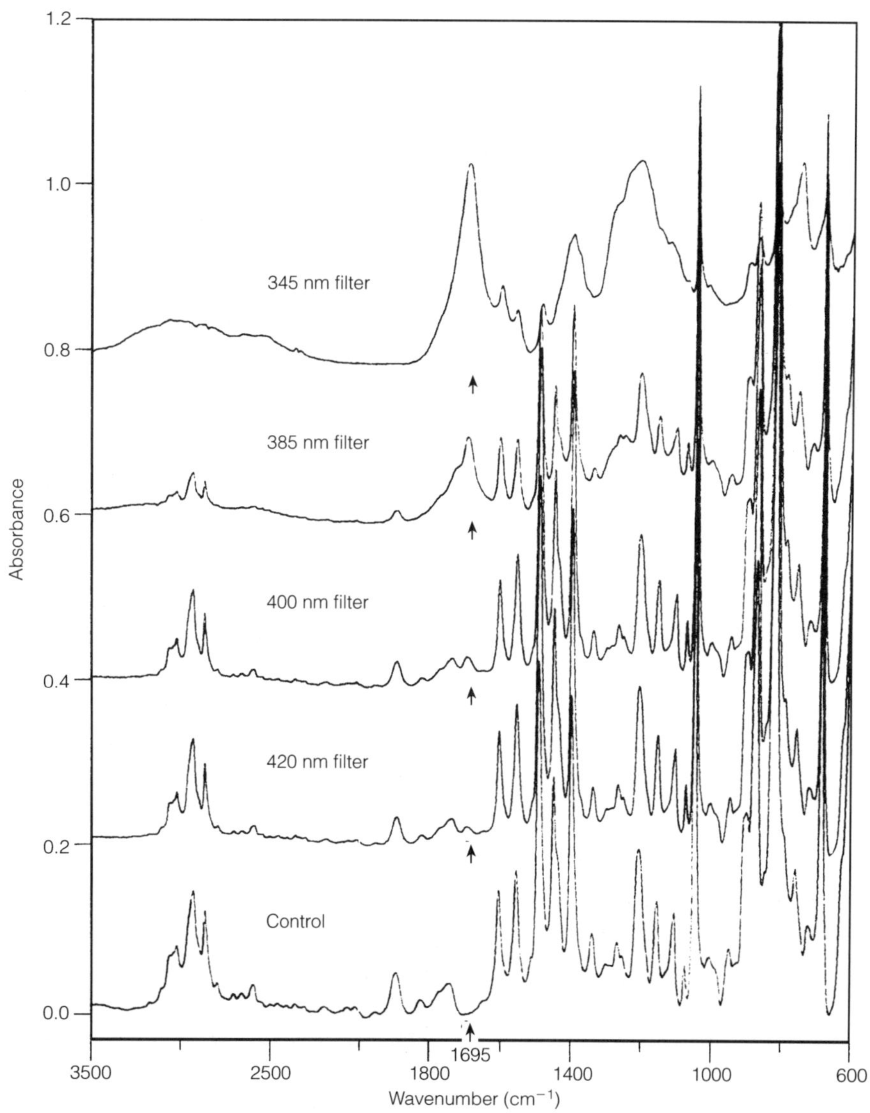

Figure 6.19
IR absorbance spectra for Parylene-C films after 750 hours of simulated sunlight exposure with UV cutoff filters (345, 385, 400, and 420 nm), along with the spectrum of an unexposed control. The carbonyl band at 1695 cm^{-1} is an indicator of photodegradation.

3400 cm^{-1}), due to an increase in the number of alcohols and acids; and (3) a decrease in the hydrocarbon stretching and bending frequencies (sharp bands at 3100–2800 cm^{-1}), which indicates a loss of carbon-hydrogen bonds due to oxidation. Overall, this spectral series shows that with an increasing amount of UV in the light, the quantity and number of oxidation products increases. A closer examination of the carbonyl region shows that the appearance and growth in the band at 1695 cm^{-1} (shown by the arrow in Fig. 6.19) is the first easily recognizable indicator of the Parylene-C oxidation process. Nowlin and coworkers found that this band corresponded directly to the oxygen uptake by parylenes in their thermal degradation studies (Nowlin, Smith, and Cieloszyk 1980).

Table 6.1 lists the total area under the carbonyl stretching band region from 1720 to 1660 cm^{-1} for each sample in the exposure matrix. Since the sample size was not uniform for every spectral measurement, the areas of the carbonyl bands were normalized to an internal reference band. Chosen for this purpose was a small band at 1880 cm^{-1} (C-H wag on the aromatic ring), because it would be a relatively stable indicator of the amount of polymer analyzed. This was also the band that Nowlin and coworkers used for normalization in their study (Nowlin, Smith, and Cieloszyk 1980).

Table 6.1 shows that there is a corresponding increase in the size of the carbonyl band versus the amount of UV light received by the sample. Additionally, there is an increase in the size of the carbonyl band with increasing time of exposure from 0 to 750 hours. Two exceptions to these trends occur for the 500 hour exposures of the unfiltered sample and the 305 nm filter sample; these have lower measurements than the sample exposed under a 345 nm filter. For the unfiltered sample, the measurement at 500 hours is lower than the measurement at 155 hours. These lower results are due to further degradation of the oxidation products. The samples for both of these measurements were extremely brittle and difficult to analyze. In fact, after the 750 hour exposure, the corresponding samples (unfiltered and with a 305 nm filter) were deteriorated to the point that they could not be analyzed.

Since the growth of the carbonyl band (presented in Table 6.1) is an early indication of the photodegradation of the polymer, it is significant to note that the area under the carbonyl band for the sample

Table 6.1
The total area under the carbonyl band (integrated from 1720 to 1660 cm^{-1}) for Parylene-C samples exposed to filtered xenon light. Each spectrum was normalized with the intensity of the absorption band at 1880 cm^{-1} at 1.0.

Time (hr)	Cutoff filter (nm)					
	none	305	345	385	400	420
0			average of 6 controls = 0.49 =/– 0.78			
35[1]	39.8	12.0	5.42	2.43	2.68	3.08
155[1]	181	91.0	24.8	9.17	6.71	5.41
500[1]	83.8	142	277	12.1	7.61	*
750[2]	*	*	442	34.1	8.74	5.98

*Not measured.
[1]Exposure study 1 (up to 500 hours).
[2]Exposure study 2 (up to 750 hours).

filtered with a 420 nm filter increased with exposure time. The cutoff filters theoretically allow zero throughput of light from wavelengths shorter than the number on the filter (i.e., higher energies), while allowing 100% transmission of the light at wavelengths above the filter number (i.e., lower energies). In reality, the cutoff lines are not sharp but are, instead, approximations. In this study, a radiometer was used to determine the exact photodistribution characteristics of each filter. The spectral distribution obtained for the xenon lamp through the 420 nm cutoff filter indicates that some radiation from 395 nm to 420 nm was present. This small portion of higher-energy radiation may be responsible for the photooxidation of the material.

Comparison of the physical changes (yellowing) with the chemical changes (formation of the carbonyl band in the IR spectra) indicated that oxidative changes in the IR spectra were apparent earlier (i.e., at lower radiation levels) than were the color changes. This finding was consistent with the formation of a colorless monobenzoyl prior to the formation of deep yellow dibenzoyl.

Resolution

This example used ATR IR spectroscopy to monitor the minor changes in the chemical composition of Parylene-C film surfaces exposed to light. Two significant points in this study were (1) that oxidation of Parylene-C occurred in light filtered to the visible region, and (2) that this oxidation was measurable prior to visual discoloration, or yellowing, of the films. When a 400 nm cutoff filter (typically found in museum environments) was used, a carbonyl band indicating oxidation was readily apparent after an exposure of 155 hours. Filtering of the violet portion of the spectrum with a wavelength minima of 420 nm still resulted in a measurable amount of chemical change in Parylene-C.

Case Study 9: Cellulose nitrate sculptures

IR analysis is often used to study degradation products, as well as to examine materials during different phases of degradation. While any analysis technique can only identify the materials currently present in a sample, information about the materials, their history, and their current condition can sometimes be used to gain insight into the degradation process.

Background

Cellulose esters, the modern plastics of the early 1900s, furnished a lightweight material that could be transformed into a myriad of shapes and colors. Because of this versatility, the plastic found many uses in everyday products (buttons, billiard balls, cosmetic cases), in new products (movie film), and in works of art (sculptures). However, it was eventually discovered that plastic products based on cellulose nitrate were inherently unstable. This polymer initially undergoes a very slow, spontaneous decomposition, at normal room conditions, that can progress to faster, autocatalytic degradation in the presence of high humidity, high temperature, or UV radiation (for a detailed explanation of this chemistry, see Selwitz 1988).

The Museum of Modern Art (MoMA) in New York has three early polymer sculptures created in the 1920s by Naum Gabo and Antoine Pevsner; the artworks exemplify the Constructivist style. These sculptures were shown by IR spectroscopy to be composed of cellulose nitrate. Examination showed that they exhibit varying degrees of deterioration and range from very good to poor in overall condition. The types of deterioration include crazing, cracking, and discoloration of the cellulose nitrate, as well as corrosion of some metal components that are in contact with the plastic. When two of the pieces began exhibiting drops of clear-to-light-brown liquid on the surface, MoMA and the GCI began an examination of these cellulose nitrate sculptures, in order to obtain a better understanding of the relation between various deterioration mechanisms and the material composition (Derrick, Stulik and Ordonez 1993).

Analysis questions

What is the composition of the droplets? Are they due to cellulose nitrate degradation?

Method

Samples were obtained from the cellulose nitrate sculptures at MoMA and from other collections. When possible, multiple samples were taken from several areas of each sculpture. The samples were acquired as broken fragments or removed as scrapings, cross sections, and exudates. They were analyzed by IR microspectroscopy at the GCI and by SEM-EDS, XRD, and optical microscopy at MoMA, to identify additives in the polymer, as well as to identify the principal component of the plastic. The application presented below will be limited to the IR microanalysis.

Analysis

The samples were characterized by FT-IR microanalysis; the analysis conditions were the same as those specified in Case Study 1.

Results

IR analysis of the liquid exudate, or "sweat," showed that it contained an inorganic nitrate. Elemental analysis by energy dispersive spectroscopy showed that the exudate contained zinc. Crystals formed when the liquid was placed in a sealed container and liquefied again upon exposure to air. This behavior corresponds to that of reference samples of nitrate salts—for example, calcium and zinc—which are hygroscopic and become hydrated to form a liquid at room conditions. In cellulose nitrate, the nitrate salts can be formed by the reaction of nitric acid (a degradation product) with fillers, stabilizers, or colorants in the piece. Zinc oxide was commonly used as a filler and opacifier. It also acts as a stabilizer by rapidly consuming nitric acid to form zinc nitrate.

The samples primarily contained camphor as a plasticizing agent, in addition to an unidentified oil. The presence of camphor was confirmed by isolating it from the sample. This was accomplished by placing a tiny sample in the center of a 0.6 cm ($^1/_4$ in.) diameter glass tube 10 cm (4 in.) long. The area of the sample was gently heated with a flame until droplet condensation was noted on the cooler portion of the tube. The tube was then quickly broken open, and a droplet was transferred to a BaF_2 pellet for analysis by IR microspectroscopy. It was imperative to work fast, since the camphor volatilized quickly. An alternate method for the analysis of a volatile liquid such as camphor would be to place the sample in a microliquid cell or between two salt pellets.

Figure 6.20 shows IR transmittance spectra of three samples of cellulose nitrate: one in good condition, one in moderate condition, and one in poor condition; also shown are spectra of the camphor and liquid nitrate salt. Noticeable in the spectra of the more degraded samples is the increase in the intensity of the band at 1340 cm^{-1}, due to the presence of nitrate salts in the sample. There is also a corresponding decrease in the intensity of the carbonyl band at 1735 cm^{-1} that is due to decreasing amounts of camphor.

The rate of deterioration of cellulose nitrate is dependent on the availability of oxygen and water and on exposure to UV radiation. This fact makes the surface of the polymer the most vulnerable area. In order to obtain a depth profile of the degradation, a cross section sample obtained from broken fragments of the sculpture was analyzed. The sample was embedded in a polyester resin and then polished to a flat, shiny surface. IR spectra were collected in reflection mode. Figure 6.21 shows three IR absorbance spectra in the region of 1900–1500 cm^{-1}; the first was collected at the surface of the cellulose nitrate sample, while the second and third were collected 1 and 2 mm, respectively, deeper into the sample. All three spectra were normalized to the height of the band at 1650 cm^{-1}, which corresponds to the nitrate ester groups in the cellulose nitrate. The spectra show that the surface of the sculpture contains cellulose nitrate with a diminished amount of camphor (based on carbonyl at 1740 cm^{-1}) compared to the composition at a depth of 2 mm into the sample.

The diminished amounts of camphor in the deteriorated areas of the plastic can be due to sublimation. Since the deteriorated surfaces often have cracks, more area is exposed, which may allow more camphor to sublime. Conversely, the loss of camphor may have a degenerative effect on the plastic. Camphor was preferred as a plasticizer because it stabilized the cellulose nitrate as well as made it less brittle (for more information on cellulose nitrate, see Reilly 1991 and Edge et al. 1990).

Resolution

IR microspectroscopy, along with other analysis methods, was used to increase understanding of the various types and degrees of degradation exhibited by each sculpture. The droplets were determined to be a hydrated nitrate salt resulting from deterioration of cellulose nitrate. At present, there are no established treatments for reversing some of the degradation processes that these objects have undergone. Since the deterioration of the polymer is caused by heat, light, acid impurities, and high humidities, preventive conservation practices may minimize further degradation by direct control of the environment.

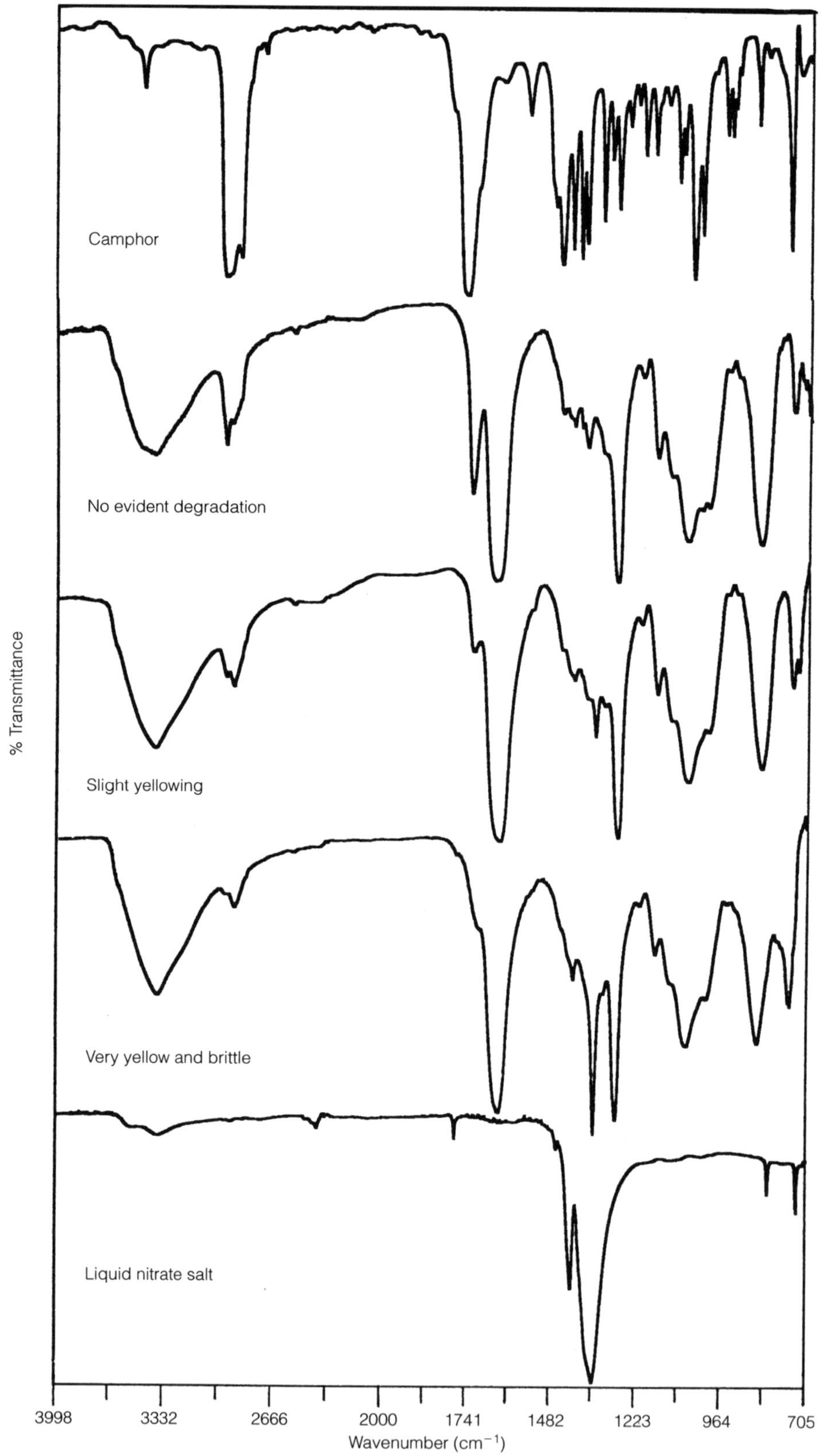

Figure 6.20
IR transmittance spectra for camphor (top) and a liquid nitrate salt (bottom), along with spectra for three samples of cellulose nitrate (plasticized with camphor) that visually exhibit varying degrees of degradation.

Case Study 10: Dead Sea Scrolls

In 1988 the GCI began an evaluation of the optimum storage and display conditions for the Dead Sea Scrolls. After the determination that control of the environment is critical to the stabilization of severely degraded

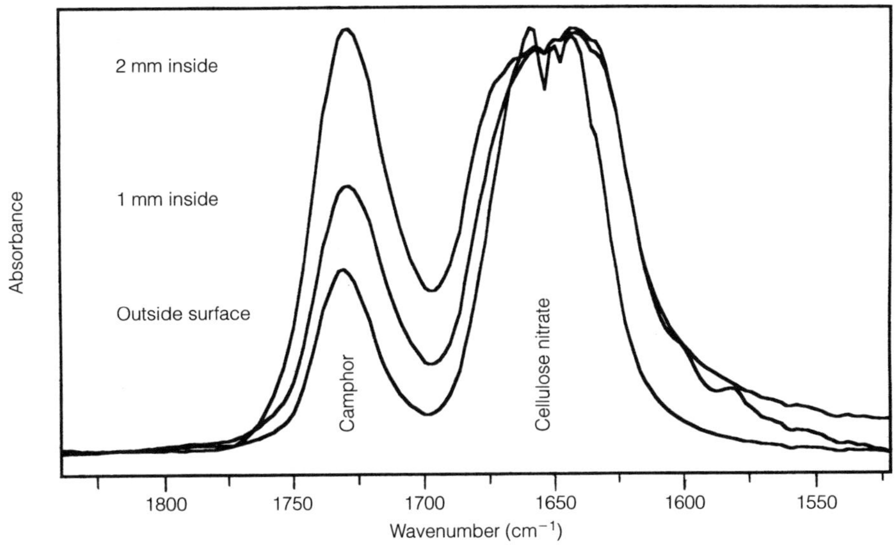

Figure 6.21
An expanded view of IR absorbance spectra for a cellulose nitrate cross section sample measured at the outer surface, 1 mm inside, and 2 mm inside. The camphor is depleted near the surface. The spectra are normalized to the height of the nitrate ester band at 1650 cm^{-1}.

parchment samples (Hansen 1993), several methods were used to evaluate the state of degradation of the scroll fragments. These included XRD, liquid chromatography, and IR spectroscopy.

Background

The collective group of documents known as the Dead Sea Scrolls were written on parchment, papyrus, and copper scrolls. The parchment scrolls were prepared from the skin of sheep and goats about two thousand years ago. Ancient preparation techniques for parchment involved abrasive dehairing of the skin and/or use of a vegetable matter and enzymatic dehairing bath. After removal from the baths, the skins were stretched tightly on a frame, and the water was removed from the skins by scraping with a half-moon knife. When dry, the skins were smoothed by abrasion, usually with pumice stones. Additionally, there is some evidence that the parchment surface was treated with small amounts of vegetable tannage or cedar oil (Poole and Reed 1962).

Collagen, a protein, is the primary component of parchment and can be irreversibly converted to gelatin. The tropocollagen molecule is held in a triple-helical structure by hydrogen bonds that, upon the addition of water and/or heat, can become structurally disorganized, or gelatinized (Weiner et al. 1980). Additionally, the protein polypeptide chain may degrade through hydrolysis or oxidation. Collagen denaturation to gelatin has been characterized by IR spectrophotometry by Brodsky-Doyle, Bendit, and Blout (1975); Susi, Ard, and Carroll (1975); and Warren, Smith, and Tillman (1969), who have shown that the most noticeable change in an IR spectrum during denaturation is an amide II band position shift from 1550 to 1530 cm^{-1} when the collagen structure is converted to the disordered form found in gelatin.

Analysis question

What is the state of degradation for nine Dead Sea Scroll parchment samples? These samples consist of fragments removed from intact scrolls, along with fragments that cannot be assigned to any specific scroll.

Method

Initially ATR analysis was used to characterize the front and back surface composition of each of the samples. Later, because it was important to determine the depth of degradation, some of the samples were prepared as thin sections for linear step-scan analysis by IR microspectroscopy.

Analysis

ATR IR analysis was used to examine front and back surface degradation on nonembedded scroll fragments and for comparison to three modern reference parchment samples. The analysis conditions for ATR measurements are given in Case Study 8.

Thin sections of the samples were characterized by FT-IR microanalysis; the analysis conditions were the same as those specified in Case Study 1. A linear spectral map of the Cave IV 9A3 scroll sample was collected with an analysis window of 20 × 100 μm stepped across the width of the sample in 15 μm increments.

Results

The surface degradation of the samples was examined by ATR by use of the method presented. With the use of Brodsky-Doyle's method, the results of the ATR analysis of this set of samples compared the relative differences in position of the amide I and II bands in the spectra to the changes in height for the two bands (Fig. 6.22; Brodsky-Doyle, Bendit, and Blout 1975). These changes in position and height indicate the degree of denaturation of the collagen to gelatin. All of the scroll samples exhibit greater degrees of denaturation than do the modern parchment reference samples.

Examination of the spectra collected across the width of several thin sections (e.g., Fig. 6.23) shows that the shift in amide II band position, in addition to the change in relative intensities of both amide I and II bands, is localized in the exterior portions of the samples. Thus, any denaturation and hydrolysis of the collagen is primarily in the outer 20–50 μm, depending on the condition of the sample. In general, the skin side of the parchment (with a smooth upper surface) exhibited degradation to a greater depth than did the flesh side (the reverse).

Photomicrographs presented in Figure 6.24, taken in normal and cross-polarized light, show that the inorganic compounds in the sample exist only near the edges, and that the inorganic material appears to be embedded in the parchment rather than being just a surface encrustation. IR absorption bands corresponding to nonproteinaceous materials or additives were found in the spectra of the surface areas of all of the scroll samples. None were detected in the interior of the samples. Three types of compounds—carbonates, silicates (pumice, talc, or dirt), and alum (aluminum ammonium sulfate)—were found as mixtures or individual components in the samples. Alum, which is used in tawing parchment, was found in two samples. Silicates, possibly from pumice used as an abrasive, were found in the exterior surfaces of all the scroll pieces. Carbonates, possibly used to fill pores or as an abrasive, were present in large amounts in one sample and in very small or nondetectable amounts

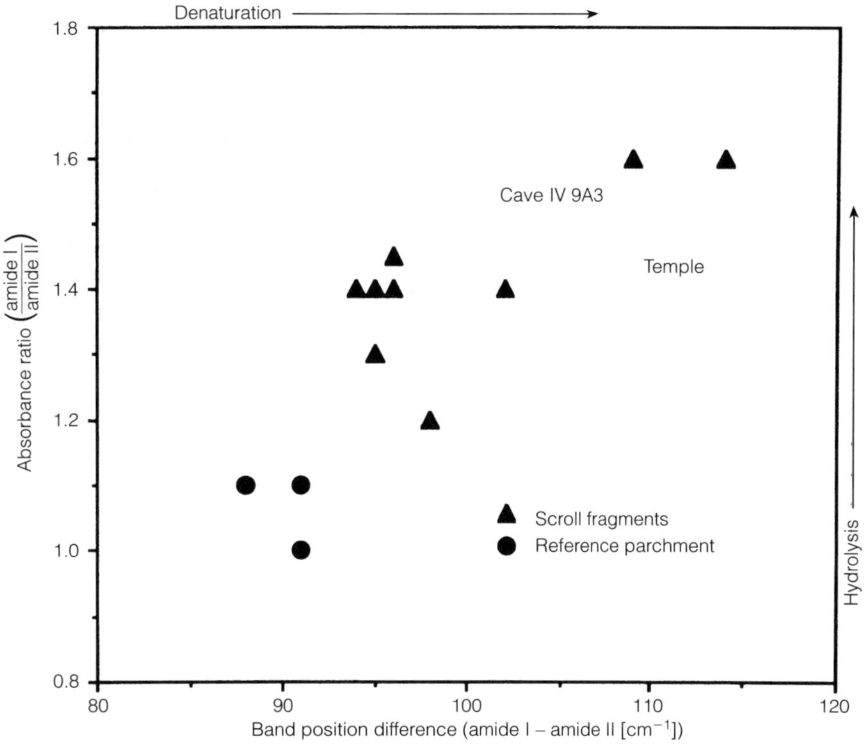

Figure 6.22
The difference in absorption band position (amide I: 1650 cm^{-1}; amide II: 1550 cm^{-1}) versus the ratio of their intensities (amide I/amide II). All scroll fragments exhibited more degradation of the collagen than did the reference parchment samples.

in the remaining scroll samples. There could be other nonproteinaceous compounds, such as organic tanning agents, present in the samples in amounts too small to be detected by IR spectroscopy.

Resolution

Microanalysis showed that the samples were nonhomogeneous and that the spectral results varied from area to area within a sample. Thus,

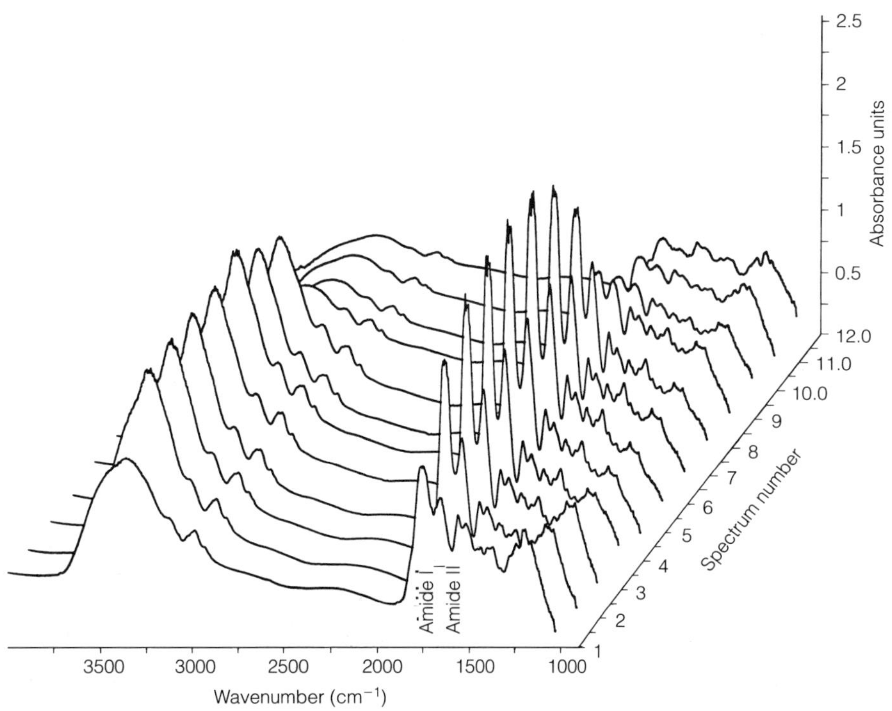

Figure 6.23
A linear IR absorption band spectral map of the Dead Sea Scroll cross section sample shown in Figure 6.24. The analysis window was 20 × 100 μm; the step size was 15 μm. The window was stepped from the flesh side of the parchment (scan 1) to the skin side of the parchment (scan 12). The relative intensities of the amide I and amide II bands change; this change indicates that the surfaces of the sample are more denatured than are the interior.

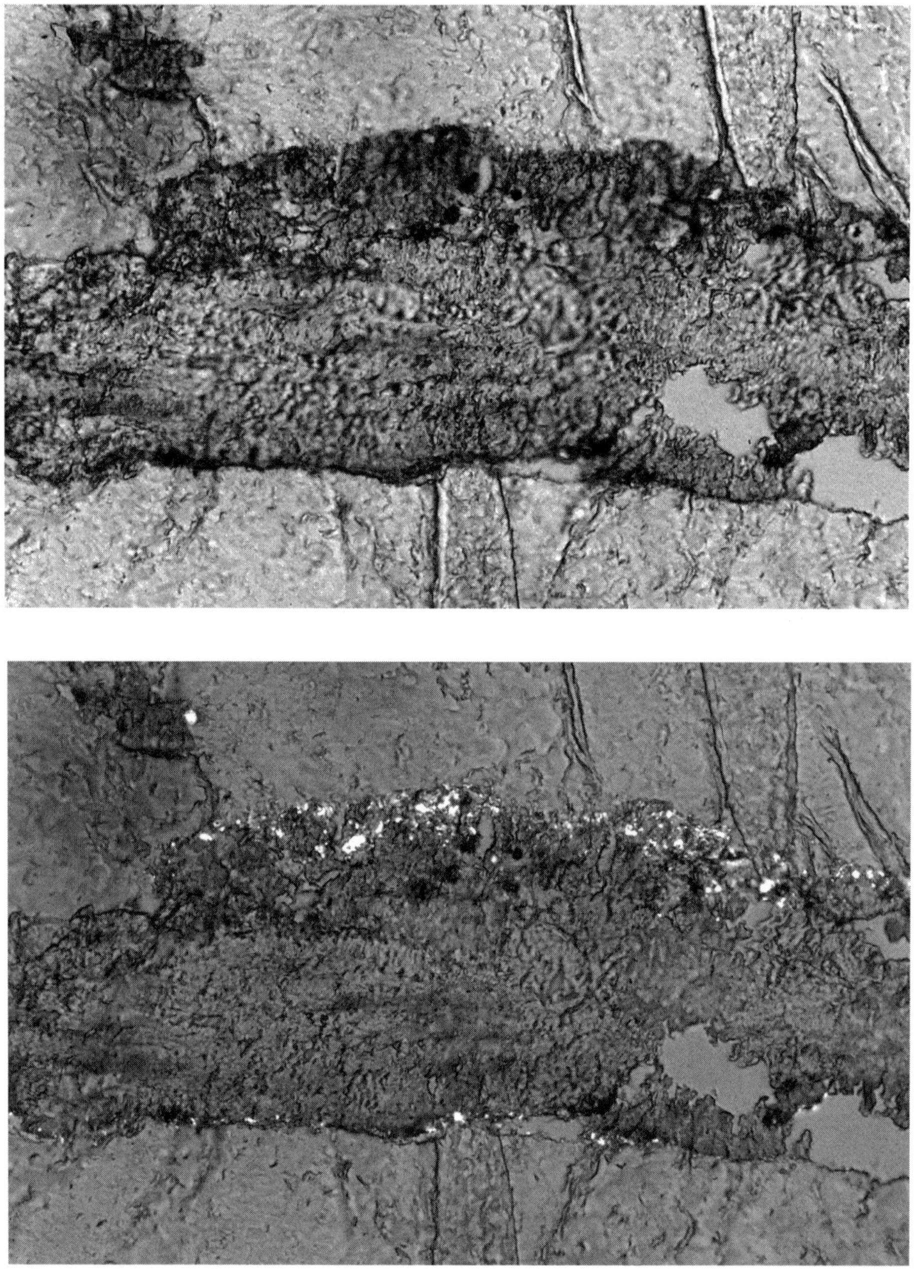

Figure 6.24
Photomicrographs of a thin section of a sample from a Dead Sea Scroll (cave IV 9A3) at ×250. Normal illumination was used for the top photo; the bottom photo was taken in cross-polarized light.

while the results themselves were reproducible (as determined by replicate analysis) and were characteristic of the area analyzed, they were not necessarily representative of the entire scroll from which the sample was taken. However, since several samples covering a wide range of types were analyzed, the total set of the IR results on the pieces were exemplary of many areas on the scrolls. In addition, since some areas of the scrolls were shown to be severely degraded, the recommended storage conditions should be based on the worst—that is, the most gelatinized scroll.

Summary

The characterization of materials is one of the most important functions of IR spectroscopy in the art conservation field. It is often used as the first analytical method for classifying the major components in a sample submitted for analysis. Almost any type or form of sample, except for metals, can be analyzed. In most cases, it is necessary to remove small, barely visible particles from the object.

Examples in this chapter illustrate the potential for stratigraphic microanalysis of multilayered materials. The optimal sample form is a thin section prepared by microtoming an embedded cross section. Also explored has been the use of IR reflection techniques for the analysis of polished cross sections for the production of molecular maps. Currently the reflection techniques are most useful when supplemental analyses by other methods are available.

In addition to its use in the identification of materials, IR spectroscopy can be used to evaluate the condition of a material and to monitor chemical reactions. Both of these functions are significant to the evaluation and monitoring of deterioration. IR analysis is a very practical method for comparative studies of chemical changes in relation to time and conditions.

Appendix I

Selected Infrared Spectra Collections and Digitized Libraries

All instrument manufacturers sell similar collections of digitized spectral libraries that work with their search software. However, to minimize repetition, only a few were selected for this appendix. Please see the Suppliers list for the names and addresses of other instrument manufacturers.

Aldrich Library of Infrared Spectra, C. J. Pouchert, ed.
Aldrich Chemical Co., 1001 W. St. Paul Ave., Milwaukee, WI 53233
Classical hard-copy reference of IR grating spectra for pure chemicals. Also includes polymers, organometallics, and carboxylic acid salts.

Aldrich-Nicolet Digitized Libraries
Nicolet Instrument Corp., 5225-1 Verona Road, Madison, WI 53711
Digitized versions of the classical Aldrich collection, along with digitized spectra for chemicals sold by Sigma Chemical. Aldrich-Nicolet collection; Sigma-Nicolet collection.

Bio-Rad Sadtler Div.—see **Sadtler Research Laboratories.**

Coblentz Society, Inc., P.O. Box 9952, Kirkwood, MO 63122
Special collection books of high-quality grating spectra contributed by government and industrial labs. Books are edited by Clara Craver and contain text sections with good information on sample preparation and spectral interpretation. *The Desk Book of Infrared Spectra;* regulated and major industrial chemicals; gases and vapors; halogenated hydrocarbons; plasticizers and other additives.

Galactic Industries Corp., 395 Main St., Salem, NH 03079
Search, data conversion, and data processing spectroscopic software (Spectra Calc); data acquisition and processing for data generated by spectroscopy and chromatography instruments (Lab Calc); designed to integrate laboratory data from several types of instruments.

Infrared Data Committee of Japan (IRDC), Sanyo Shuppan Boeki Co., Hoyu Bldg., 8, 2-Chrome, Takaru-cho, Chuo-ku, Tokyo, Japan
Search software based on wavenumbers and intensities; possibly changing to fully digitized spectra (14,000 spectra).

National Chemical Laboratory for Industry (NCLI) collection, Japan
An integrated system with a digitized collection of electron spin resonance, nuclear magnetic resonance, IR, and mass spectra. This database is available online in Japan; it is described in O. Yamamoto, K. Someno, N. Wasada, J. Hiraishi, K. Hayamizu, K. Tanabe, T. Tamura, and M. Yanagisawa, "An Integrated Spectral Database System including IR, MS, proton-NMR, carbon-13-NMR, ESR, and Raman Spectra," *Analytical Sciences* 4(1988):233–39.

Sadtler Research Laboratories, Bio-Rad Sadtler Div., 3316 Spring Garden St., Philadelphia, PA
Search software (IR Mentor); hard-copy and digitized versions of spectral libraries (libraries contain FT-IR and grating IR spectra). Condensed-phase and vapor-phase standards; adhesives and sealants; attenuated total reflectance of polymers; coating chemicals; automobile paint chips; commonly abused drugs; controlled pyrolysates of polymers; dyes, pigments, and stains; fats, waxes, and derivatives; fiber and textile chemicals; flame retardants; flavors and fragrances; food additives; inorganics; lubricants; manufacturing starting materials and intermediates; minerals and clays; monomers and polymers; organometallics; pesticides and agricultural chemicals; petroleum chemicals; pharmaceuticals; plasticizers; polymer additives; prepared and prescription drugs; priority pollutants; rubber chemicals; solvents; steroids; surface-active agents; water-treatment chemicals.

Also suppliers for automobile paint chips database; Canadian Forensic Package Library; enhanced EPA Vapor-Phase Package Library; Georgia State Crime Lab Package Library; Merck/Sadtler Library; Hummel/Sadtler Polymer Library; Sadtler/Scholl Polymer Processing Library; University Package Library of Pure Compounds.

Sprouse Scientific Systems, Inc., 1801 Crossbeam Dr., Charlotte, NC 28217
Search software (Micro-search), hard-copy and digitized versions of spectral libraries (libraries contain FT-IR spectra; books are edited by D. L. Hansen). *Polymers; Solvents (Transmission Spectra); Solvents (CIRCLE Cell Spectra); Surface-Active Agents, Solvents: Condensed-Phase, Vapor-Phase, and Mass Spectra.* In addition to these four collections, digitized IR spectral libraries on the following topics are available: the Canada Collection: coal, shale, and clay minerals; coatings and resins; environmental toxins; EPA vapor-phase (corrected); epoxy resins, curing agents, and additives; fragrances and essential oils; fibers by IR microscope; gas-phase IR of environmental chemicals; small-molecule gases and environmental pollutants; general chemical compounds; Georgia State Crime Lab forensic library; Georgia State Crime Lab automotive paint; inorganics on KBr beam splitter; inorganics on CsI beam splitter; lubricants, additives, and raw materials; minerals, U.S. Geological Survey collection; polymers by transmission; polymers by attenuated total reflectance; polymer additives; general organic compounds; solvents; solvents by cylindrical internal reflectance; solvents, vapor-phase; surface-active agents.

Other Collections

Afremow, L. C., and J. T. Vandeberg
1966. High resolution spectra of inorganic pigments and extenders in the mid-infrared region from 1500 cm^{-1} to 200 cm^{-1}. *Journal of Painting Technology* 38(495):169–202. (78 spectra of pigments with known composition.)

American Society for Testing Materials
1969. *ASTM-Wyandotte Index: Alphabetical List of Compound Names, Formulas and References to Published Infrared Spectra; an Index to 92,000 Published Infrared Spectra.* Philadelphia: ASTM. (Previous versions: ASTM special publications 131[1962], 131-A[1963].)

Association of Official Analytical Chemists
1975. *Infrared and Ultraviolet Spectra of Some Compounds of Pharmaceutical Interest.* Rev. ed. Washington, D.C.: Association of Official Analytical Chemists.

Bellamy, L. J.
1975. *Advances in Infrared Group Frequencies.* London: Chapman and Hall.

1980. *The Infrared Spectra of Complex Molecules.* 2d ed. London: Chapman and Hall.

Bellanto, J., and A. Hidalgo
1971. *Infrared Analysis of Essential Oils.* London: Heyden and Sons.

Boldyrev, A. I.
1976. *Infrared Spectra of Minerals.* Moscow: Nedra.

British Pharmacopoeia Commission
1980. *Infrared Reference Spectra.* (First supplement 1981, second supplement 1982, third supplement 1984.) London: H.M. Stationery Office.

Cain, D. S., and S. S. Stimler
1967. *Infrared Spectra of Plastics and Resins.* Part 3. Naval Research Laboratory, Report 6503, Feb. 28. Washington, D.C.: U.S. Department of Commerce.

Carroll, G. R., W. C. LaLonde, B. D. Gaudette, S. L. Hawley, and Hubert
1988. A computerized database for forensic textile fibres. *Journal of the Canadian Society of Forensic Scientists* 21:1.

Chicago Society for Coating Technology
1980. *An Infrared Spectroscopy Atlas for the Coatings Industry.* Philadelphia: Federation of Societies for Paint Technology. (1433 spectra.)

Coates, J. P., and L. C. Setti
1983. *Oils, Lubricants, and Petroleum Products—a Library Containing Spectra from Fuels, Lubricants, Additives and Related Materials.* Norwalk, Conn.: Perkin-Elmer Corp.

Colthup, N. B., L. H. Daly, and S. E. Wiberley
1975. *Introduction to Infrared and Raman Spectroscopy.* 2d ed. New York: Academic Press. (About 600 spectra.)

Dobriner, K., E. R. Katzenellenbogen, and R. N. Jones
1953. *Infrared Absorption Spectra of Steroids—an Atlas.* Vol. 1. New York: Wiley-Interscience.

Dolphin, D., and A. E. Wick
1977. *Tabulation of Infrared Spectral Data.* New York: Wiley.

Elliott, A.
1969. *Infrared Spectra and Structure of Organic Long-Chain Polymers.* London: Arnold.

Farmer, V. C., ed.
1974. *The Infrared Spectra of Minerals.* Mineralogical Society Monograph 4. Surrey, U.K.: Adlard and Son. (Contains detailed information on the relationship of mineral structures to their infrared spectra.)

Flett, M.
1969. *Characteristic Frequencies of Chemical Groups in the Infrared.* New York: American Elsevier.

Fox, R. H., and H. I. Schuetzman
1968. The infrared identification of microscopic samples of man-made fibers. *Journal of Forensic Science* 13(3):397–406. (Contains 13 spectra of synthetic fibers.)

Gadsen, J. A.
1975. *Infrared Spectra of Minerals and Related Inorganic Compounds.* London: Butterworth.

Gore, R. C., R. W. Hannah, S. C. Pattacini, and T. J. Porro
1971. Infrared and ultraviolet spectra of seventy-six pesticides. *Journal of the Association of Official Analytical Chemists* 54(5):1028–40.

Hershenson, H. H.
1959, 1964. *Infrared Absorption Spectra Index.* New York: Academic Press. (Two volumes cover 1945–62.)

Hummel, D. O., and F. Scholl
1981. *Atlas of Polymer and Plastics Analysis.* (Vol. 1, *Polymers: Structures and Spectra;* vol. 2, *Plastics, Fibres, Rubbers, Resins;* vol. 3, *Additives and Processing Aids.*) New York: Verlag Chemie International. (Over 7000 spectra.)

Infrared Users' Group (IRUG)
1995. *Art and Conservation Materials Infrared Spectral Library.* (Vol. 1, *Natural and Synthetic Products;* vol. 2, *Colorants* [limited distribution].) Marina del Rey, Calif.: Scientific Program of the Getty Conservation Institute.

Jakes, K. A., L. R. Sibley, and R. Yerkes
1994. A comparative collection for the study of fibres used in prehistoric textiles from eastern North America. *Journal of Archaeological Science* 21:641–50.

Kagarise, R. E., and L. A. Weinberger
1954. *Infrared Spectra of Plastics and Resins.* Part 1. U.S. Naval Research Lab report #4369, Washington, D.C.: U.S. Department of Commerce.

Karcher, W., ed.
1983, 1988. *Spectra Atlas of Polycyclic Aromatic Compounds.* Vols. 1, 2. Boston: Kluwer Academic Publishers.

Keller, R. J.
1986. *The Sigma Library of FT-IR Spectra.* Vols. 1, 2. St. Louis, Mo.: Sigma Chemical Collection.

Langenheim, J. H., and C. W. Beck
1968. Catalogue of infrared spectra of fossil resins (ambers). I. North and South America. *Botanical Museum Leaflets* 22(3):65–120. (45 spectra.)

McClure, A., J. Thomson, and J. Tannahill
1968. Infrared spectra of ninety-six organic pigments. *Journal of Oil and Colour Chemists' Association* 51:580.

Merck, E., ed.
1988. *Merck IR Atlas—A Collection of FT-IR Spectra.* New York: VCH Publishers.

Miller, F. A., and C. H. Wilkins
1952. Infrared spectra and characteristic frequencies of inorganic ions. *Analytical Chemistry* 24:1253–94.

Ministry of Aviation Technical Information and Library Services
1960. *An Index of Published Infrared Spectra.* Vols. 1, 2. London: Her Majesty's Stationery Office.

Mitzner, B. M., E. T. Theimer, and S. K. Freeman
1965. The infrared spectra of monoterpenes and related compounds. *Applied Spectroscopy* 19(6):169–85.

Morris, W. W.
1973. High resolution infrared spectra of fragrance and flavor compounds. *Journal of the Association of Official Analytical Chemists* 56:1027.

Nakamoto, K.
1978. *Infrared and Raman Spectra of Inorganic and Coordination Compounds.* 3d ed. New York: John Wiley and Sons.

Nyquist, R. A., and R. O. Kagel
1971. *Infrared Spectra of Inorganic Compounds.* New York: Academic Press. (892 spectra.)

Plyusnina, I. I.
1977. *Infrared Spectra of Minerals.* Moscow: Izd. Mosk. Univ.

Polchlopek, S. E., and R. L. Harris
1963. *ATR Spectra of Surface Coatings.* Stamford, Conn.: Barnes Engineering Co.

Pouchert, C. J., ed.
1981. *The Aldrich Library of Infrared Spectra.* 3d ed. Milwaukee, Wis.: Aldrich Chemical Co. (About 11,000 spectra.)

1985–89. *The Aldrich Library of FT-IR Spectra.* (Vol. 1, 1985; vol. 2, 1988; vol. 3, 1989.) Milwaukee, Wis.: Aldrich Chemical Co.

Price, B., and J. Carlson
Forthcoming. *Infrared Spectra of Naturally Occurring Minerals.* Newark, N.J.: University of Delaware.

Rouen, R. A., and V. C. Reeve
1970. A comparison and evaluation of techniques for identification of synthetic fibers. *Journal of Forensic Science* 15(3):410–32. (Contains 20 spectra of synthetic fibers.)

Siesler, H. W., and K. Holland-Moritz
1980. *Infrared and Raman Spectroscopy of Polymers.* New York: Marcel Dekker.

Snodgrass, A., and B. Price
1993. *Infrared Spectra of the Gettens Collection.* Cambridge, Mass.: Harvard Art Museums.

Stimler, S. S., and R. E. Kagarise
1966. *Infrared Spectra of Plastics and Resins 2: Materials Developed since 1954.* U.S. Naval Research Lab, #6392, Washington, D.C.: U.S. Department of Commerce.

Szymanski, H. A., and R. E. Erickson
1970. *Infrared Band Handbook.* Rev. 2d ed. (2 vols.). New York: Plenum.

Thermodynamics Research Center Hydrocarbon Project
1990. *Selected Infrared Spectral Data.* College Station, Tex.: Thermodynamics Research Center, Texas Agricultural and Mechanical University.

Thompson, B.
1974. *Hazardous Gases and Vapors: Infrared Spectra and Physical Constants.* Technical Report no. 595 (August). Fullerton, Calif.: Beckman Instruments.

Tipson, R. S.
1968. *Infrared Spectroscopy of Carbohydrates.* Washington, D.C.: National Bureau of Standards.

Tungol, M. W., E. G. Bartick, and A. Montaser
1990a. The development of a spectral data base for the identification of fibers by infrared microscopy. *Applied Spectroscopy* 44(4):543–49. (Describes the development of a reference library for polymer fibers.)

1990b. Spectral data base for fibers by infrared microscopy. *Spectrochimica Acta* 46B:1535E.

Van der Marel, H. W., and H. Beutelspacher
1976. *Atlas of Infrared Spectroscopy of Clay Minerals and Their Admixtures.* New York: Elsevier.

Welti, D.
1970. *Infrared Vapour Spectra.* New York: Heyden and Sons.

Wenninger, J. A., and R. L. Yates
1967. High resolution infrared spectra of some naturally occurring sesquiterpene hydrocarbons, part 1. *Journal of the Association of Official Analytical Chemists* 50(6):1313–45.

1970. High resolution infrared spectra of some naturally occurring sesquiterpene hydrocarbons, 2: Second series. *Journal of the Association of Official Analytical Chemists* 53(5):949–61.

White, R. G.
1964. *Handbook of Industrial Infrared Analysis.* New York: Plenum Press.

Wilks, P. A., and M. R. Iszard
1964. *The Identification of Fibers and Fabrics by Internal Reflection Spectroscopy.* Presented at the 15th Mid-America Spectroscopy Symposium, Chicago. South Norwalk, Conn.: Wilks Scientific Corp. (Contains 47 spectra of fibers.)

Yamaguchi, K.
1968. *Spectral Data of Natural Products.* Vols. 1, 2. New York: Elsevier.

Zeller, M. V., and M. R. Grabowski
1978. *A Collection of Reference Spectra for Flame Retardant Reagents.* Perkin-Elmer Infrared Bulletin no. 66. Norwalk, Conn.: Perkin-Elmer Corp. (Contains 30 spectra of flame retardants for fabrics.)

Zeller, M. V., and M. P. Juszli
1975. *Reference Spectra of Minerals.* Norwalk, Conn.: Perkin-Elmer Corp.

Zeller, M. V., and S. C. Pattacini
1973. *The Infrared Grating Spectra of Polymers.* Perkin-Elmer Infrared Applications Study no. 13. Norwalk, Conn.: Perkin-Elmer Corp. (Contains spectra for 29 polymers along with a flowchart to aid in the identification of polymers.)

Appendix II
Infrared Reference Spectra

An important factor for IR spectral identification is access to reference spectra corresponding to appropriate materials. While Appendix I supplied numerous commercial and literature sources for spectra, this appendix gives a few examples of spectra for art and conservation materials, because commercial libraries often do not include materials encountered in cultural artifacts.

The spectra are organized by their material classifications. For example, gelatin and casein are both classified as proteins. Materials in the same classification have similar chemical compositions and thus similar spectra. The spectra sheets also provide physical data about the sample, as well as its analysis conditions. Accompanying descriptions supply information about the material and its use in art and conservation. An alphabetical listing of the materials is given below.

Alphabetical Listing of IR Reference Spectra

Gum Arabic

PROVENANCE	Los Angeles County Museum of Art; RBG, 66.1848
SOURCE	Sudan
APPEARANCE	Amber color, transparent, chunks

CHARACTERISTIC IR ABSORPTION BANDS

3600–3200 cm^{-1}	O-H	stretching band
3000–2800 cm^{-1}	C-H	stretching bands
1650 cm^{-1}	O-H	bending band
1480–1300 cm^{-1}	C-H	bending band
1300–900 cm^{-1}	C-O	stretching bands

Gum Arabic

Gum arabic is the most commonly used gum in preparation of paints. It is a dried, amorphous exudate from the stem of several species of *Acacia* trees *(Acacia senegal)* in tropical and subtropical areas of the world. Most of the current output of gum arabic is from the sub-Sahara region in Africa. The gum is sold in the form of colorless, round lumps, as granules, as thin flakes, or as a powder; all of these may be white or slightly yellowish. Gum arabic is completely soluble in hot and cold water, yielding a viscous solution. It is insoluble in alcohol. Gum arabic is used in watercolor, paints, and inks and for textile sizing. The earliest known inks consisted of gum arabic and lamp black.

SYNONYMS: Kordofan, picked turkey, white Sennar, Senegal gum, Ghezineh gum, *gomme blonde, gomme blanche,* gum acacia, East India gum, kami.

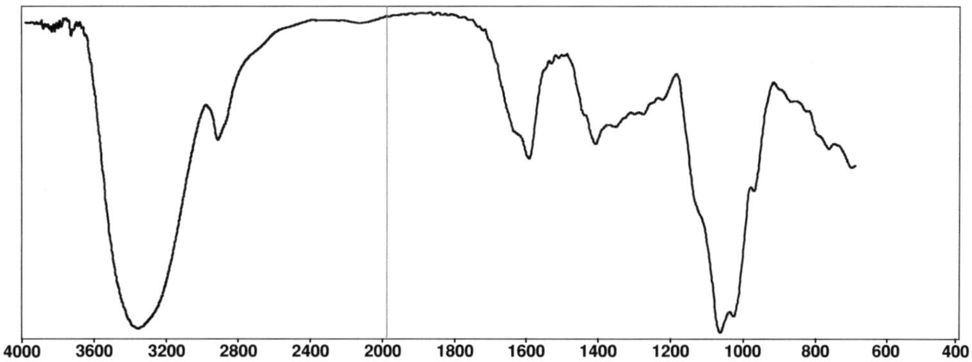

IR ANALYSIS LAB

Loyola Marymount University, 7/6/89

IR ANALYSIS CONDITIONS

P-E 1700
Microscope
Cast from solvent on BaF_2
Resolution = 4 cm^{-1}
Scans = 120
Range = 4000–700 cm^{-1}

Gum Arabic

Gum / Sugar

Gum Tragacanth *(Astragalus gummifer)*

PROVENANCE	Los Angeles County Museum of Art; RBG, 66.1848
SOURCE	Western Persia
APPEARANCE	Brown, opaque, sheets

CHARACTERISTIC IR ABSORPTION BANDS

3600–3200 cm^{-1}	O-H	stretching band
3000–2800 cm^{-1}	C-H	stretching bands
1750–1600 cm^{-1}	C=O	stretching band
1650 cm^{-1}	O-H	bending band
1480–1300 cm^{-1}	C-H	bending bands
1300–900 cm^{-1}	C-O	stretching bands

Gum Tragacanth

Gum tragacanth is an exudate from several species of shrubs of the genus *Astragalus* found in the dry regions of Iran, Syria, and Turkey. It is available in the form of dull white, translucent plates or as a yellowish powder. It is insoluble in alcohol but soluble in alkaline solutions and solutions of hydrogen peroxide. A soluble fraction, tragacanthin, dissolves when added to water, whereas an insoluble fraction, bassorin (60–70% by wt.) swells to a gel-like state. A solution is prepared by wetting the powder with alcohol, then adding water and shaking. Gum tragacanth is used for textile sizing and printing, pastel crayon production, leather curing, and furniture polishes.

SYNONYMS: Gum tragacanth, gum dragon, *gomme adragante,* Smyrna tragacanth, Anatolian tragacanth, Persian tragacanth.

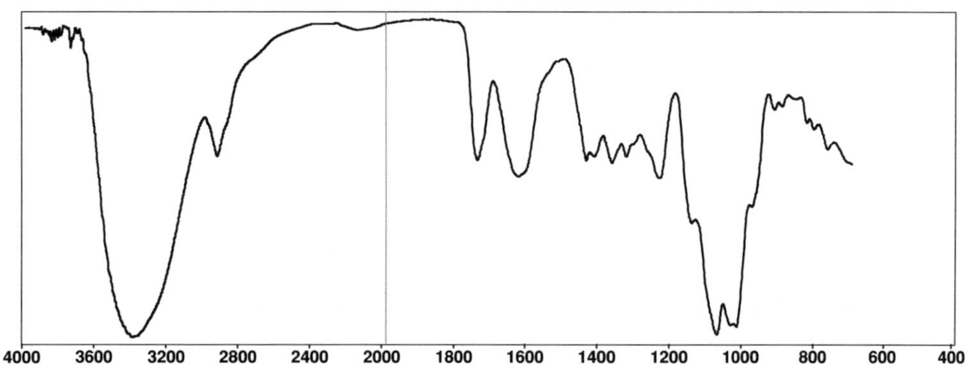

IR ANALYSIS LAB

Loyola Marymount University, 7/11/89

IR ANALYSIS CONDITIONS

P-E 1700
Microscope
Cast from solvent on BaF_2
Resolution = 4 cm^{-1}
Scans = 120
Range = 4000–700 cm^{-1}

Gum Tragacanth

Gum / Sugar

Rice Starch

PROVENANCE U.S. Customs Lab, Long Beach, Calif.

SOURCE Unknown

APPEARANCE White, opaque, powder

CHARACTERISTIC IR ABSORPTION BANDS

3600–3200 cm^{-1}	O-H	stretching band
3000–2800 cm^{-1}	C-H	stretching bands
1650 cm^{-1}	O-H	bending band
1480–1300 cm^{-1}	C-H	bending bands
1300–900 cm^{-1}	C-O	stretching bands

Starch

Starches occur as granules of varying size in the roots, bulbs, and seeds of most plants. Chemically, starch is a carbohydrate in the same family as cellulose, gums, and sugars. Starch contains about 20% of water soluble amylose and 80% of a water-insoluble fraction called amylopectin. Both fractions correspond to different carbohydrates of high molecular weight and formula $(C_6H_{10}O_5)_n$. Starch turns iodine blue. Starch is a fairly poor adhesive but has been used in some cases for lining paintings. As a paint binder, it was most popular for fingerpaints and cheap house paints.

VARIETIES: Wheat, corn, rice, dextrin, dextran, British gum, mucilage, sorghum, potato, tapioca, arrowroot, sago palm.

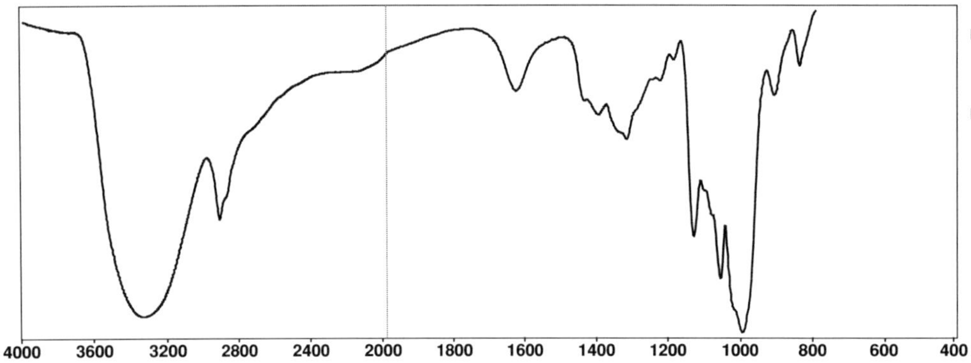

IR ANALYSIS LAB

Getty Conservation Institute, 1/19/93

IR ANALYSIS CONDITIONS

Spectra-Tech IRµS
Microscope
Bulk sample on BaF_2
Resolution = 4 cm^{-1}
Scans = 200
Range = 4000–800 cm^{-1}

Starch **Gum / Sugar**

Honey, Clover

PROVENANCE J. Paul Getty Museum Antiquities Conservation

SOURCE USA

APPEARANCE Golden-yellow, highly viscous, liquid

CHARACTERISTIC IR ABSORPTION BANDS

3600–3200 cm^{-1}	O-H	stretching band
3000–2800 cm^{-1}	C-H	stretching bands
1650 cm^{-1}	O-H	bending band
1480–1300 cm^{-1}	C-H	bending bands
1300–900 cm^{-1}	C-O	stretching bands

Honey

A sweet, viscous fluid produced by bees from the nectar of flowers. Honey is a mixture of fructose, glucose, dextrose, and water (~20%), with trace amounts of enzymes and oils. Its composition varies slightly depending on the source of nectar. Honey was used since early times as a plasticizing additive to watercolors, tempera, size, and glair.

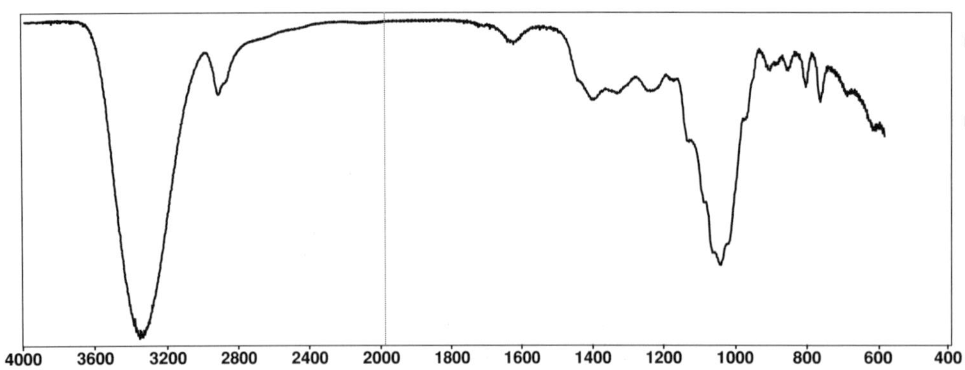

IR ANALYSIS LAB

Getty Conservation Institute, 9/21/84

IR ANALYSIS CONDITIONS

Digilab 15E/80
KBr pellet
Ground with KBr
Resolution = 4 cm^{-1}
Scans = 200
Range = 4000–600 cm^{-1}

Honey **Gum / Sugar**

Hide Glue (pearls)

PROVENANCE J. Paul Getty Museum Decorative Arts Conservation; Gleck & Co.

SOURCE USA

APPEARANCE Light amber, translucent, chunks

CHARACTERISTIC IR ABSORPTION BANDS

3400–3200 cm^{-1}	N-H	stretching band
3100–2800 cm^{-1}	C-H	stretching bands
1660–1600 cm^{-1}	C=O	stretching band
1565–1500 cm^{-1}	C-N-H	bending band
1480–1300 cm^{-1}	C-H	bending band

Animal Glue

Animal glue is an adhesive consisting primarily of gelatin and other protein residues of collagen, keratin, or elastin. Glues may be made from bones, skins, hides, and intestines of animals (fish, goats, sheep, cattle, horses, etc.). These agglutinating materials are removed by extraction with hot water, then cooled and dried to produce gelatin or glue. Animal glues are available in the form of sheets, droplets, chips, granules, cubes, and powder. They occur in a wide variety of colors ranging from transparent to opaque and white to brown. Glue is soaked in cool water to form a turbid jelly that will become clear and thinner upon heating to 40 °C. Glue will decompose and darken when it is boiled. Animal glues are strong adhesives that have been used in furniture manufacture, gilding, gessoes, and paint binders.

VARIETIES: Animal glue, glue, gelatin, size, isinglass, fish glue, rabbit-skin glue, bone glue, hide glue, parchment glue, calfskin glue, skin glue, Nikawa, sturgeon glue, sturgeon's glue, deerskin glue.

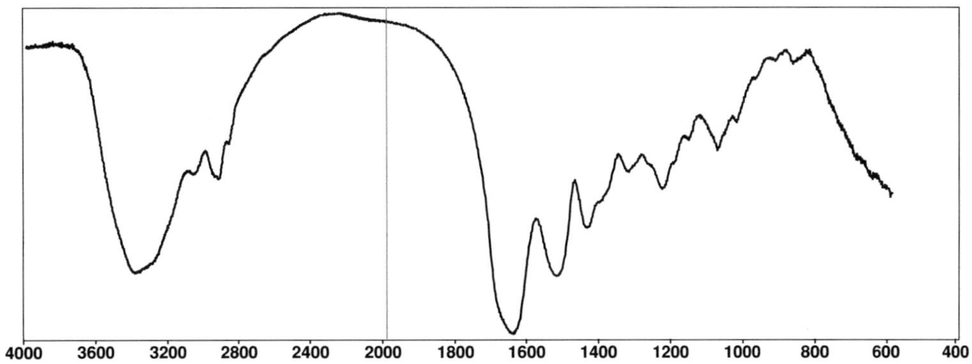

IR ANALYSIS LAB
Getty Conservation Institute, 9/14/84

IR ANALYSIS CONDITIONS
Digilab 15E/80
KBr pellet
Ground with KBr
Resolution = 4 cm^{-1}
Scans = 200
Range = 4000–600 cm^{-1}

Animal Glue

Protein

Isinglass (sheets)

PROVENANCE Zecchi

SOURCE Russia

APPEARANCE White, opaque, sheets

CHARACTERISTIC IR ABSORPTION BANDS

3400–3200 cm^{-1}	N-H	stretching band
3100–2800 cm^{-1}	C-H	stretching bands
1660–1600 cm^{-1}	C=O	stretching band
1565–1500 cm^{-1}	C-N-H	bending band
1480–1300 cm^{-1}	C-H	bending band

Fish Glue

Fish glue is made from unsorted fish waste and is an adhesive consisting primarily of gelatin and other protein residues of collagen, keratin, or elastin. These agglutinating agents are removed by extraction with hot water, then cooled and dried to produce gelatin or glue. Varied production techniques can produce poor-quality fish glue. The highest quality is made from the bladders of sturgeons. It is clear, bluish white, and very flexible. It is solidified into flat disks that are usually broken into smaller bits for sale. Isinglass is a particularly fine glue made from a specific type of sturgeon. It is generally sold in narrow, soft, translucent strips. Isinglass is very expensive. Glue is to be soaked in cool water to form a turbid jelly that will become clear and thinner upon heating to 40°C. Glue will decompose and darken when it is boiled. High-quality fish glues are used for paintings and gilding.

VARIETIES: Gelatin, size, isinglass, fish glue, sturgeon glue, sturgeon's glue, ichthycoll.

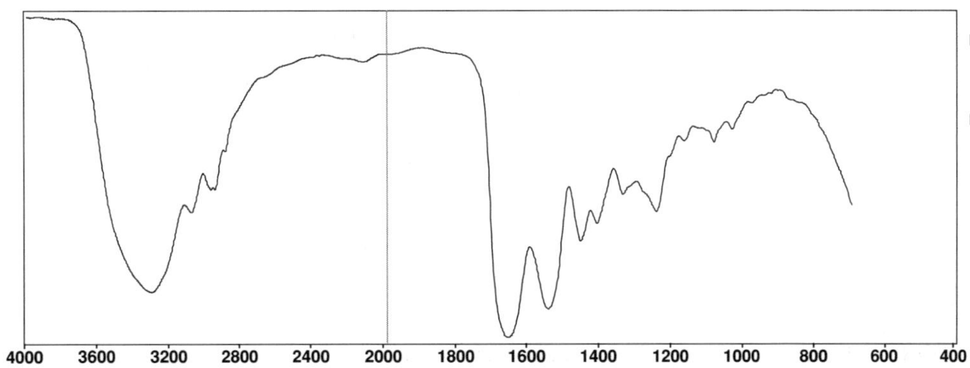

IR ANALYSIS LAB
Getty Conservation Institute, 1/28/94

IR ANALYSIS CONDITIONS
Spectra-Tech IRμS
Microscope
Bulk sample on BaF_2
Resolution = 4 cm^{-1}
Scans = 50
Range = 4000–700 cm^{-1}

Fish Glue

Protein

Gelatin

PROVENANCE Knox Gelatine, Inc.

SOURCE USA

APPEARANCE Light yellow or white, opaque, powder

CHARACTERISTIC IR ABSORPTION BANDS

3400–3200 cm^{-1}	N-H	stretching band
3100–2800 cm^{-1}	C-H	stretching bands
1660–1600 cm^{-1}	C=O	stretching band
1565–1500 cm^{-1}	C-N-H	bending band
1480–1300 cm^{-1}	C-H	bending band

Gelatin

Gelatin is a mixture of proteins prepared by hydrolyzing, via boiling, collagen obtained from skin, ligaments, and tendons. Its production differs from that of animal glue in that raw materials are selected, cleaned, and treated with special care, so that the product is cleaner and purer than glue. Gelatin is strongly hydrophilic and can absorb up to ten times its weight of water. It is sold as colorless sheets or as a fine powder and is more elastic than most animal glues. Gelatin is used for photographic film emulsions, sizing, adhesive, ink, encapsulation, and food products.

SYNONYMS: Gelatin, gelatine, size.

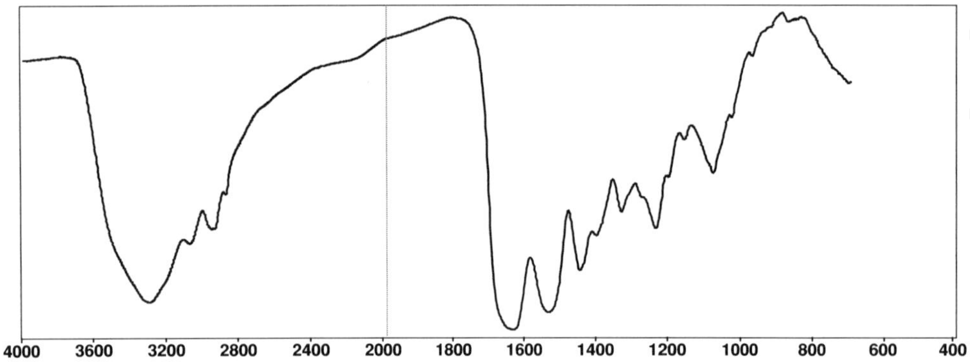

IR ANALYSIS LAB

Getty Conservation Institute, 1/28/94

IR ANALYSIS CONDITIONS

Spectra-Tech IRμS
Microscope
Bulk sample on BaF_2
Resolution = 4 cm^{-1}
Scans = 50
Range = 4000–700 cm^{-1}

Gelatin **Protein**

Casein

PROVENANCE Harvard Art Museums; Eimer & Amend

SOURCE USA

APPEARANCE Light beige, opaque, powder

CHARACTERISTIC IR ABSORPTION BANDS

3400–3200 cm^{-1}	N-H	stretching band
3100–2800 cm^{-1}	C-H	stretching bands
1660–1600 cm^{-1}	C=O	stretching band
1565–1500 cm^{-1}	C-N-H	bending band
1480–1300 cm^{-1}	C-H	bending band

Casein

Casein has been used as a glue and binder since earliest recorded periods. It is a proteinaceous, dried precipitate produced from milk. It contains sulfur and phosphorus. Casein can be prepared in different ways, but it is best used when fresh. One preparation method is to add dilute hydrochloric acid to hot skim milk. The precipitate is then collected, washed, and dried. It is a white to yellowish powder that is insoluble in water and alcohol but is soluble in carbonates and other alkaline solutions. For use, casein is soaked overnight and a weak alkali (lime, borax) is added to increase solubility. Casein provides strong adhesion and is insoluble in water when dried.

SYNONYMS: Casein, caseinate, whey glue, Casco glue, milk acid powder, *Kasein.*

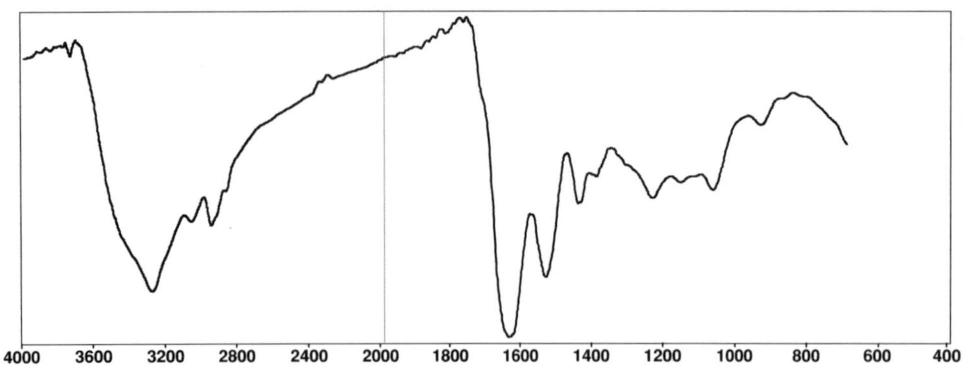

IR ANALYSIS LAB

Loyola Marymount University, 10/4/90

IR ANALYSIS CONDITIONS

P-E 1700
Microscope
Dissolved in DI H_2O, dried on BaF_2
Resolution = 4 cm^{-1}
Scans = 120
Range = 4000–700 cm^{-1}

Casein **Protein**

Egg Yolk, Hen (dried film on glass)

PROVENANCE A. Parker, Getty Conservation Institute

SOURCE USA

APPEARANCE Yellowish, oily, powder

CHARACTERISTIC IR ABSORPTION BANDS

3400–3200 cm^{-1}	N-H	stretching band
3100–2800 cm^{-1}	C-H	stretching bands
1750–1600 cm^{-1}	C=O	stretching bands
1565–1500 cm^{-1}	C-N-H	bending band
1480–1300 cm^{-1}	C-H	bending band

Egg

The whole egg, yolk, or white may be used as a tempera medium. The egg yolk is a stable emulsion of an aqueous liquid with an oily, proteinaceous medium that dries quickly into a hard, insoluble film. It is the traditional tempera medium and may be mixed with oil and/or resin for painting. The white of the egg, or glair, has been used as a medium for illuminated manuscripts. It is also used as a size for attaching gold leaf. Albumen is the adhesive substance of egg white. As a pure film, albumen is clear, brittle, and water soluble. Water solubility can be decreased by heating or adding tannin.

VARIETIES: Egg, yolk, whole egg, egg white, glair, egg tempera, hen, duck, chicken, goose, pheasant, pigeon, quail, albumen, albumine.

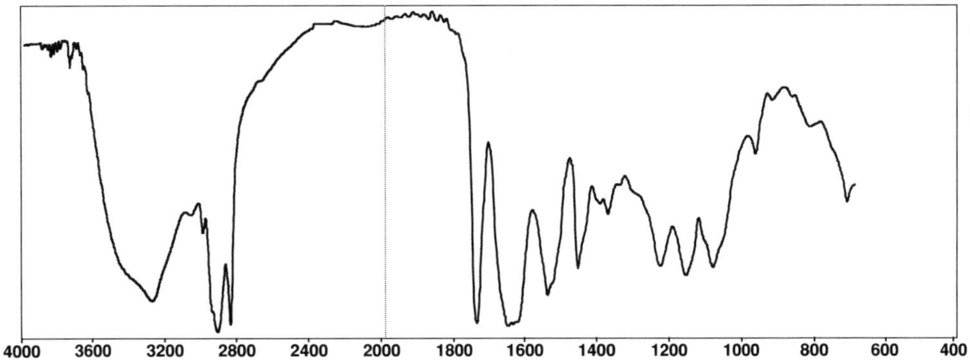

IR ANALYSIS LAB

Loyola Marymount University, 5/17/89

IR ANALYSIS CONDITIONS

P-E 1700
Microscope
Dried film on KBr
Resolution = 4 cm^{-1}
Scans = 120
Range = 4000–700 cm^{-1}

Egg **Protein**

Microcrystalline Wax

PROVENANCE A. F. Suter & Co.

SOURCE England

APPEARANCE White, opaque, chunks

CHARACTERISTIC IR ABSORPTION BANDS

3000–2800 cm^{-1}	C-H	stretching bands
1480–1300 cm^{-1}	C-H	bending bands
750–700 cm^{-1}	C-H	torsion bands

Mineral Wax

Mineral waxes are relatively pure materials that consist of a hydrocarbon series with little to none of the alcohols or esters found in plant waxes and beeswax. Mineral waxes are generally obtained from the fractional distillation of shale oil, lignite, or petroleum. They are soluble in mineral oil, chloroform, naphtha, benzene, and ether. Mineral waxes are very stable and nonreactive. Paraffin and microcrystalline waxes are white, translucent materials.

VARIETIES: Mineral wax, paraffin, ozokerite, ceresin, ceresine, microcrystalline wax, earth wax, cerin, cerosin, ozocerite.

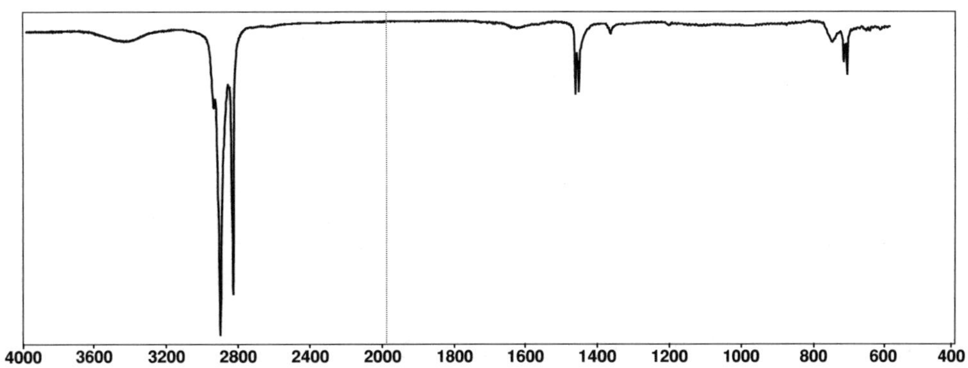

IR ANALYSIS LAB

Getty Conservation Institute, 6/15/84

IR ANALYSIS CONDITIONS

Digilab 15E/80
KBr pellet
Ground with KBr
Resolution = 4 cm^{-1}
Scans = 200
Range = 4000–600 cm^{-1}

Mineral Wax **Wax**

Beeswax, Crude

PROVENANCE A. F. Suter & Co.

SOURCE England

APPEARANCE Brownish yellow, opaque, chunks

CHARACTERISTIC IR ABSORPTION BANDS

3600–3200 cm^{-1}	O-H	stretching band
3000–2800 cm^{-1}	C-H	stretching bands
1780–1700 cm^{-1}	C=O	stretching band
1480–1300 cm^{-1}	C-H	bending bands
1300–900 cm^{-1}	C-O	stretching bands
750–700 cm^{-1}	C-H	torsion bands

Beeswax

Beeswax is produced by many species of bees; the most common is *Apis mellifica*. It is secreted from the organs on the underside of the abdomen of the worker bees and is used in forming the cells of the honeycomb. Wax may be obtained by melting the combs in hot water and straining to remove impurities that may contain resins, sugars, and other plant materials. The waxes from different localities vary considerably in color and texture and chemical composition. The color ranges from light yellow to dark brown. The darker varieties are often bleached by exposure to light and air or with ozone or hydrogen peroxide. Beeswax contains about 10% hydrocarbons in addition to alcohols, acids, and ester. Punic wax is refined beeswax.

VARIETIES: Beeswax, punic wax, crude beeswax, bleached beeswax, yellow beeswax, white beeswax, virgin beeswax.

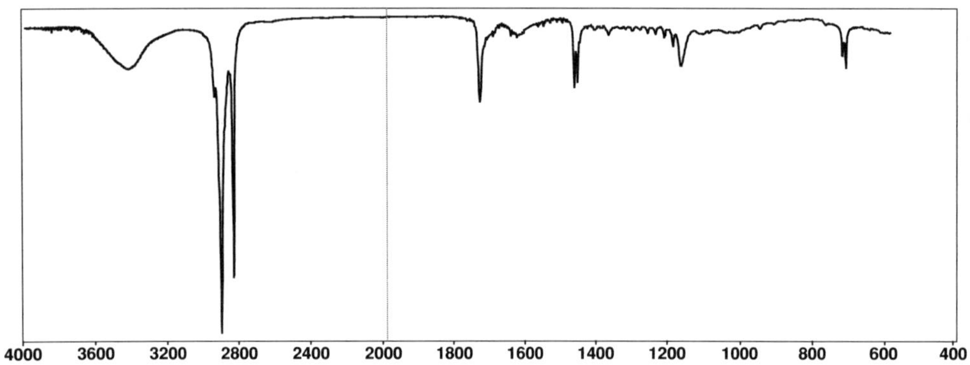

IR ANALYSIS LAB

Getty Conservation Institute, 6/14/84

IR ANALYSIS CONDITIONS

Digilab 15E/80
KBr pellet
Ground with KBr
Resolution = 4 cm^{-1}
Scans = 200
Range = 4000–600 cm^{-1}

Beeswax

Wax

Carnauba Wax

PROVENANCE A. F. Suter & Co.

SOURCE Northern Brazil

APPEARANCE Yellow, opaque, chunks

CHARACTERISTIC IR ABSORPTION BANDS

3600–3200 cm^{-1}	O-H	stretching band
3000–2800 cm^{-1}	C-H	stretching bands
1780–1700 cm^{-1}	C=O	stretching band
1480–1300 cm^{-1}	C-H	bending bands
1300–900 cm^{-1}	C-O	stretching bands
750–700 cm^{-1}	C-H	torsion bands

Vegetable Wax

Vegetable waxes are low-melting mixtures of long-chain hydrocarbon compounds found in or on plants. Their properties range widely from the soft white of Japan wax to the hard yellow of carnauba wax to the brownish black of bitumen wax. The hard waxes, such as carnauba, are often added to softer waxes, such as beeswax, for stiffening. Vegetable waxes generally contain fatty acids or alcohols along with the hydrocarbon series. Carnauba, one the hardest waxes, is obtained from the leaves of the wax palm, *Copernicia cerifera*, in Brazil.

VARIETIES: Vegetable wax, bitumen wax, montan wax, candelilla wax, candilla wax, carnauba wax, Japan wax, rice wax, Brazil wax.

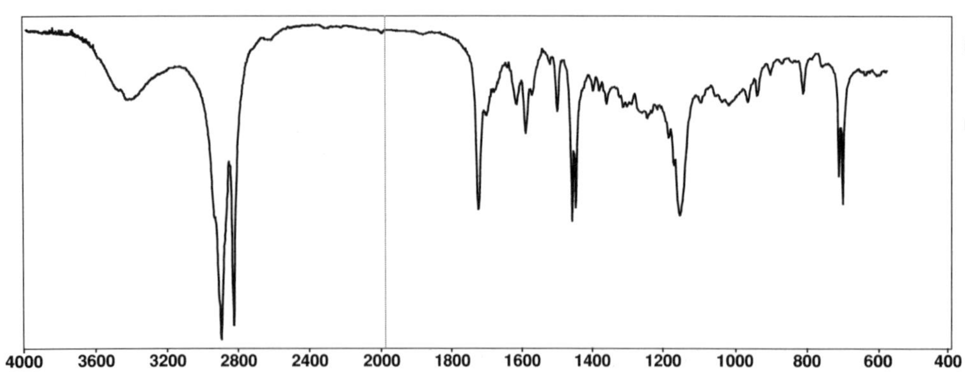

IR ANALYSIS LAB

Getty Conservation Institute, 6/15/84

IR ANALYSIS CONDITIONS

Digilab 15E/80
KBr pellet
Ground with KBr
Resolution = 4 cm^{-1}
Scans = 200
Range = 4000–600 cm^{-1}

Vegetable Wax

Wax

Linseed Stand Oil

PROVENANCE Ashley's Art; Grumbacher

SOURCE USA

APPEARANCE Yellow, transparent, liquid

CHARACTERISTIC IR ABSORPTION BANDS

3600–3200 cm^{-1}	O-H	stretching band
3000–2800 cm^{-1}	C-H	stretching bands
1750–1730 cm^{-1}	C=O	stretching band
1480–1300 cm^{-1}	C-H	bending bands
1300–900 cm^{-1}	C-O	stretching bands
750–700 cm^{-1}	C-H	torsion band

Linseed Oil

Linseed oil is obtained from the seeds of the flax *(Linum usitatissimum)* plant. It is the most important drying oil in artists' media. The oil is commercially extracted from the crushed seed by hot water and steam. Cold pressing is a less efficient method for extraction, but it produces a higher-quality artist paint. Many types of aging, refining, and bleaching procedures have been used to purify the oil and make it dry faster. In the past, one method to refine linseed oil was to let it stand over time. Any mucilage or impurities settled out so that a clarified oil was produced. Modern stand oil is prepared by steam treatment, in the absence of oxygen, to produce a thicker and glossier oil.

VARIETIES: Linseed oil, raw, cold-pressed, refined, stand oil, blown, bodied, boiled, sun-refined, sun-bleached, flaxseed oil.

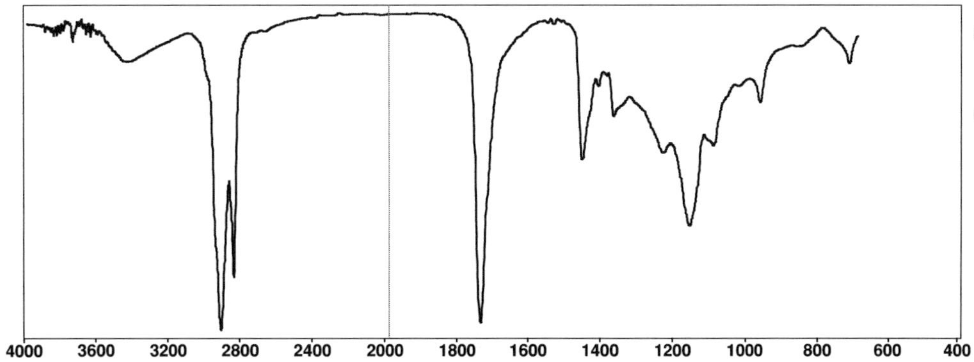

IR ANALYSIS LAB

Loyola Marymount University, 5/3/89

IR ANALYSIS CONDITIONS

P-E 1700
Microscope
Dried film on KBr
Resolution = 4 cm^{-1}
Scans = 120
Range = 4000–700 cm^{-1}

Linseed Oil

Oil

Poppyseed Oil (medium/slow drying)

PROVENANCE Ashley's Art; Grumbacher

SOURCE USA

APPEARANCE Very light yellow, transparent, liquid

CHARACTERISTIC IR ABSORPTION BANDS

3600–3200 cm^{-1}	O-H	stretching band
3000–2800 cm^{-1}	C-H	stretching bands
1750–1730 cm^{-1}	C=O	stretching band
1480–1300 cm^{-1}	C-H	bending bands
1300–900 cm^{-1}	C-O	stretching bands
750–700 cm^{-1}	C-H	torsion band

Poppyseed Oil

Poppy oil is a naturally colorless transparent oil obtained from the seed of the opium poppy *(Papaver somniferum)*. It comes primarily from India, Russia, France, and Asia Minor. Poppy oil dries slower than linseed oil and yellows less, so it was sometimes used with white pigments. It produced a soft, rubbery paint film with a long wet-in-wet work time that was popular with Impressionist painters. Thick layers of poppy oil paint films tend to wrinkle and crack upon aging.

SYNONYMS: Poppyseed oil, poppy oil.

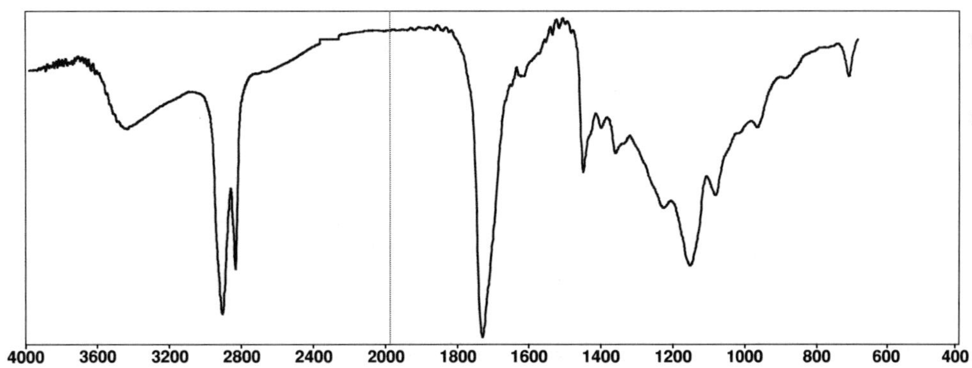

IR ANALYSIS LAB

Loyola Marymount University, 4/26/89

IR ANALYSIS CONDITIONS

P-E 1700
Microscope
Dried film on KBr
Resolution = 4 cm^{-1}
Scans = 120
Range = 4000–700 cm^{-1}

Poppyseed Oil

Oil

Walnut Oil

PROVENANCE Spectrum Naturals

SOURCE USA

APPEARANCE Slightly yellow, transparent, liquid

CHARACTERISTIC IR ABSORPTION BANDS

3600–3200 cm^{-1}	O-H	stretching band
3000–2800 cm^{-1}	C-H	stretching bands
1750–1730 cm^{-1}	C=O	stretching band
1480–1300 cm^{-1}	C-H	bending bands
1300–900 cm^{-1}	C-O	stretching bands
750–700 cm^{-1}	C-H	torsion band

Walnut Oil

Walnut oil, pressed from the seeds of a walnut tree *(Juglans regia)*, is pale in color and dries slower than does linseed oil. It also yellows and cracks less than linseed oil and dries faster than poppyseed oil. Nut oil was popular in Italy, the Netherlands, and Germany. Perhaps the reason that it has now fallen into disuse is its high cost and its tendency to turn rancid and putrid on storage.

SYNONYMS: Walnut oil, nut oil.

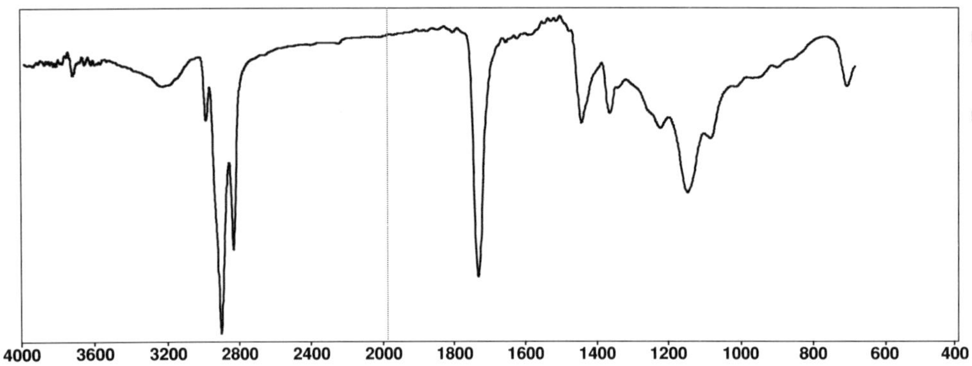

IR ANALYSIS LAB

Loyola Marymount University, 6/28/89

IR ANALYSIS CONDITIONS

P-E 1700
Microscope
Dried film on KBr
Resolution = 4 cm^{-1}
Scans = 120
Range = 4000–700 cm^{-1}

Walnut Oil

Oil

Rosin (colophony)

PROVENANCE A. F. Suter & Co.

SOURCE England

APPEARANCE Golden-yellow, transparent, chunks

CHARACTERISTIC IR ABSORPTION BANDS

3600–3200 cm^{-1}	O-H	stretching band
3100–2800 cm^{-1}	C-H	stretching bands
1740–1640 cm^{-1}	C=O	stretching bands
1650–1600 cm^{-1}	C-C	stretching band
1480–1300 cm^{-1}	C-H	bending bands
1300–900 cm^{-1}	C-O	stretching bands

Balsam

Balsam is a general term used to designate the resinous exudate from *Coniferae* trees. These softer resins generally contain a large amount of essential oils that come from trees that grow in sandy soil near the sea. Balsam is a soft semiliquid consisting of terpenes of resinous character. Upon distillation, a liquid portion called turpentine and a solid residue called colophony, or rosin, are produced. Balsams are often used in varnishes or as paint media; however, they deteriorate easily unless a harder resin is mixed with them. Rosin is often used in recipes for oil varnishes.

VARIETIES: Balsam, colophony, rosin, Greek pitch, Venice turpentine, Strasbourg turpentine, Canada balsam, copaiba balsam, pine resin, larch balsam, larch turpentine, tolu balsam, natural balsam, oleoresin, ester gum, Bordeaux turpentine, Burgundy pitch, Burgundy resin, gum thus, *olio d'Abezzo,* Jura turpentine, copaiva balsam, silver fir turpentine.

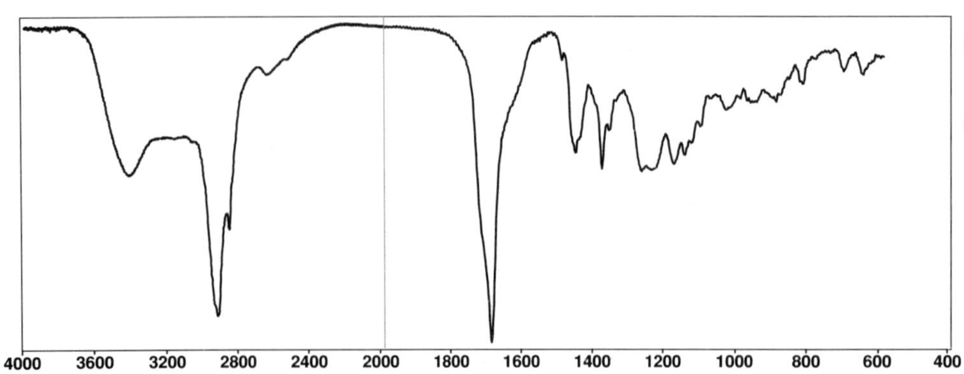

IR ANALYSIS LAB

Getty Conservation Institute, 3/22/85

IR ANALYSIS CONDITIONS

Digilab 15E/80
KBr pellet
0.8 mg ground with KBr
Resolution = 4 cm^{-1}
Scans = 200
Range = 4000–600 cm^{-1}

Balsam

Natural Resin

Pine Resin (*Pinus edulis,* piñon)

PROVENANCE U.S. National Park Service, Colorado

SOURCE Las Animas County, Colorado, USA

APPEARANCE Yellow tears, sticky

CHARACTERISTIC IR ABSORPTION BANDS

3600–3200 cm^{-1}	O-H	stretching band
3100–2800 cm^{-1}	C-H	stretching bands
1740–1640 cm^{-1}	C=O	stretching band
1650–1600 cm^{-1}	C-C	stretching bands
1480–1300 cm^{-1}	C-H	bending bands
1300–900 cm^{-1}	C-O	stretching bands

Balsam

Balsam is a general term used to designate the resinous exudate from *Coniferae* trees. These softer resins generally contain a large amount of essential oils that come from trees that grow in sandy soil near the sea. Balsam is a soft semiliquid consisting of terpenes of resinous character. Upon distillation, a liquid portion called turpentine and a solid residue called colophony, or rosin, are produced. Balsams are often used in varnishes or as paint media; however, they deteriorate easily unless a harder resin is mixed with them. The United States, France, and Spain are the largest producers of balsams.

VARIETIES: Balsam, colophony, rosin, Greek pitch, Venice turpentine, Strasbourg turpentine, Canada balsam, copaiba balsam, pine resin, larch balsam, larch turpentine, tolu balsam, natural balsam, oleoresin, ester gum, Bordeaux turpentine, Burgundy pitch, Burgundy resin, gum thus, *olio d'Abezzo,* Jura turpentine, copaiva balsam, silver fir turpentine, pine resin.

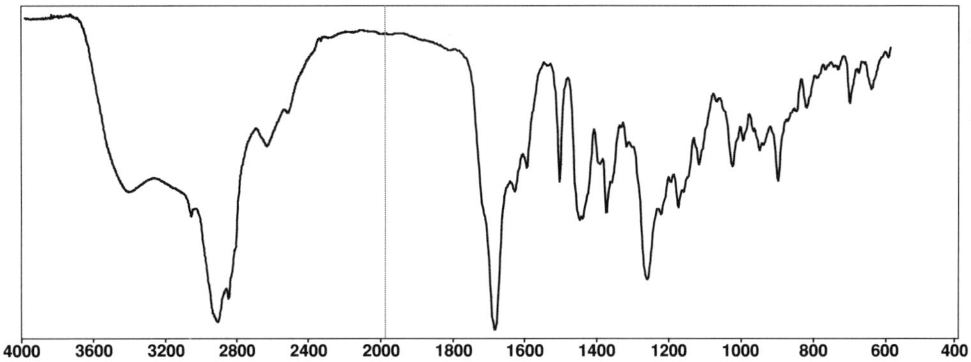

IR ANALYSIS LAB

Getty Conservation Institute, 7/24/90

IR ANALYSIS CONDITIONS

Digilab 15E/80
KBr micropellet
Ground with KBr
Resolution = 4 cm^{-1}
Scans = 200
Range = 4000–600 cm^{-1}

Balsam

Natural Resin

Copal, Manila

PROVENANCE Kremer-Pigmente

SOURCE Unknown

APPEARANCE Light amber, opaque, chunks

CHARACTERISTIC IR ABSORPTION BANDS

3600–3200 cm^{-1}	O-H	stretching band
3100–2800 cm^{-1}	C-H	stretching bands
1740–1640 cm^{-1}	C=O	stretching band
1650–1600 cm^{-1}	C-C	stretching band
1480–1300 cm^{-1}	C-H	bending bands
1300–900 cm^{-1}	C-O	stretching bands

Copal

Copal is a term given to a large variety of hard natural resins obtained directly from trees or as fossil resins. They have a large range of solubility and color (from colorless to a bright yellow-brown). The hardest copal resin is Zanzibar; Sierra Leone, kauri, and Congo are of medium hardness; Manila and Borneo are soft copals. The oldest resins are the hardest. Copal resins may be purchased as large lumps or small tears. Congo copal is often used in commercial spirit varnish manufacture today. Copal resins have also been used as oil varnishes. They tend to darken and become insoluble with age.

VARIETIES: Zanzibar, Demerara, Benguela, Sierra Leone, Mozambique, red Angola, white Angola, Congo, kauri, Manila, Pontianak, Madagascar, Accra, Loango, Gaboon, bastard Angola, Borneo, Singapore, South American, Cochin, Brazilian, Benin, swamp gum, kauri gum, bush gum, Manila nubs, old bold Pontianak.

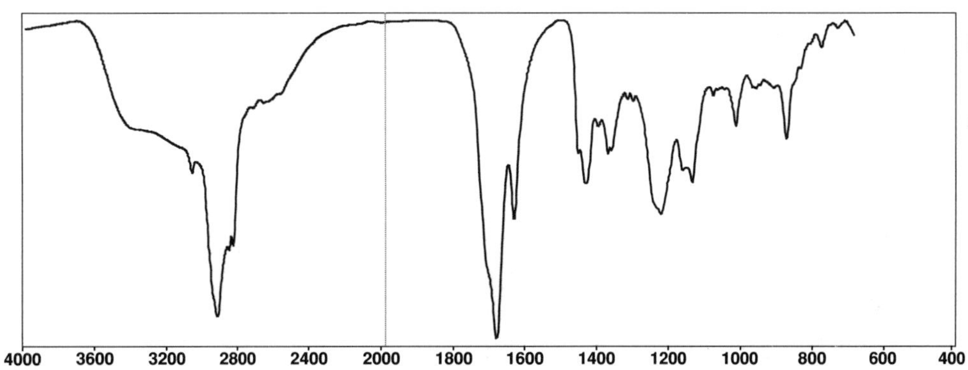

IR ANALYSIS LAB

Getty Conservation Institute, 1/28/94

IR ANALYSIS CONDITIONS

Spectra-Tech IRμS
Microscope
Bulk sample on BaF_2
Resolution = 4 cm^{-1}
Scans = 50
Range = 4000–700 cm^{-1}

Copal

Natural Resin

Dammar

PROVENANCE Kremer-Pigmente

SOURCE Unknown

APPEARANCE Beige, translucent, chunks

CHARACTERISTIC IR ABSORPTION BANDS

3600–3200 cm^{-1}	O-H	stretching band
3100–2800 cm^{-1}	C-H	stretching bands
1740–1640 cm^{-1}	C=O	stretching band
1650–1600 cm^{-1}	C-C	stretching band
1480–1300 cm^{-1}	C-H	bending bands
1300–900 cm^{-1}	C-O	stretching bands

Dammar

Dammar, the palest natural resin, is obtained from Dipterocarpaceae (genus *Shorea* or *Hopea*) trees growing in Malaysia and Indonesia. The soft, viscous, highly aromatic resin oozes readily from incisions in the bark and dries to become transparent, brittle, odorless lumps that are sorted into the following grades: pale (A), yellow (B), amber (C), and dust. To prepare as a varnish, dammar pieces are placed in a cheesecloth bag partially submersed in turpentine. After a few hours, the dammar is dissolved and any residual material remaining in the bag is thrown out. This produces a high-quality, clear varnish for paintings.

SYNONYMS: Damar, Malay dammar, Mata Kuching, cat's eye dammar, penak, gum batu, hitam, Batavian dammar, Singapore dammar, Borneo dammar, Pontianak dammar, black dammar, Pedong, East India dammar, grade A Batavia, no.1 Singapore, Bata gum.

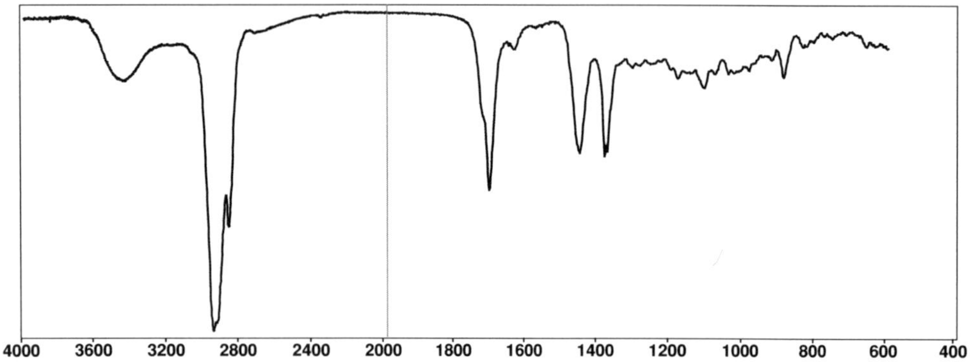

IR ANALYSIS LAB

Getty Conservation Institute

IR ANALYSIS CONDITIONS

Digilab 15E/80
KBr pellet
Ground with KBr
Resolution = 4 cm^{-1}
Scans = 200
Range = 4000–600 cm^{-1}

Dammar **Natural Resin**

Elemi

PROVENANCE Kremer-Pigmente

SOURCE Unknown

APPEARANCE Light yellow, opaque, paste

CHARACTERISTIC IR ABSORPTION BANDS

3600–3200 cm^{-1}	O-H	stretching band
3100–2800 cm^{-1}	C-H	stretching bands
1740–1640 cm^{-1}	C=O	stretching band
1650–1600 cm^{-1}	C-C	stretching band
1480–1300 cm^{-1}	C-H	bending bands
1300–900 cm^{-1}	C-O	stretching bands

Elemi

Elemi is a resin derived from trees of the family Burseraceae. Because of the high oil content of the elemis, the term was used to describe oleoresins in the seventeenth and eighteenth centuries. Now the term usually describes Manila elemi, which originates in the Philippines and is gathered from *Canarium communis*. This resin is extremely soft and has a very pungent odor. Elemi has been used in varnishes, but the components responsible for its initial malleability (mono- and sesquiterpenoids) evaporate, and it eventually hardens. Elemi has been used in printing inks, textile coatings, paper coatings, perfume bases, and waterproofing.

SYNONYMS: Elemi gum, Luzon, Manila elemi, Nauli elemi, Canarium.

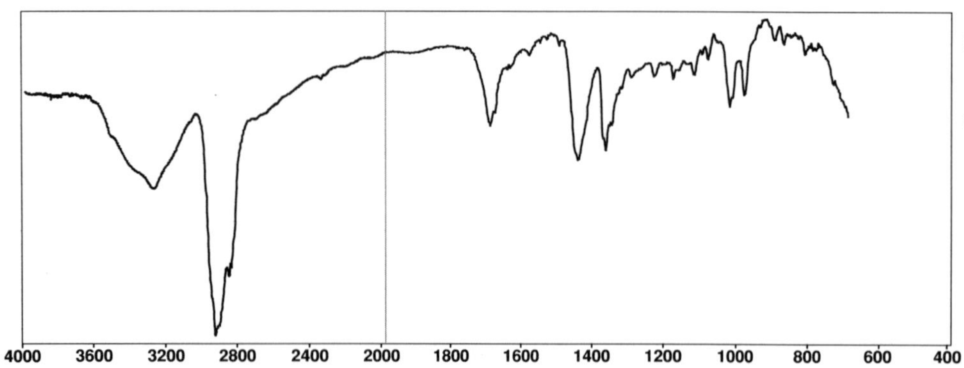

IR ANALYSIS LAB

Getty Conservation Institute, 1/28/94

IR ANALYSIS CONDITIONS

Spectra-Tech IRμS
Microscope
Bulk sample on BaF_2
Resolution = 4 cm^{-1}
Scans = 50
Range = 4000–700 cm^{-1}

Elemi **Natural Resin**

Mastic

PROVENANCE A. F. Suter & Co.

SOURCE Chios, Greece

APPEARANCE Hard, pale yellow, tears

CHARACTERISTIC IR ABSORPTION BANDS

3600–3200 cm^{-1}	O-H	stretching band
3100–2800 cm^{-1}	C-H	stretching bands
1740–1640 cm^{-1}	C=O	stretching band
1650–1600 cm^{-1}	C-C	stretching band
1480–1300 cm^{-1}	C-H	bending bands
1300–900 cm^{-1}	C-O	stretching bands

Mastic

Mastic is produced by a *Pistacia lentiscus* tree, which grows in southern Europe and northern Africa. The resin collected from the island of Chios has a reputation for highest quality. Mastic is sold commercially in small, transparent "tears" of a pale straw color. The resin is soluble in alcohol but insoluble in petroleum ethers. It is used as a varnish for oil paintings and as an additive in an oil medium called megilp. As with dammar, mastic varnish is prepared by placing the resin bits in a gauze bag suspended in solvent. Mastic varnishes yellow and become insoluble with time.

SYNONYMS: Chios mastic, Indian mastic, khinjak, Turkish mastic, pistacia galls, Bombay mastic.

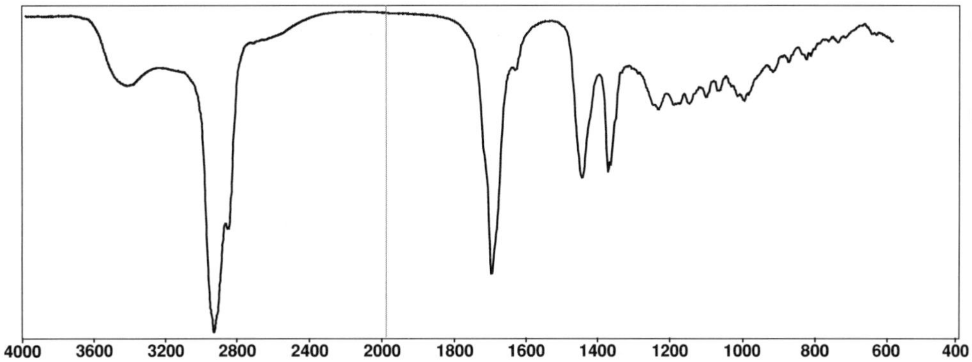

IR ANALYSIS LAB

Getty Conservation Institute, 11/5/90

IR ANALYSIS CONDITIONS

Digilab 15E/80
KBr pellet
Inside portion ground with KBr
Resolution = 4 cm^{-1}
Scans = 200
Range = 4000–600 cm^{-1}

Mastic **Natural Resin**

Sandarac

PROVENANCE A. F. Suter & Co.

SOURCE Northern Africa

APPEARANCE Hard, yellow, lumps with powdery surface

CHARACTERISTIC IR ABSORPTION BANDS

3600–3200 cm^{-1}	O-H	stretching band
3100–2800 cm^{-1}	C-H	stretching bands
1740–1640 cm^{-1}	C=O	stretching band
1650–1600 cm^{-1}	C-C	stretching band
1480–1300 cm^{-1}	C-H	bending bands
1300–900 cm^{-1}	C-O	stretching bands

Sandarac

Sandarac comes from *Callitris quadrivalvis*, a small tree that grows in northern Africa and Australia. The resin comes in pale yellow lumps or sticks that are hard, brittle, and powdery on the surface because of oxidation. It is soluble in alcohol and hot turpentine and forms a hard white film that becomes darker and redder with age. Sandarac spirit varnishes are often sold as retouching varnishes because they dry very quickly.

SYNONYMS: Sandarac, Berenice, Mogador, gum sandarac, gum juniper, white gum, Cyprus pine, Australian pine gum.

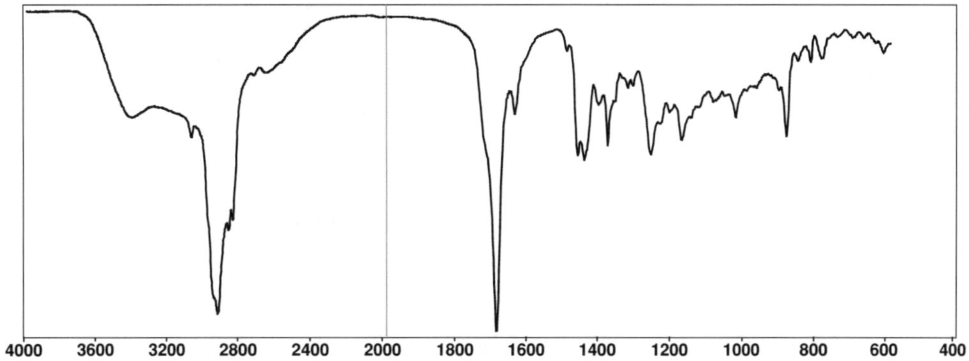

IR ANALYSIS LAB

Getty Conservation Institute, 3/22/85

IR ANALYSIS CONDITIONS

Digilab 15E/80
KBr micropellet
Ground with KBr
Resolution = 4 cm^{-1}
Scans = 200
Range = 4000–600 cm^{-1}

Sandarac **Natural Resin**

Shellac (medium, flake-button)

PROVENANCE J. Paul Getty Museum Paintings Conservation

SOURCE N/A

APPEARANCE Amber, transparent, flakes

CHARACTERISTIC IR ABSORPTION BANDS

3600–3200 cm^{-1}	O-H	stretching band
3100–2800 cm^{-1}	C-H	stretching bands
1740–1640 cm^{-1}	C=O	stretching band
1650–1600 cm^{-1}	C-C	stretching band
1480–1300 cm^{-1}	C-H	bending bands
1300–900 cm^{-1}	C-O	stretching bands

Shellac

Shellac is the resinous secretion of the lac insect *Tachardia lacca*. The insect feeds on a plant and converts the plant juice into the resin and red dye exudate. It comes almost entirely from India. The crude lac, seed lac, is gathered from the trees and crushed as graded. The lac is then washed, heated, and drawn into thin sheets. When cool, the sheets are broken into fragments for sale as flake shellac. Shellac colors range from a deep red to a pale gold. Shellac is soluble in alcohol and is used to obtain the high gloss on French polished furniture.

VARIETIES: Shellac, baisakhi, jethwi, seed lac, stick lac, button lac, black button lac, kiri, garnet lac, orange flakes, lemon flakes, bleached shellac, refined shellac.

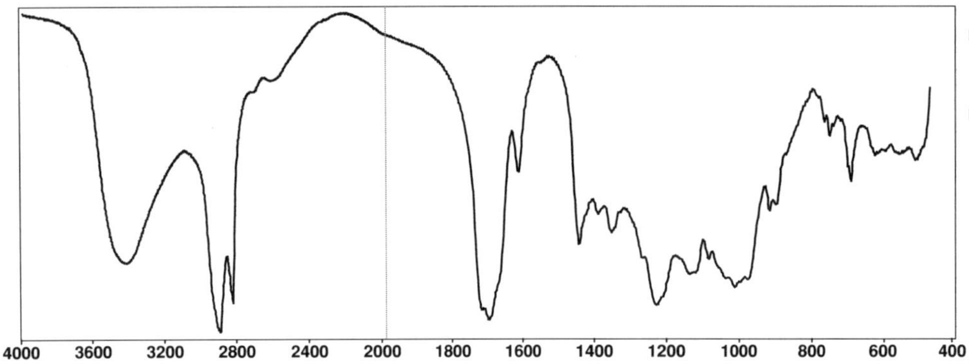

IR ANALYSIS LAB

Getty Conservation Institute

IR ANALYSIS CONDITIONS

Digilab 15E/80
KBr pellet
Ground with KBr
Resolution = 4 cm^{-1}
Scans = 200
Range = 4000–500 cm^{-1}

Shellac **Natural Resin**

Cellulose Nitrate (11.8% N)

PROVENANCE Lawrence Livermore Labs

SOURCE USA

APPEARANCE Colorless, transparent

CHARACTERISTIC IR ABSORPTION BANDS

3600–3200 cm^{-1}	O-H	stretching band
3100–2800 cm^{-1}	C-H	stretching bands
1660–1625 cm^{-1}	N-O	stretching band
1285–1270 cm^{-1}	N-O	stretching band
1480–1300 cm^{-1}	C-H	bending bands
1300–900 cm^{-1}	C-O	bending bands
890–800 cm^{-1}	N-O	bending band

Cellulose Nitrate

Some of the earliest synthetic resins were made from cellulose fibers. Cellulose nitrate was first made as a substitute for ivory and later was used for photographic film and as clear lacquers, adhesives, and high-gloss paints. Celluloid is a proprietary product of cellulose nitrate mixed with camphor as a plasticizer. Cellulose nitrate is inherently unstable and slowly decomposes at room temperature. UV light, heat, and/or high humidities can hasten its decomposition. Cellulose nitrate is still commercially available and is used as adhesives and coatings.

VARIETIES: Celluloid, pyroxylin, airplane wing dope, nitrocellulose, guncotton, collodion, Duco, celloidin, celluidine, photoxylin.

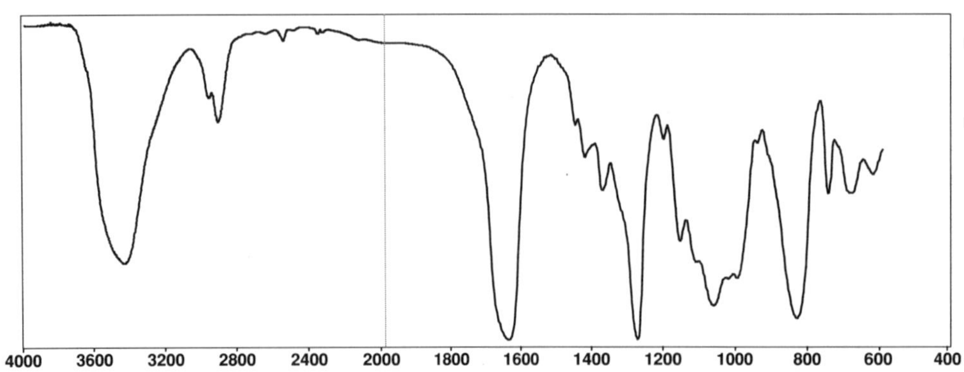

IR ANALYSIS LAB

Getty Conservation Institute, 2/23/87

IR ANALYSIS CONDITIONS

Digilab 15E/80
KBr micropellet
Ground with KBr
Resolution = 4 cm^{-1}
Scans = 200
Range = 4000–600 cm^{-1}

Cellulose Nitrate **Synthetic Resin**

Acryloid B-72: methyl acrylate/ethyl methacrylate copolymer

PROVENANCE Conservation Materials Ltd.; Rohm & Haas

SOURCE USA

APPEARANCE Transparent, colorless, pellets

CHARACTERISTIC IR ABSORPTION BANDS

3100–2800 cm^{-1}	C-H	stretching bands
1740–1640 cm^{-1}	C=O	stretching band
1480–1300 cm^{-1}	C-H	bending bands
1300–900 cm^{-1}	C-O	stretching bands

Acrylic

Acrylic resins are a commercially important family of polymers that were first made in 1901 and sold by Rohm & Haas and Du Pont in the United States since the 1930s. Acrylics have many popular uses. They are sold in solid form as glass substitutes under the names of Plexiglas and Lucite; they are also used as adhesives, varnishes, and paint media (Acryloid F-10, Lucite 44, Acryloid B-72). Bocour Artists Colors began selling Magna acrylic-based artist paints in 1949. While the resins are generally soluble in mineral spirits and turpentine, they may also be dispersed in water to form acrylic emulsions such as Rhoplex AC-234. Liquitex, an acrylic emulsion paint, was first marketed in 1954.

VARIETIES: Plexiglas, Lucite, Acryloid, Paraloid, Rhoplex, Liquitex.

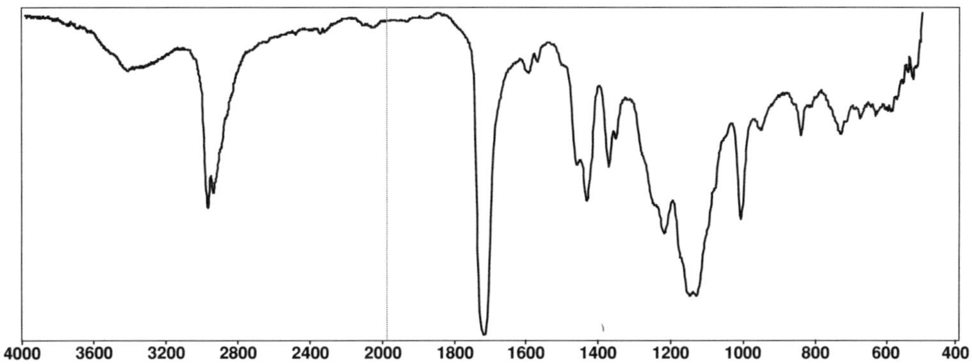

IR ANALYSIS LAB

Getty Conservation Institute, 4/30/90

IR ANALYSIS CONDITIONS

Digilab 15E/80
KBr pellet
Ground with KBr
Resolution = 4 cm^{-1}
Scans = 200
Range = 4000–500 cm^{-1}

Acrylic **Synthetic Resin**

Polyester 12F

PROVENANCE American Hoechst Corp.

SOURCE USA

APPEARANCE White, translucent, fibers

CHARACTERISTIC IR ABSORPTION BANDS

3100–2800 cm^{-1}	C-H	stretching bands
1740–1640 cm^{-1}	C=O	stretching band
1620–1420 cm^{-1}	aromatic bands	
1480–1300 cm^{-1}	C-H	bending bands
1300–900 cm^{-1}	C-O	stretching bands

Polyester

Polyester resins are a special type of alkyd resins. When catalyzed, they can harden at room temperature and pressure with very little shrinkage to produce a clear, colorless filament, block, or film. They are often used for encapsulating and embedding samples and objects. One type of polyester, polyethylene terephthalate, is used to make Mylar and other strong, moisture-resistant films, as well as to make Dacron, an important textile fiber.

SYNONYMS: Polyester, Dacron, Mylar, Bio-Plastic, Caroplastic, Castolite, Vestopal, Terelene, Cronar.

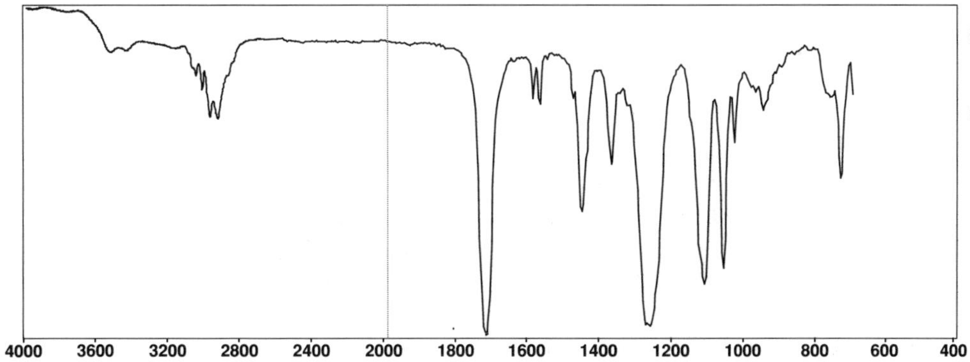

IR ANALYSIS LAB

Getty Conservation Institute

IR ANALYSIS CONDITIONS

Spectra-Tech IRµS
Microscope
Bulk sample on BaF_2
Resolution = 4 cm^{-1}
Scans = 100
Range = 4000–700 cm^{-1}

Polyester **Synthetic Resin**

Poly(vinyl acetate)(PVAC)

PROVENANCE Conservation Materials Ltd.

SOURCE USA

APPEARANCE Transparent, colorless, pellets

CHARACTERISTIC IR ABSORPTION BANDS

3100–2800 cm^{-1}	C-H	stretching bands
1750–1650 cm^{-1}	C=O	stretching band
1480–1300 cm^{-1}	C-H	bending bands
1300–900 cm^{-1}	C-O	stretching bands
750–700 cm^{-1}	C-H	torsion band

Poly(vinyl acetate)

Poly(vinyl acetate) (PVAC) was first produced in 1912 and was used as an artists' medium in 1938. Water-based emulsions, or latex, paints have been used as house paints as well as artists' media. Vinyl polymer resins produce clear, hard films and are also used as coatings, hot melts, and adhesives. Other types of vinyl polymers include polyvinyl butyral, polyvinyl chloride, polyvinylidene chloride, and polyvinyl alcohol.

VARIETIES: Poly(vinyl acetate), Vinylite, Vinylac, Elmer's glue, Vinamul, Mowilith, AYAT.

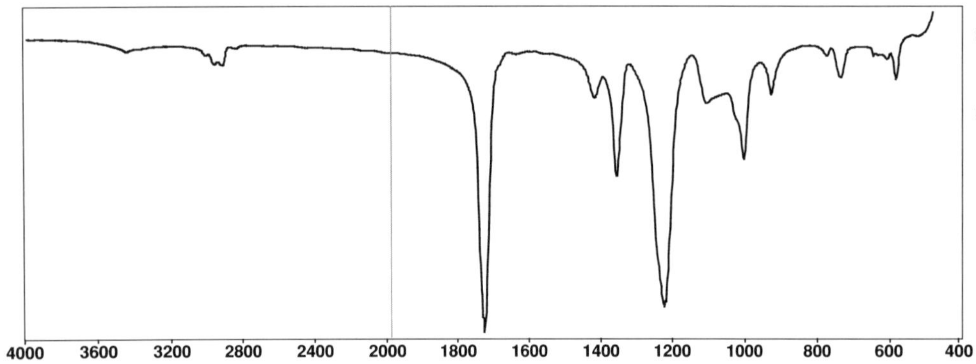

IR ANALYSIS LAB

Getty Conservation Institute

IR ANALYSIS CONDITIONS

Digilab 15E/80
KBr pellet
Ground with KBr
Resolution = 4 cm^{-1}
Scans = 200
Range = 4000–500 cm^{-1}

Poly(vinyl acetate)

Synthetic Resin

BEVA 371: ethylene/vinyl acetate copolymer mixed with polycyclohexanone

PROVENANCE Australian Museum; Adam Chemical Co.

SOURCE Australia

APPEARANCE White, opaque gel, dried to film

CHARACTERISTIC IR ABSORPTION BANDS

3100–2800 cm^{-1}	C-H	stretching bands
1750–1650 cm^{-1}	C=O	stretching band
1480–1300 cm^{-1}	C-H	bending bands
1300–900 cm^{-1}	C-O	stretching bands
750–700 cm^{-1}	C-H	torsion band

BEVA 371

BEVA 371 is a thermoplastic, elastomeric polymer mixture. It is composed of Elvax (ethylene/vinyl acetate [EVA] copolymer), Ketone Resin N (polycyclohexanone), A-C copolymer (EVA), Cellolyn 21 (phthalate ester of hydroabietyl alcohol), and paraffin. It is an opaque gel at room temperature and has a melting point of 50–55 °C. It is soluble in naphtha, toluene, acetone, and alcohol. BEVA produces a matte, waxy finish and is used as a consolidant for paintings and textiles.

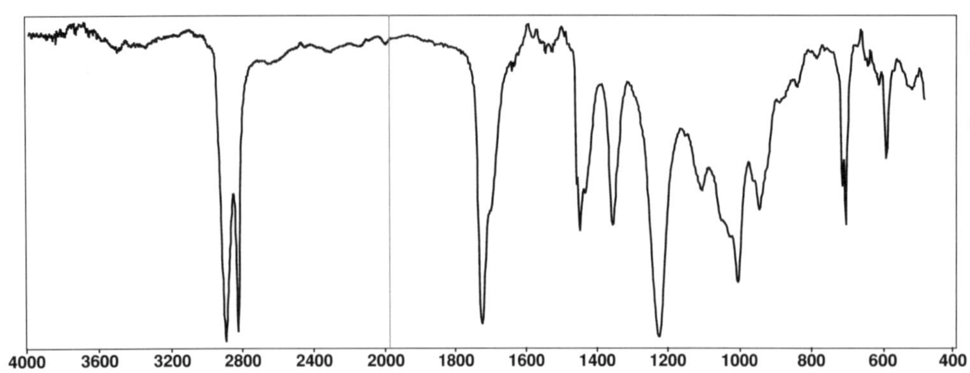

IR ANALYSIS LAB

Getty Conservation Institute, 4/30/90

IR ANALYSIS CONDITIONS

Digilab 15E/80
Attenuated total reflectance (ATR)
Free film
Resolution = 4 cm^{-1}
Scans = 200
Range = 4000–500 cm^{-1}

BEVA 371

Synthetic Resin

Polycyclohexanone

PROVENANCE H. Lanke

SOURCE Unknown

APPEARANCE White, translucent, chunks

CHARACTERISTIC IR ABSORPTION BANDS

3600–3200 cm^{-1}	O-H	stretching band
3100–2800 cm^{-1}	C-H	stretching bands
1750–1650 cm^{-1}	C=O	stretching band
1480–1300 cm^{-1}	C-H	bending bands
1300–900 cm^{-1}	C-O	stretching bands
750–700 cm^{-1}	C-H	torsion band

Polycyclohexanone

Polycyclohexanone resins are important because they are soluble in turpentine. This allows them to be mixed with or used instead of natural resins as varnishes. They resemble dammar closely but are harder and remain practically colorless. Some studies have shown, however, that solubility may be lost over time.

OTHER VARIETIES: AW-2, MS2A, ketone N, BASF N, Laropal K-80.

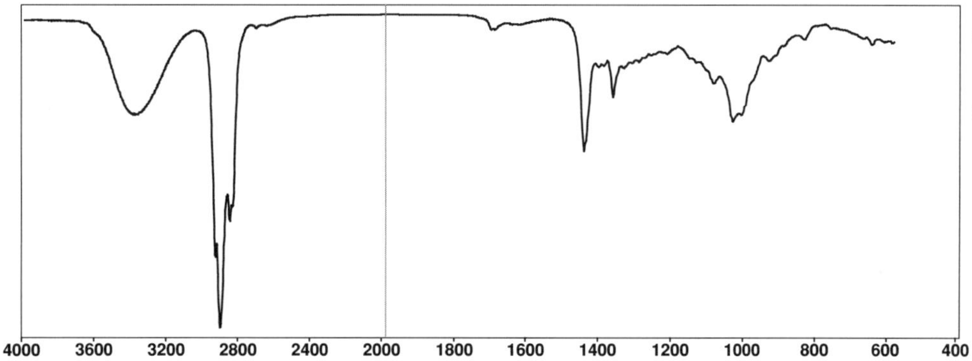

IR ANALYSIS LAB

Getty Conservation Institute, 10/8/84

IR ANALYSIS CONDITIONS

Digilab 15E/80
KBr pellet
Ground with KBr
Resolution = 4 cm^{-1}
Scans = 200
Range = 4000–600 cm^{-1}

Polycyclohexanone

Synthetic Resin

Polyamide

PROVENANCE Scientific Polymer Products, Inc. (catalog no. 385)

SOURCE USA

APPEARANCE White, powder

CHARACTERISTIC IR ABSORPTION BANDS

3400–3250 cm^{-1}	N-H	stretching band
3100–2800 cm^{-1}	C-H	stretching bands
1700–1630 cm^{-1}	C=O	stretching band
1620–1550 cm^{-1}	N-H	bending band
1480–1300 cm^{-1}	C-H	bending bands

Polyamide

Polyamides can be thought of as synthetic proteins because they are made by the polymerization of amino acids or lactams. Polyamides are thermoplastic resins that are characterized by their high degree of toughness, strength, and durability, along with their resistance to chemicals and heat. They are manufactured as bristles, fibers, molding powders, sutures, adhesives, and coatings. The most important examples of polyamides are the various kinds of nylon.

VARIETIES: Nylon, Versamid, Nylon-6, soluble nylon.

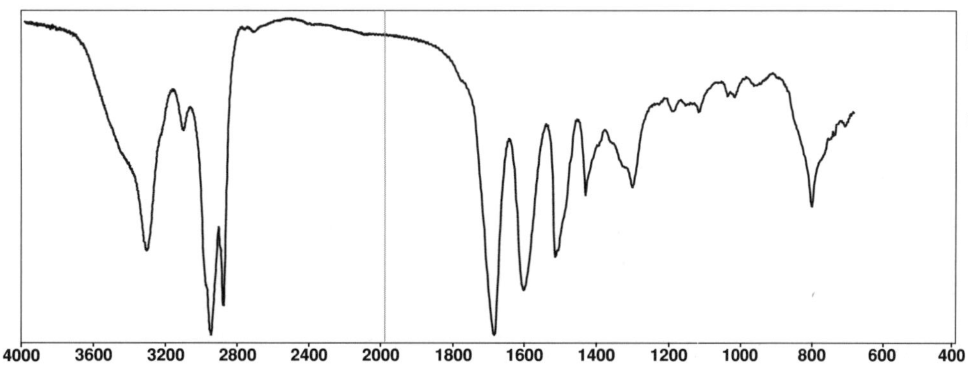

IR ANALYSIS LAB

Getty Conservation Institute, 10/1/84

IR ANALYSIS CONDITIONS

Digilab 15E/80
KBr pellet
Ground with KBr
Resolution = 4 cm^{-1}
Scans = 200
Range = 4000–600 cm^{-1}

Polyamide

Synthetic Resin

Chalk: calcium carbonate, $CaCO_3$

PROVENANCE Kremer-Pigmente

SOURCE France

APPEARANCE White, opaque, powder

CHARACTERISTIC IR ABSORPTION BANDS
1490–1370 cm^{-1} CO_3^{2-} stretching band
910–850 cm^{-1} O-C-O bending band

Calcium Carbonate

Calcium carbonate is found in many natural forms such as chalk, limestone, marble, and seashells. It can be found worldwide and ranges in color (because of impurities) from white to gray to yellow. The pigment is prepared by grinding the stone or shell with water and by levigating to separate the coarser material. Artificial chalk is known as precipitated chalk and is whiter and more homogeneous than natural chalk. Pearl white is made from calcined oyster shells. Calcium carbonate reacts with acids to evolve carbon dioxide.

SYNONYMS: Calcium carbonate, chalk, pearl white, oystershell white, marble, limestone, whiting, lime white, aragonite, calcite, marl, travertine, CI Pigment White 18.

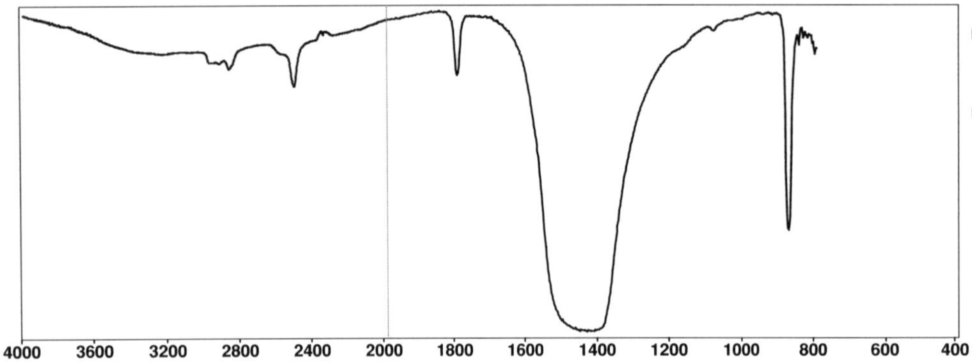

IR ANALYSIS LAB
Getty Conservation Institute, 10/15/92

IR ANALYSIS CONDITIONS
Spectra-Tech IRμS
Microscope
Bulk sample on BaF_2
Resolution = 4 cm^{-1}
Scans = 200
Range = 4000–800 cm^{-1}
Comment: baseline corrected

Calcium Carbonate

White

Gypsum: calcium sulfate, dihydrate, $CaSO_4 \cdot 2H_2O$

PROVENANCE Kremer-Pigmente

SOURCE Unknown

APPEARANCE White, powder

CHARACTERISTIC IR ABSORPTION BANDS
1140–1080 cm^{-1} asymmetric SO_4^{2-} stretching band
~620 cm^{-1} SO_4^{2-} bending band (not shown)
3700–3200 cm^{-1} antisymmetric and symmetric O-H stretching bands

Calcium Sulfate

Calcium sulfate can be commonly found in three forms: anhydrous (anhydrite), dihydrate (gypsum), and hemihydrate (plaster of Paris). Anhydrite is a colorless, inert pigment that is often a component in *gesso grosso,* while pure calcium sulfate dihydrate is found in *gesso sottile.* Gypsum is also used as a filler and as a base for lake pigments.

VARIETIES: Calcium sulfate, anhydrite, gypsum, plaster, terra alba, alabaster, hydrated calcium sulfate, mineral white, stucco, Keene's cement, Martin's cement, Mack's cement.

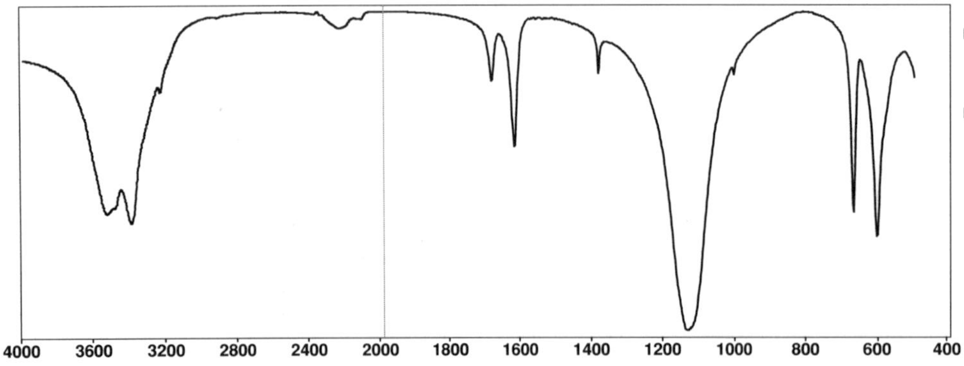

IR ANALYSIS LAB
Getty Conservation Institute

IR ANALYSIS CONDITIONS
Digilab 15E/80
KBr pellet
Ground with KBr
Resolution = 4 cm^{-1}
Scans = 200
Range = 4000–500 cm^{-1}

Calcium Sulfate

White

Plaster: calcium sulfate, hemihydrate, $CaSO_4 \cdot 1/2H_2O$

PROVENANCE Kremer-Pigmente

SOURCE Germany

APPEARANCE White

CHARACTERISTIC IR ABSORPTION BANDS
1140–1080 cm^{-1} asymmetric SO_4^{2-} stretching band
~620 cm^{-1} SO_4^{2-} bending band (not shown)
3700–3200 cm^{-1} antisymmetric and symmetric O-H stretching bands

Calcium Sulfate

Calcium sulfate can be commonly found in three forms: anhydrous (anhydrite), dihydrate (gypsum), and hemihydrate (plaster of Paris). Anhydrite is a colorless, inert pigment that is often a component in *gesso grosso,* while pure calcium sulfate dihydrate is found in *gesso sottile.* Gypsum is also used as a filler and as a base for lake pigments.

VARIETIES: Calcium sulfate, anhydrite, gypsum, plaster, terra alba, alabaster, hydrated calcium sulfate, mineral white, stucco, Keene's cement, Martin's cement, Mack's cement.

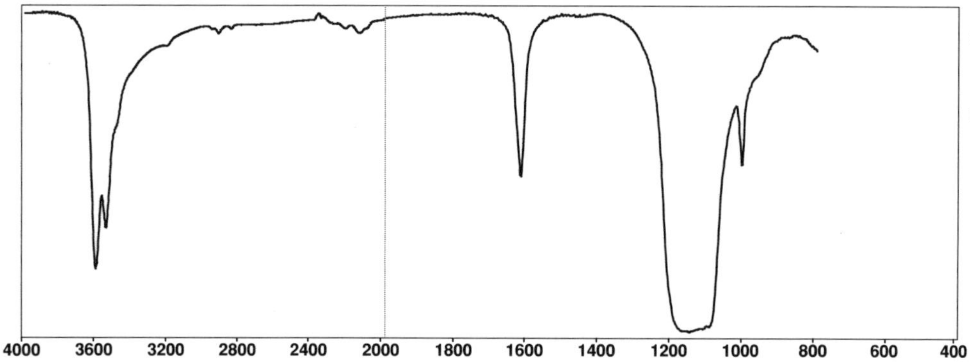

IR ANALYSIS LAB
Getty Conservation Institute, 10/15/92

IR ANALYSIS CONDITIONS
Spectra-Tech IRμS
Microscope
Bulk sample on BaF_2
Resolution = 4 cm^{-1}
Scans = 200
Range = 4000–800 cm^{-1}
Comment: baseline corrected

Calcium Sulfate

White

Kaolin: $Al_2Si_2O_5(OH)_4$

PROVENANCE Kremer-Pigmente

SOURCE Germany

APPEARANCE White

CHARACTERISTIC IR ABSORPTION BANDS
3700–3200 cm^{-1} O-H stretching bands
1100–1000 cm^{-1} asymmetric Si-O-Si stretching bands
910–830 cm^{-1} Si-O stretching bands

Clay

Many types of clay naturally occur around the world. They are principally composed of hydrated aluminum silicate. Small amounts of other minerals can change the color (white, yellow, brown, or red) and texture of the clays. Clay is usually formed by the weathering of aluminum-bearing rocks, such as granite. When pure, china clay (kaolin) is a fine, white, amorphous powder that becomes very plastic when water is added. When heated to high temperatures, clays became hard because of the loss of water and are used to make pottery, porcelain, and bricks. Clay is also used as a filler and whiting in paints and grounds.

VARIETIES: China clay, kaolin, modeling clay, porcelain clay, Tonerde, fuller's earth, white bole, bolus alba, terra alba, pipe clay, Bouvigal white, Rouen white, Spanish white, feldspar, ball clay, bentonite, stoneware clay, kaolinite, illite, montmorillonite, halloysite, argilla.

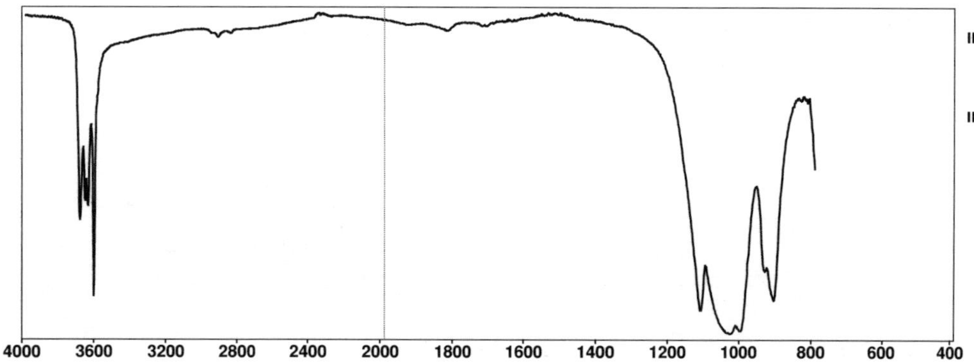

IR ANALYSIS LAB
Getty Conservation Institute, 10/15/92

IR ANALYSIS CONDITIONS
Spectra-Tech IRμS
Microscope
Bulk sample on BaF_2
Resolution = 4 cm^{-1}
Scans = 200
Range = 4000–800 cm^{-1}
Comment: baseline corrected

Clay

White

Silica: SiO_2

PROVENANCE Kremer-Pigmente

SOURCE USA

APPEARANCE White

CHARACTERISTIC IR ABSORPTION BANDS 1100–1000 cm^{-1} asymmetric Si-O-Si stretching band

Silica

Silica is widely available because it makes up one of the largest portions of the earth's crust. In its purest form, silica, or silicon dioxide, occurs as quartz. The more common, but less pure, forms are quartzite, sandstone, and sand. The fossil form of silica is diatomaceous earth. All forms of silica are inert, unaffected by heat, insoluble in strong acids (except hydrofluoric), and slowly attacked by strong alkalis. Silica is not commonly used as a pigment; however, it is found in grounds, primers, and wood fillers. Silica is used in the manufacture of glass, ceramics, and enamelware.

VARIETIES: Silica, silicon dioxide, quartz, silex, diatomaceous earth, sand, flint, chalcedony, opal, agate, diatomite.

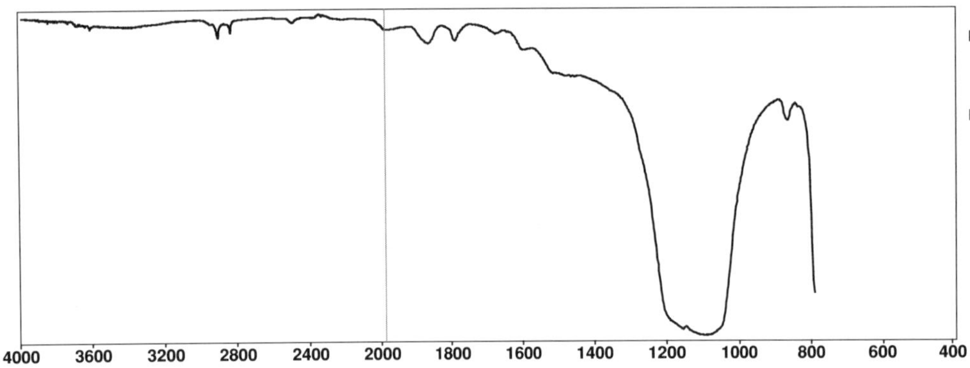

IR ANALYSIS LAB

Getty Conservation Institute, 10/16/92

IR ANALYSIS CONDITIONS

Spectra-Tech IRμS
Microscope
Bulk sample on BaF_2
Resolution = 4 cm^{-1}
Scans = 200
Range = 4000–800 cm^{-1}
Comment: baseline corrected

Silica White

Barium Sulfate: $BaSO_4$

PROVENANCE Kremer-Pigmente

SOURCE Germany

APPEARANCE White

CHARACTERISTIC IR ABSORPTION BANDS 1200–1050 cm^{-1} asymmetric SO_4^{2-} stretching bands
3700–3200 cm^{-1} O-H stretching bands

Barium Sulfate

Barium white is obtained naturally from the mineral barite. It can also be made artificially by a process discovered in the late nineteenth century. The artificially prepared barium sulfate is called blanc fixe. It is a white, opaque pigment composed of zinc sulfide and barium sulfate. The mixture of the two components is so intimate that it is hard to distinguish microscopically. It is an inert, transparent pigment that is often used as a filler or as a base for lake pigments.

SYNONYMS: Barium sulfate, baryte, barite, terra ponderosa, blanc fixe, permanent white, silver white, Albilith, Becton white, Orr's white, Permalba, baryta white, Constant white, *Schwerspatweiss,* heavy spar, CI Pigment White 22.

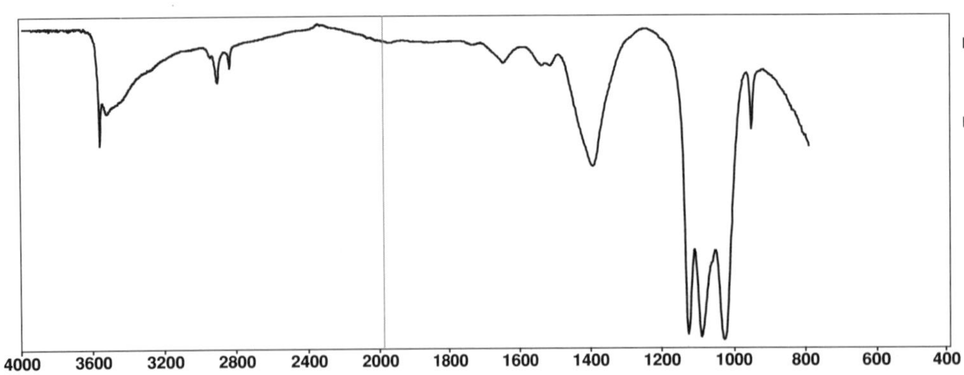

IR ANALYSIS LAB

Getty Conservation Institute, 10/15/92

IR ANALYSIS CONDITIONS

Spectra-Tech IRμS
Microscope
Bulk sample on BaF_2
Resolution = 4 cm^{-1}
Scans = 200
Range = 4000–800 cm^{-1}
Comment: baseline corrected, contains carbonate (1450 cm^{-1}) and hydrocarbon (3000–2800 cm^{-1}) impurities

Barium Sulfate White

Indigo, Natural (Naturindigo): $C_{16}H_{10}N_2O_2$

PROVENANCE H. Schweppe

SOURCE Java

APPEARANCE Blue, powder

CHARACTERISTIC IR ABSORPTION BANDS

3400–3200 cm^{-1}	N-H	stretching band
3100–2800 cm^{-1}	C-H	stretching bands
1700–1550 cm^{-1}	C=O	stretching band
1620–1420 cm^{-1}	aromatic bands	

Indigo

Indigo is a natural, dark blue dye obtained from the *Indigofera tinctoria* plants native to India, Java, and other tropical areas. Synthetic indigo, first produced in 1880, has almost entirely replaced the natural dyestuff. The natural material is collected as a precipitate from a fermented solution of the plant. It is a fine powder that may be used directly as a pigment in oil, tempera, or watercolor media, but it is more commonly used as a textile dye. The exposed material can fade rapidly in strong sunlight.

SYNONYMS: Indigo carmine, intense blue, indico, indican, anil, anneill, blue ynde, CI 73000.

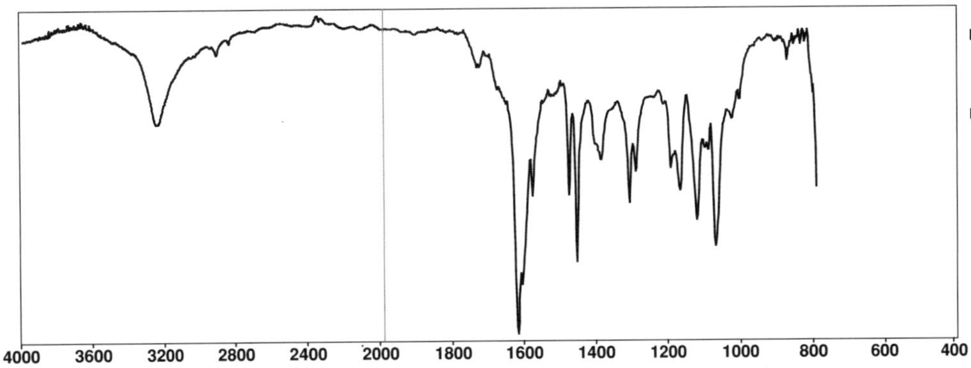

IR ANALYSIS LAB

Getty Conservation Institute, 4/29/93

IR ANALYSIS CONDITIONS

Spectra-Tech IRμS
Microscope
Bulk sample on BaF_2
Resolution = 4 cm^{-1}
Scans = 200
Range = 4000–800 cm^{-1}
Comment: baseline corrected

Indigo **Blue**

Phthalocyanine Blue (royal blue PB153)

PROVENANCE Kremer-Pigmente

SOURCE Germany

APPEARANCE Blue, powder

CHARACTERISTIC IR ABSORPTION BANDS

3100–2800 cm^{-1}	C-H	stretching bands
1700–1550 cm^{-1}	C=N	stretching band
1620–1420 cm^{-1}	aromatic bands	
1480–1300 cm^{-1}	C-H	bending bands
1330–1100 cm^{-1}	C-N	stretching bands

Phthalocyanine Blue

Copper phthalocyanine (phthalocyanine blue) is a synthetic organic pigment that was first introduced in 1935. It is usually adsorbed on an aluminum hydrate base to form a deep blue color. Other colors are achieved by varying the formulation—i.e., chlorinated copper phthalocyanine produces a green colorant. Phthalocyanine colors are important commercial pigments because of their light and chemical stability. They are used in enamels, automotive paints, plastics, and inks.

SYNONYMS: Heliogen blue, Monastral blue, phthalocyanine green, Winsor blue, CI Pigment Blue 15.

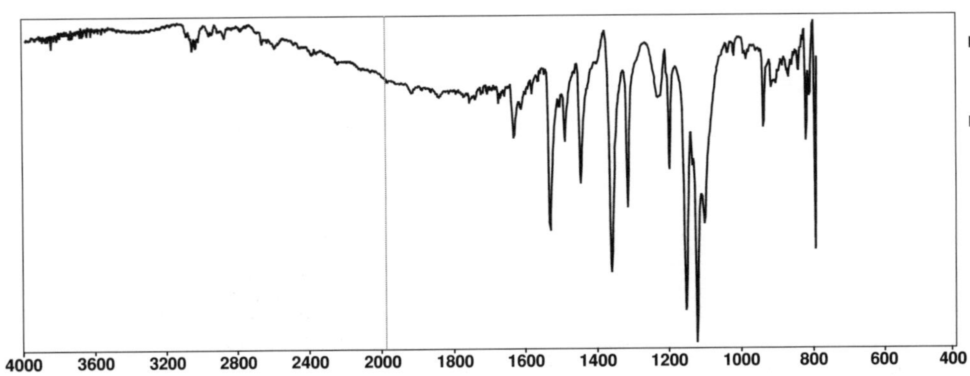

IR ANALYSIS LAB

Getty Conservation Institute, 10/13/92

IR ANALYSIS CONDITIONS

Spectra-Tech IRμS
Microscope
Bulk sample on BaF_2
Resolution = 4 cm^{-1}
Scans = 200
Range = 4000–800 cm^{-1}
Comment: baseline corrected

Phthalocyanine Blue **Blue**

Prussian Blue: $Fe_4[Fe(CN)_6]_3$

PROVENANCE Kremer-Pigmente

SOURCE Germany

APPEARANCE Blue

CHARACTERISTIC IR ABSORPTION BANDS ~2100 cm^{-1} $[Fe(C{\equiv}N)_6]^{3-}$ ion stretching band

Prussian Blue

Prussian blue, synthetically produced ferric ferrocyanide, was discovered in 1704. Its finely divided particles are a deep blue. It is transparent, has high tinting strength, and is stable to light and high temperatures; but it turns brown in the presence of alkalis. It is used in paints and printing inks.

SYNONYMS: Prussian blue, Turnbull's blue, Paris blue, Saxon blue, Milori blue, Chinese blue, bronze blue, Berlin blue, American blue, Antwerp blue, steel blue, mineral blue, iron blue.

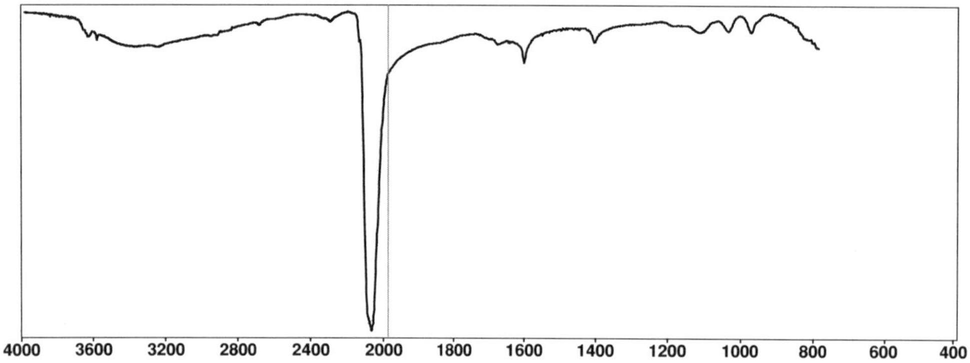

IR ANALYSIS LAB

Getty Conservation Institute, 10/13/92

IR ANALYSIS CONDITIONS

Spectra-Tech IRμS
Microscope
Bulk sample on BaF_2
Resolution = 4 cm^{-1}
Scans = 200
Range = 4000–800 cm^{-1}
Comment: baseline corrected

Prussian Blue

Blue

Ultramarine: $Na_8(S_2)(Al_6Si_6O_{24})$

PROVENANCE Kremer-Pigmente

SOURCE Unknown

APPEARANCE Blue, opaque, powder

CHARACTERISTIC IR ABSORPTION BANDS
1150–950 cm^{-1} overlapping stretching bands for Si-O-Si, and Si-O-Al
2340 cm^{-1} sulfur ion stretching band that occurs in some natural ultramarines

Ultramarine

The pigment ultramarine can be prepared from a natural semiprecious stone, lapis lazuli. The pigment is composed of silicon, aluminum, sodium, sulfur, and oxygen. A commercial process for synthetic ultramarine, developed in the 1820s, produces a very pure, deep, fine-particle material. Variations in the process give a wide range of shades of blues, violets, and reds. The pigment is stable to light but is affected by acids.

SYNONYMS: Lapis lazuli blue, lazurite, ultramarine blue, *bleu d'azur,* ultramarine ash, French ultramarine, ultramarine red, ultramarine violet, ultramarine green, French blue, new blue, permanent blue, oriental blue, Gmelin's blue, Guimet's blue.

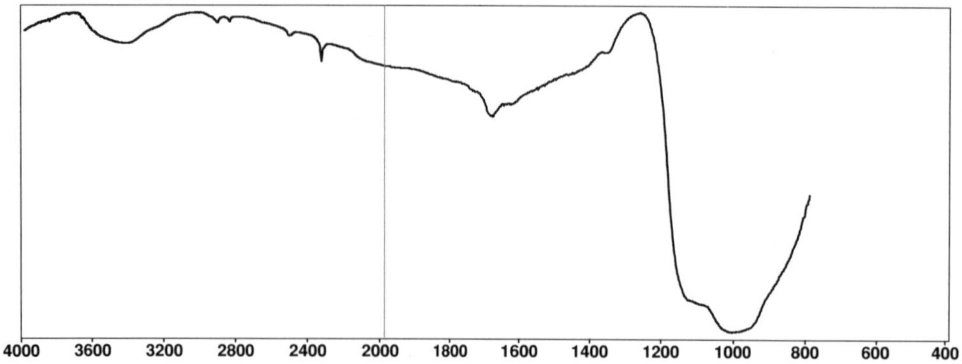

IR ANALYSIS LAB

Getty Conservation Institute

IR ANALYSIS CONDITIONS

Spectra-Tech IRμS
Microscope
Bulk sample on BaF_2
Resolution = 4 cm^{-1}
Scans = 200
Range = 4000–800 cm^{-1}

Ultramarine

Blue

Malachite: basic copper carbonate, $CuCO_3 \cdot Cu(OH)_2$

PROVENANCE Forbes Collection

SOURCE Unknown

APPEARANCE Green, opaque, powder

CHARACTERISTIC IR ABSORPTION BANDS
1530–1350 cm^{-1} CO_3^{2-} stretching bands
900–650 cm^{-1} O-C-O bending bands
3700–3100 cm^{-1} O-H stretching bands
1100–1000 cm^{-1} O-H bending bands

Basic Copper Carbonate

Basic copper carbonate naturally occurs in two mineral forms, azurite (blue: $2CuCO_3 \cdot Cu(OH)_2$) and malachite (green: $CuCO_3 \cdot Cu(OH)_2$). The two minerals usually occur together. Pigments are prepared by careful selection, grinding, washing, and levigation. Both are sensitive to acids. The synthetic pigment is called green verditer and is a pale greenish blue.

SYNONYMS: Basic copper carbonate, azurite, mountain blue, malachite, mountain green, blue verditer, blue bice, green verditer, green bice, mineral green.

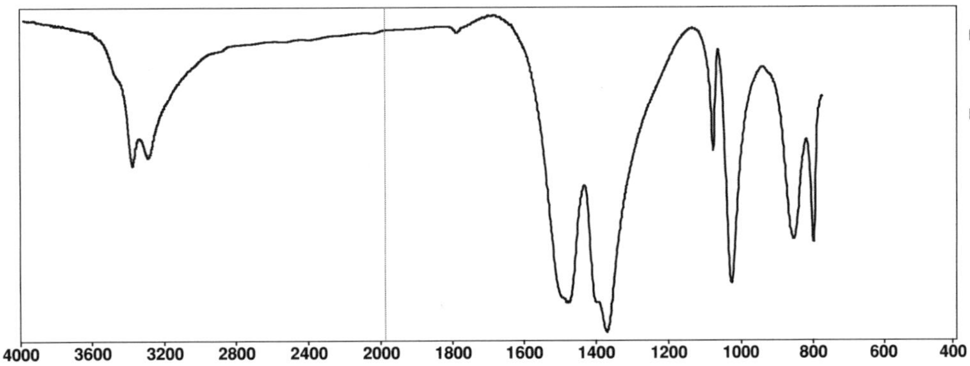

IR ANALYSIS LAB
Analytical Answers, Inc., 12/96

IR ANALYSIS CONDITIONS
Bio-Rad FTS 40
Microscope
Bulk sample on BaF_2
Resolution = 4 cm^{-1}
Scans = 50
Range = 4000–750 cm^{-1}

Basic Copper Carbonate

Green

Verdigris: $Cu(C_2H_3O_2)_2 \cdot H_2O$

PROVENANCE Kremer-Pigmente

SOURCE Germany

APPEARANCE Green

CHARACTERISTIC IR ABSORPTION BANDS
1660–1550,1450–1400,~1050,~1020,~925 cm^{-1} acetate ion stretching and bending bands
3550–3200 cm^{-1} antisymmetric and symmetric O-H stretching bands

Copper Acetate

Basic copper acetate is a greenish blue crystalline powder that is called verdigris. Its preparation, known since ancient times, involves exposing copper to vapors of fermenting solutions. The crystals are soluble in water and acids and are toxic. The pigment is unstable and can leave a black residue upon decomposition. When combined with terpenoid resins, such as Venice turpentine, verdigris forms copper resinate. Verdigris is used as a pigment and a textile dye.

SYNONYMS: Copper acetate, verdigris, copper resinate, crystals of Venus, cupric acetate.

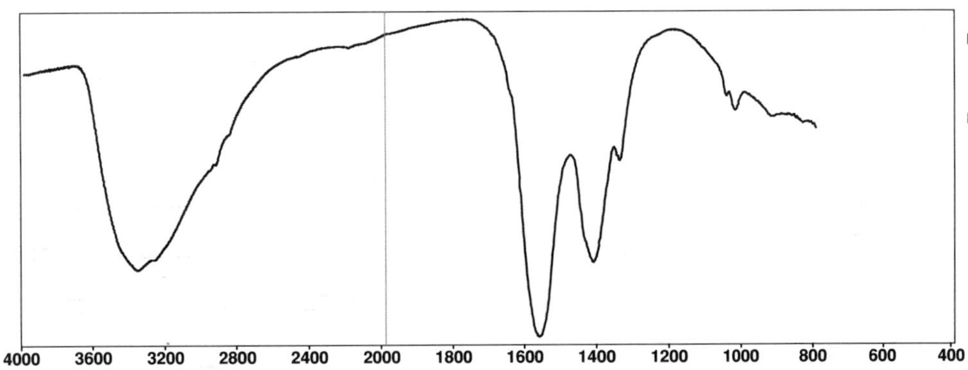

IR ANALYSIS LAB
Getty Conservation Institute, 10/14/92

IR ANALYSIS CONDITIONS
Spectra-Tech IRμS
Microscope
Bulk sample on BaF_2
Resolution = 4 cm^{-1}
Scans = 200
Range = 4000–800 cm^{-1}

Copper Acetate

Green

Madder (CI Natural Red 8)

PROVENANCE H. Schweppe

SOURCE Germany

APPEARANCE Red, powder

CHARACTERISTIC IR ABSORPTION BANDS

3600–3200 cm^{-1}	O-H	stretching band
3100–2800 cm^{-1}	C-H	stretching bands
1740–1640 cm^{-1}	C=O	stretching bands
1620–1420 cm^{-1}	aromatic bands	
1480–1300 cm^{-1}	C-H	bending bands
1300–900 cm^{-1}	C-O	stretching bands

Madder

Madder is a natural red dye obtained from the root of the *Rubia tinctorium* plant, which is cultivated in Europe and Asia Minor. The dye contains three principal colorants: alizarin (red), purpurin (red), and xanthine (yellow). The dye is extracted from the dried, powdered root as a precipitate when it is boiled in water. Alum lakes of madder, madder lake, and rose madder were used as artists' pigments. Madder was commonly used to dye textiles until synthetic alizarin was developed in 1868.

SYNONYMS: Madder lake, alizarin (natural), CI Pigment Red 83.

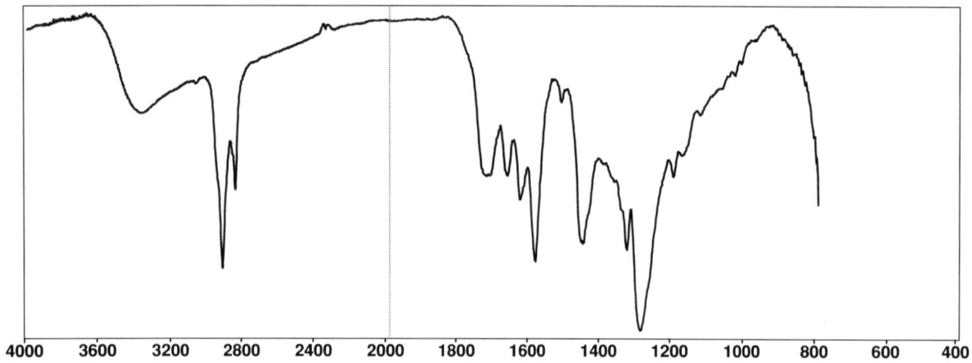

IR ANALYSIS LAB

Getty Conservation Institute, 2/24/92

IR ANALYSIS CONDITIONS

Spectra-Tech IRμS
Microscope
Bulk sample on BaF_2
Resolution = 4 cm^{-1}
Scans = 200
Range = 4000–800 cm^{-1}
Comment: baseline corrected

Madder **Red**

Dragon's Blood

PROVENANCE A. F. Suter & Co.

SOURCE Unknown

APPEARANCE Red, resinous, powder

CHARACTERISTIC IR ABSORPTION BANDS

3600–3200 cm^{-1}	O-H	stretching band
3100–2800 cm^{-1}	C-H	stretching bands
1740–1640 cm^{-1}	C=O	stretching band
1650–1600 cm^{-1}	C=C	stretching band
1620–1420 cm^{-1}	aromatic bands	
1480–1300 cm^{-1}	C-H	bending bands
1300–900 cm^{-1}	C-O	stretching bands

Dragon's Blood

Dragon's blood is a transparent, red, resinous exudation from the rattan cane palm, *Calamus draco,* that grows in eastern Asia. It is soluble in alcohol but insoluble in water. The color is fugitive when exposed to air but remains when protected by a resin film. It is used in stain varnishes, marble coloring, paints, and lithography.

SYNONYMS: Called *cinnabaris* by the ancients and by Pliny.

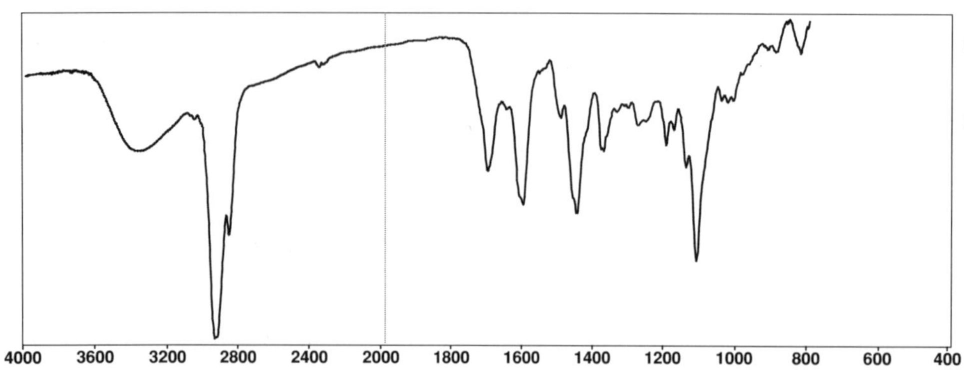

IR ANALYSIS LAB

Getty Conservation Institute

IR ANALYSIS CONDITIONS

Spectra-Tech IRμS
Microscope
Bulk sample on BaF_2
Resolution = 4 cm^{-1}
Scans = 200
Range = 4000–800 cm^{-1}

Dragon's Blood **Red**

Glossary

absorbance The amount of light absorbed by the sample at any specific wavelength. The intensity of a band in *absorbance* units is directly proportional to the concentration of sample responsible for the absorption band at that wavelength. It is useful to plot the y-axis of IR spectra in *absorbance* units (absorption bands ascending, or peaks) for quantitative analysis methods, spectral searching, and subtraction routines.

absorption The absorption of specific wavelengths of energy by an atom or molecule that results in an electronic, vibrational, translational, or rotational motion.

absorption bands (or peaks) A spectral absorption recorded as a band (transmittance units) or as a peak (absorbance units).

amplitude The intensity, volume, or magnitude of a wave.

attenuated total reflection (ATR) An older, although still-used, name for internal reflection spectroscopy. See **internal reflection**.

baseline The position of a spectral curve when the sample is not absorbing any radiation. Under optimized conditions, the baseline occurs at 100% transmittance, or 0.0 absorbance.

blackbody radiation A radiator that emits a continuous spectrum of wavelengths with no missing frequencies. The term is based on the thermodynamic principle that any material that emits all wavelengths when it is hot must absorb all wavelengths when it is cold. The color of a totally absorbing material is black, and thus the perfect emitter was given the name *blackbody* (Crooks 1978).

combination band A weak absorption band that is the product of the combination of two, or even three, strong vibrational quanta. The frequency of the band is equal to the sum of the frequencies of the originating bands. A combination band for a strong absorption band at 1600 cm^{-1} and one at 1250 cm^{-1} would occur at 2850 cm^{-1}.

computer search routine A computerized library search program typically based on the comparison of peak position and intensity of an unknown spectrum versus those in thousands of reference spectra.

Conne's advantage An advantage of FT-IR spectroscopy over dispersive IR, due to FT-IR's wavelength accuracy. Because in FT-IR a laser is used to calibrate the wavelengths in the spectrum, accuracy is ensured without any external calibration. This accuracy is important for the comparison of spectra to one another, as is done in library searching and spectral subtraction routines.

correlation chart A chart showing most common IR group frequencies and their positions, ranges, and relative intensities.

deconvolution A mathematical routine that operates on two or more overlapping bands, reducing the line width of individual components, thereby improving the spectral resolution.

degrees of freedom Allowed molecular motions—translational, rotational, and vibrational—within a molecule. The degrees of freedom depend on the number of atoms in a molecule and its symmetry.

detector An instrument or cell capable of converting radiant energy into an electrical signal.

diffraction A physical phenomenon in which radiation is dispersed or scattered when it passes by a sharp edge. In IR microspectroscopy, diffraction may occur when the beam passes through the apertures.

diffuse reflection (DRIFTS) An IR analysis technique in which the IR beam is diffusely reflected from the surface of a sample. It may be used with rough or porous samples, such as samples collected on silicon carbide paper.

dispersion Separation of radiant energy into discrete wavelengths.

electromagnetic spectrum The total range of electromagnetic radiation. The IR region is one small part of the electromagnetic spectrum.

electronic transition A change in the energy levels or spin directions of electrons with the absorption or emission of energy.

far infrared (far-IR or FIR) The region of the electromagnetic spectrum (500–20 cm^{-1}; 20–500 μm) that falls between the mid-IR and the microwave region. The far-IR region is well suited to the study of organometallic or inorganic compounds and is useful in the identification and differentiation of many minerals and colorants.

Fourier transform A complex mathematical function that converts an interferogram into a spectrum.

frequency (ν) The number of oscillations, or waves, per unit time—i.e., cycles per second. The term is also used to designate the specific wavenumber of an absorption band.

functional group A group of two or more atoms within a molecule that impart a given characteristic to that molecule (e.g., hydroxyl, carbonyl). Most functional groups absorb IR radiation at reproducible frequencies, known as group frequencies.

fundamental The frequency corresponding to a first-order vibration of the molecule. The fundamental vibration will produce the strongest absorption band for a given transition of a functional group.

group frequency A frequency that is associated with a particular functional group (i.e., carbonyl, C=O) and is common to all molecules contained in that group. Group frequencies are used to determine the chemical class of a material.

interferogram A plot of intensity versus displacement of an optical light path. This plot is the measured result of radiation that has passed through an interferometer; it may be converted to a spectrum using a Fourier transform mathematical process.

interferometer An apparatus with a moving mirror and a fixed mirror that can modulate a spectrum into an interferogram, thus allowing all frequencies to be measured at one time.

internal reflection spectroscopy (IRS) An IR analysis technique in which the sample is placed in optical contact with a high-refractive-index element. When an IR beam is passed through the element, a slight penetration of the beam into the sample occurs. This technique is useful for surface analysis.

microspectrophotometer An IR microscope coupled to an IR spectrometer, to compose a system capable of acquiring IR spectra of microscopic size samples.

mid-infrared (mid-IR or MIR) The region in which most fundamental vibrations for organic molecules occur (4000–500 cm^{-1}; 2.5–20 μm). This region can be subdivided into the group frequency region (4000–1300 cm^{-1}; 2.5–8.0 μm) and the fingerprint region (1300–500 cm^{-1}; 8.0–20 μm).

modulate To change a wave form. An interferometer modulates, or changes, the wave form of light from a sine pattern to an interferogram pattern.

monochromator A device that separates light into its spectrum of frequencies. In dispersive IR spectrometers, a monochromator is used to disperse the radiation prior to passing each wavelength through the sample.

multiplex advantage (Fellgett's advantage) The major advantage of FT-IR spectrometers over dispersive spectrometers. FT-IRs measure all wavelengths of IR radiation simultaneously, while dispersive spectrophotometers measure one resolution element at a time. Thus, an FT-IR can acquire a whole spectrum in the time it takes a dispersive spectrometer to collect one resolution element.

near infrared (near-IR or NIR) The IR region from 14,000 to 4000 cm^{-1} (0.7–2.5 μm). Spectra generated in the near-IR region consist entirely of overtones, combinations, and combinations of overtones of fundamental vibration modes from the mid-IR region.

noise Small, rapid fluctuations usually observed in the baseline of the spectrum. Noise is associated with instrument conditions rather than being specific to a sample.

overtone A weak vibrational transition occurring at approximately twice the frequency of a strong fundamental absorption frequency. An overtone band for a strong absorption band at 1700 cm^{-1} would occur at 3400 cm^{-1}.

pathlength The thickness of a measured medium. The absorption intensity is proportional to the pathlength of measurement. This relationship is often observed when a sample is too thick and thus absorbs too strongly, producing saturated absorption bands.

qualitative Analysis used to determine what kind of material is present.

quantitative Analysis used to determine how much of a material is present.

reflectance The amount of radiation reflected from a sample. An IR spectrum plotted in reflectance units can usually be converted to appear as a transmittance spectrum with the Kubelka-Munk or Kramers-Kronig transformations for diffuse reflection and specular reflection, respectively.

reflection Radiation that bounces off the surface of a material after contact.

refraction The change in angle of a light path as it passes from one medium into another with a different refractive index.

resolution The ability of an instrument to separate two closely occurring absorptions. This ability is a function of instrumental and spectral parameters such as detector sensitivity, absorption intensity, absorption frequency, interferometer scan distance, and signal-to-noise ratio.

rotational transition The rotation of a molecule around its center of mass with the absorption or emission of energy.

selection rule A rule that determines whether a given vibration will be seen in the spectrum, based on the symmetry of the molecule. The primary selection rule, or requirement, for active IR absorptions is that the vibration must result in a change in the dipole moment of the molecule.

signal-to-noise ratio The intensity of the recorded absorption band versus the intensity of noise.

source A high-temperature body used to generate radiant energy.

spectral line Absorption related specifically to a quantum energy change. This absorption is never actually seen as a line but rather as a band or peak.

spectrophotometer An instrument for recording the intensity and frequency of spectral absorptions. Also sometimes referred to as a spectrometer.

spectroscopy The study of the interaction of light and matter. The term is also used to specify the technique of recording and studying spectra.

spectrum (pl., spectra) A recorded tracing of the amount of radiation absorbed at each frequency over a given spectral range of interest.

specular reflection (external reflection) A type of reflection in which the reflection angle of a beam is equal to the incident angle, as in a mirror.

subtraction A mathematical routine for comparing two spectra by subtracting the absorption bands of one from the other. It is useful for elucidating small changes between two similar spectra that may otherwise be overlooked.

throughput advantage (Jacquinot's advantage) An advantage of FT-IR over dispersive instruments. A greater amount of radiation reaches an FT-IR detector than reaches a detector for a dispersive instrument, because of the limitations of the monochromator slit. This greater energy throughput gives FT-IR more sensitivity, which is often needed for reflection techniques.

translational transition The movement of an entire molecule to a new position in space due to the absorption or emission of energy.

transmission The passage of radiation through a material without its being absorbed or reflected.

transmittance The amount of radiation transmitted through a sample. Transmittance is equivalent to the logarithmic reciprocal of absorbance. IR spectra are often plotted with the y-axis giving percent transmittance (absorption bands descending) to increase the size of smaller absorption bands in relation to more intense absorption bands.

vibrational transitions The change in bond angles or bond lengths within a molecule due to the absorption or emission of energy. Molecular vibrations account for the strongest absorption bands in an IR spectrum.

wavelength (λ) The distance between two successive maxima or minima of a wave—i.e., the length of one wave.

wavenumber ($\bar{\nu}$) The number of waves per unit length. Wavenumber units are commonly used in IR spectroscopy and are expressed in cm^{-1}.

Suppliers

Becton Dickinson, Acute Care, 1 Becton Dr., Franklin Lakes, NJ 07417; (201) 847-6800. Sells Beaver blades and other microsurgical tools.

Bio-Rad, Digilab Div., 237 Putnam Ave., Cambridge, MA 02139; (617) 868-4330. Sells FT-IR spectrophotometers, microscopes and other accessories, thermal and chromatographic IR attachments.

Bomem International, 7800 Quincy St., Willowbrook, IL 60521; (708) 986-1090. Sells dispersive and FT-IR spectrophotometers along with many accessories, including microscopes.

Bruker Instruments, Inc., 19 Fortune Dr., Billerica, MA 01821; (508) 667-9580. Sells dispersive and FT-IR spectrophotometers along with many accessories, including microscopes.

Buck Scientific, Inc., 58 Fort Point St., East Norwalk, CT 06855-1097; (203) 853-9444. Sells dispersive and FT-IR spectrophotometers along with many accessories.

Carolina Biological Supply Co., 2700 York Road, Burlington, NC 27215; (800) 334-5551. Sells Caroplastic polyester for embedding.

Castolite Co., P.O. Box 391, Woodstock, IL 60098; (815) 338-4670. Sells Castolite polyester for embedding.

Conservation Materials, 1275 Kleppe Lane, Suite 10, Sparks, NV 89431; (702) 331-0582. Sells conservation products, including Rhoplex AC-33.

Digilab—see **Bio-Rad**.

Ernest E. Fullam, Inc., 900 Albany Shaker Road, Dept. AL, Latham, NY 12110-1491; (518) 785-5533. Sells microanalysis tools and containers, silicone molds for preparing cross sections.

Galactic Industries Corp., 395 Main St., Salem, NH 03079; (603) 898-7600. Sells spectral conversion and search software, data processing routines, including three-dimensional spectral processing routines.

Graseby Specac, 301 Commerce Dr., Fairfield, CT 06430; (800) 447-2558. Sells accessories for IR spectrometers, gas cells, optical crystals, sample preparation equipment.

Harrick Scientific, 88 Broadway, P.O. Box 1288, Ossining, NY 10562; (914) 762-0020. Sells accessories for IR spectrometers, optical crystals, sample preparation equipment, custom accessories (such as ATR and DRIFTS units).

High Pressure Diamond Optics, 231 Gianconda Way, Suite 103, Tucson, AZ; (520) 544-9338. Sells high- and low-pressure diamond anvil cells; diamond microtoming knives.

International Crystal Laboratories, 11 Erie St., Garfield, NJ 07026; (201) 478-8944. Sells accessories for IR spectrometers, optical crystals, sample preparation equipment.

JASCO, 8649 Commerce Dr., Easton, MD 21601; (800) 333-5272. Sells dispersive and FT-IR spectrophotometers, accessories including microscopes.

KVB/Analect, 17819 Gillette Ave., Irvine, CA 92714; (714) 660-8801. Sells dispersive and FT-IR spectrophotometers, accessories including microscopes.

Ladd Research Industries, Inc., 13 Dorset Lane, Williston, VT 05495; (802) 658-4961. Sells general microscope supplies and acrylic, epoxy, and wax materials for embedding.

Mattson Instruments, 1001 Fourier Dr., Madison, WI 53717; (608) 831-5515. Sells dispersive and FT-IR spectrophotometers, accessories including microscopes.

McCarthy Scientific Co., P.O. Box 5332, Fullerton, CA 92635; (714) 526-2742. Sells optical crystals, sample preparation equipment.

Midac Corp., 17911 Fitch Ave., Irvine, CA 92714; (714) 660-8558. Sells FT-IR systems, portable IR air-monitoring systems.

Minitools, 634 University Ave., Los Gatos, CA 95030; (408) 395-1585. Sells microtools, micromanipulation devices, microdrills.

Nicolet Instrument Corp., 5225-5 Verona Road, Madison, WI 53711; (608) 276-6100. Sells FT-IR spectrophotometers, accessories including microscopes.

Perkin-Elmer Corp., 761 Main Ave., Norwalk, CT 06859; (800) 762-4000. Sells dispersive and FT-IR spectrophotometers, accessories including microscopes, other types of analytical instrumentation.

Photometrics, Inc., 15801 Graham St., Huntington Beach, CA 92649; (714) 895-4465. Provides searches of numerous commercial IR libraries.

Pike Technologies, 2919 Commerce Park Dr., Madison, WI 53719; (608) 274-2721. Sells IR accessories, microscope computer-controlled mapping stages.

Shimadzu Scientific Instruments, Inc., 7102 Riverwood Dr., Columbia, MD 21046; (800) 477-1227. Sells dispersive and FT-IR spectrophotometers, accessories including microscopes, other types of analytical instrumentation.

Spectra-Tech (Spectra Technology), Inc., 2 Research Dr., Shelton, CT 06484-0849; (203) 926-8998. Develops and sells innovative IR accessories, including microscopes, IRμS microspectrophotometers, compression cells that use salt plates or diamond windows, roller knives. Also sells 3M disposable IR cards containing thin polymer films for sample preparation.

Structure Probe, Inc., P.O. Box 656, West Chester, PA 19381; (215) 436-5400. Sells microanalysis tools and containers, silicone molds for preparing cross sections.

Ted Pella, Inc., P.O. Box 2318, Redding, CA 96099; (916) 243-2200. Sells acrylic, epoxy, and wax materials for embedding, along with several types of Pelco silicone embedding molds.

Union Carbide, 39 Old Ridgebury Rd., Danbury, CT 06817; (203) 794-5300. Manufactures and sells Parylene.

Ward's Natural Science, P.O. Box 92912, Rochester, NY; (800) 962-2660. Sells Bio-Plastic polyester for embedding.

Whatman, Balston Div., 100 Ames Pond Dr., Tewksbury, MA 01876; (800) 343-4048. Sells Balston dry air systems to purge IR spectrometers.

Wig-L-Bug, Crescent Dental Manufacturing Co., 7750 47th St., Lyons, IL 60534; (708) 447-8050.

References

Abbey Newsletter
1988. Parylene made available for conservation. *Abbey Newsletter* 12(6):108.

Abd El-Hady, M. M.
[1988]. Infra-red investigations on monumental limestone samples. In *6th International Congress on Deterioration and Conservation of Stone*, vol. 1, comp. J. Ciabach, 387–94. Torun, Poland: Nicholas Copernicus University Press Department.

Afremow, L. C., and J. T. Vandeberg
1966. High resolution spectra of inorganic pigments and extenders in the mid-infrared region from 1500 cm^{-1} to 200 cm^{-1}. *Journal of Painting Technology* 38(495):169–202.

Aldrich, D. S., and M. A. Smith
1995. Pharmaceutical applications of infrared microspectroscopy. In *Practical Guide to Infrared Microspectroscopy*, ed. H. Humecki, 323–76. New York: Marcel Dekker.

American Society for Testing Materials (ASTM)
1981. *Standard Practices for Internal Reflection Spectroscopy*. ASTM E 573-81. Philadelphia: ASTM.

Baer, N. S., and N. Indictor
1976. Linseed oil–metal acetylacetonate systems 3: Infrared spectroscopic observation of the drying film. *Journal of Coatings Technology* 48:58–62.

Baker, M., D. von Endt, W. Hopwood, and D. Erhardt
1988. FTIR microspectrometry: A powerful conservation analysis tool. In *Preprints of Papers Presented at the Sixteenth Annual Meeting of the AIC, New Orleans, Louisiana, June 1–5, 1988*, 1–13. Washington, D.C.: American Institute for Conservation of Historic and Artistic Works.

Baker, M. T., H. D. Burgess, N. E. Binnie, M. R. Derrick, and J. R. Druzik
1990. Laboratory investigation of the fumigant Vikane. In *ICOM Committee for Conservation Preprints, 9th Triennial Meeting, Dresden*. Paris: ICOM Committee for Conservation.

Banerjee, A., and T. Hager
1992. On some crystals of lapis lazuli. *Zeitschrift für Naturforschung* 47:1094.

Barclay, R. L.
1989. An improved method for solvent extraction of oil from a whale bone sculpture. *Journal of the International Institute for Conservation—Canadian Group* 14:18–22.

Barer, R., A. R. Cole, and H. W. Thompson
1949. Infra-red spectroscopy with the reflecting microscope in physics, chemistry and biology. *Nature* 163:198–203.

Bartick, E. G.
1987. Considerations for fiber sampling with infrared microspectroscopy. In *The Design, Sample Handling, and Applications of Infrared Microscopes: A Symposium*, ed. P. B. Roush. ASTM Special Technical Publication 949. Philadelphia: American Society for Testing Materials.

Bartick, E. G., J. C. Corbett, and G. L. McClure
1982. Computerized quantitative analysis of copolymers by IR spectroscopy. In *Computer Applications in Applied Polymer Science,* 185–96. Washington D.C.: American Chemical Society.

Bartick, E. G., M. W. Tungol, and J. A. Reffner
1994. A new approach to forensic analysis with infrared microscopy: Internal reflection spectroscopy. *Analytica Chimica Acta* 288:35–42.

Beck, C. W.
1972. Aus der Bernstein-Forschung (Amber research). *Die Naturwissenschaften* 59(7):294–98.

Beck, C. W., A. B. Adams, G. C. Southard, and C. Fellows
1971. Determination of the origin of Greek amber artifacts by computer-classification of infrared spectra. In *Science and Archaeology,* ed. R. H. Brill, 235–40. Cambridge, Mass: MIT Press.

Beck, C. W., J. Greenlie, M. Diamond, A. Macchiarulo, A. Hannenberg, and M. Hauck
1978. The chemical identification of Baltic amber at the Celtic Oppidum Stare Hradisko in Moravia. *Journal of Archaeological Science* 5:343–54.

Beck, C. W., C. J. Smart, and D. J. Ossenkop
1989. Residues and linings in ancient Mediterranean transport amphoras. In *Archaeological Chemistry.* Washington D.C.: American Chemistry Society.

Beck, C. W., E. Wilbur, and S. Meret
1964. Infrared spectra and the origin of amber. *Nature (*London) 201(4916):256–57.

Beck, C. W., E. Wilbur, S. Meret, D. Kossove, and K. Kermani
1965. The infrared analysis of amber and the identification of Baltic amber. *Archaeometry* 9:96–109.

Bellamy, L. J.
1980. *The Infrared Spectra of Complex Molecules.* 2d ed. London: Chapman and Hall.

Berni, R. J., and N. M. Morris
1984. Infrared spectroscopy. In *Analytical Methods for a Textile Laboratory,* ed. J. W. Weaver, 265–92. Research Triangle Park, N.C.: American Association of Textile Chemists and Colorists.

Besaninou, D.
1984. *Cire et cires: Étude sur la composition, l'alteration, la restauration, la conservation des oeuvres ceroplastiques.* Paris: Institut français de restauration des oeuvres d'art.

Bettembourg, J. M.
1988. Examen et analyse des processus de corrosion vitraux anciens (Examination and analyses of corrosion processes of ancient stained glass windows). In *2nd International Conference on Non-Destructive Testing: Microanalytical Methods and Environment Evaluation for Study and Conservation of Works of Art, Perugia, 17–20 April 1988,* 1.1–1.15. Rome: Comas Grafica.

Billmeyer, F. W., Jr., and M. Saltzman
1966. *Principles of Color Technology.* New York: John Wiley and Sons.

Birstein, V. J.
1975. On the technology of central Asian wall paintings: The problem of binding media. *Studies in Conservation* 20:8–19.

1978. Infra-red spectroscopic investigation of grounds and pigments of central Asian wall paintings. *Khudozestvennoe nasledie* 434:5–22.

Birstein, V. J., and V. M. Tul'Chinskii
1977. Ob identifikatsii vesestv pokrovnogo sloia egiptskikh kamennikh stel (On the identification of substance of covering layers of Egyptian stone tombs). *Khudozestvennoe nasledie* 333:8–11.

1979. Identifikatsiia nekotor'ikh materialov zivopisi ikon'i 'mucenik i mucenitsa' metodom ik-spektroskopii. *Khudozestvennoe nasledie* 535:198–201.

1981. IR-spectroscopic analysis of aged gelatins. In *ICOM Committee for Conservation 6th Triennial Meeting, Ottawa, 21–25 September 1981, Preprints,* 9.1–9.8. Paris: ICOM Committee for Conservation.

Biscontin, G., and S. Volpin
1989. Indagine sulle patine ad ossalato mediante FT-IR (FT-IR investigations on calcium oxalate films). In *Le pellicole ad ossalati: Origine e significato nella conservazione delle opere d'arte* (Oxalate films: Origin and meaning in the conservation of works of art), 151–63. Milan: Centro del C.N.R. "Gino Bozza."

Bishop, M.
1994. Conversation with author, Getty Conservation Institute, Marina del Rey, Calif.

Blanco, M. T., F. Puertas, A. Palomo, and F. Accion
1991. Infrared spectroscopy as a complementary technique to identify inorganic salts in ancient deteriorated mortars, grouts and plasters. In *Science, Technology, and European Cultural Heritage: Proceedings of the European Symposium, Bologna, Italy, 13–16 June 1989,* ed. N. S. Baer, C. Sabbioni, and A. Sors, 601–6. Oxford and Boston: Butterworth-Heinemann.

Blasius, E., H. Wagner, H. Braun, R. Krumbholz, and B. Schwartz
1983. Tile fragments as characteristic evidence for ancient Roman settlements: Scientific investigations. *Archaeometry* 25(2):165–78.

Bleck, R.-D., and R. Ziessler
1967. Chemische Untersuchungen an der Kirche in Thorey, Kreis Arnstadt (Chemical investigations at the church in Thorey near Arnstadt). *Neue Museumskunde* 10(1).

Bockoff, F. J., K.-M. Guo, G. E. Richards, and E. Bockoff
1984. Infrared studies of the kinetics of insolubilization of soluble nylon. In *Adhesives and Consolidants: Preprints of the Contributions to the IIC Paris Congress, 2–8 September 1984,* ed. N. S. Brommelle, E. M. Pye, P. Smith, and G. Thomson. London: International Institute for Conservation of Historic and Artistic Works.

Brachert, T.
1978–79. Historic transparent varnishes and furniture polishes. Translated by R. Mussey. Parts 1–5. *Maltechnik-Restauro* 1(1978):56; 2(1978):120; 3(1978):185; 4(1978):263; 2(1979):131.

Bradley, S.
1987. Techniques for the analysis of salts. *Conservation News* 34:24–25.

Brodsky-Doyle, B., E. G. Bendit, and E. R. Blout
1975. Infrared spectroscopy of collagen and collagen-like peptides. *Biopolymers* 14:937–45.

Brown, M. K., and C. A. Shelton
1994. Bayou Bend's latest acquisition: Old World inspiration for a New World masterpiece. In *42nd Annual Theta Charity Antiques Show, Houston, Texas, September 22–25, 1994.* Exhibition catalogue. Houston: Theta Charity Antiques Show.

Burnstock, A., and T. Learner
1992. Changes in the surface characteristics of artificially aged mastic varnishes after cleaning using alkaline reagents. *Studies in Conservation* 37(3):165–84.

Butler, S. A.
1988. *The Ageing of Paraloid B-72 Based Adhesives 1: Solubility and Infrared and UV-Visible Spectroscopy Studies.* London: British Museum, Department of Conservation, Conservation Research Section.

Butz, W. H.
1960. Infrared absorption analysis of powdered samples. In *Beckman Application Data Sheet.* Application note IR-8063, Feb. 1960. Fullerton, Calif.: Beckman Instruments.

Cain, C. E., and V. F. Kalasinsky
1985. Characterization of eighteenth- and nineteenth-century paper using Fourier transform infrared spectroscopy and thin-layer chromatography. In *Application of Science in Examination of Works of Art: Proceedings of the Seminar, September 7–9, 1983,* ed. P. A. England and L. van Zelst, 55–58. Boston: Research Laboratory, Museum of Fine Arts.

Camaiti, M., R. Ingoglia, G. Moggi, A. Pasetti, and A. Scala
1991. Controllo della reversibilità di trattamenti protettivi effettuati con polimeri fluorurati su materiali lapidei (Controlling the reversibility of protective treatments using fluoropolymers on stone materials). In *Le pietre nell'architettura, struttura e superfici: Atti del convegno di studi, Bressanone, 25–28 giugno 1991,* ed. Guido Biscontin and Daniela Mietto, 267–76. Padua, Italy: Libreria Progetto Editore.

Cardamone, Jeanette M.
1988. Fourier transform infrared reflectance absorbance spectroscopy (IRAS) for non-destructive evaluation of textiles. In *2nd International Conference on Non-Destructive Testing: Microanalytical Methods and Environment Evaluation for Study and Conservation of Works of Art, Perugia, 17–20 April 1988.* Rome: Comas Grafica.

Carriveau, G. W.
1984. Technical examination of oriental lacquer. In *Archaeological Chemistry 3,* ed. J. B. Lambert, 395–402. Advances in Chemistry series 205. Washington, D.C.: American Chemical Society.

Cartwright, L. J., N. S. Cartwright, and P. G. Rodgers
1977. A microtome technique for sectioning multilayer paint samples for microanalysis. *Journal of the Canadian Society for Forensic Science* 10:10–12.

Cennini, C.
1960. Il libro dell'arte. In *The Craftsman's Handbook*, ed. and trans. D. V. Thompson Jr. New York: Dover Publications.

Chess, C. A.
1995. Applications of reflectance microspectroscopy in the electronics industry. In *Practical Guide to Infrared Microspectroscopy*, ed. H. Humecki. New York: Marcel Dekker.

Chicago Society for Coating Technology
1980. *An Infrared Spectroscopy Atlas for the Coatings Industry.* Philadelphia: Federation of Societies for Paint Technology.

Clark, R. J. H.
1964. The application of diffuse reflectance spectroscopy to inorganic chemistry. *Journal of Chemical Education* 41(9):489.

Coblentz, W. W.
1905. *Investigation of Infrared Spectra.* Carnegie Institute Publications no. 35. Washington, D.C.: Carnegie Institute.

Cole, A. R. H., and R. N. Jones
1952. A reflecting microscope for infrared spectroscopy. *Journal of the Optical Society of America* 42:348.

Connor, M., and F. Girardet
1992. Étude du mode de fixation du soufre sur un gres calcareux (Study of sulfur fixation modes on a calcareous sandstone). In *Proceedings of the 7th International Congress on Deterioration and Conservation of Stone, Held in Lisbon, Portugal, 15–18 June 1992,* ed. J. D. Rodrigues, F. Henriques, and F. T. Jeremias, 407–16. Lisbon: Laboratório Nacional de Engenharia Civil.

Cook, R., and M. D. Paterson
1978. New techniques for the identification of microscopic samples of textile fibers by infrared spectroscopy. *Forensic Science International* 12:237–43.

Cooley, J. W., and J. W. Tukey
1965. An algorithm for the machine calculation of complex Fourier series. *Mathematics of Computation* 19:297–301.

Coville, F. V.
1892. The Panamint Indians of California. *American Anthropologist* 5:351–61.

Crooks, J. E.
1978. *The Spectrum in Chemistry.* London: Academic Press.

Dauphin, Y.
1993. Potential of the diffuse reflectance infrared Fourier transform (DRIFT) method in paleontological studies of bones. *Applied Spectroscopy* 47(1):52.

de Keijzer, M., and M. F. S. Karreman
1989. Een vroeg-twintigste-eeuwse pigmentcollectie: De pigmentcollectie van het Missiemuseum te Steyl-Tegelen (An early 20th-century pigment collection: The pigment collection in the Missiemuseum at Steyl-Tegelen). *Mededelingenblad (IIC Nederland)* 6(2):7–24.

Delbourgo, S.
1972. La technique de la fresque de Cennino Cennini à Botticelli: Étude physico-chimique (Fresco technique from Cennino Cennini to Botticelli: Physico-chemical study). *Annales du laboratoire de recherche des Musées de France,* 4–15.

Delbourgo, S., and M. C. Gay
1968. La physico-chimie au service de l'archéologie—examen d'une série d'échantillons provenant de fouilles effectuées au Soudan (Physico-chemistry at the service of archaeology—examination of a series of samples from excavations done in Sudan). *Bulletin du laboratoire du Musée du Louvre* 12:20–25.

Demmler, K.
1980. The determination of residual active oxygen in parts of unsaturated polyester resins and its influence on post curing and yellowing by light. *Kunstoffe* 70:786–92.

Derrick, M. R.
1989. Fourier transform infrared spectral analysis of natural resins used in furniture finishes. *Journal of the American Institute for Conservation* 28(1):43–56.

1991. Evaluation of the state of degradation of Dead Sea Scroll samples using FT-IR spectroscopy. In *Book and Paper Annual,* 49–65. Washington, D.C.: AIC.

1995. Infrared microspectroscopy in the analysis of cultural artifacts. In *Practical Guide to Infrared Microspectroscopy*, ed. H. Humecki, 287–322. New York: Marcel Dekker.

Derrick, M. R., E. Doehne, A. Parker, and D. Stulik
1994. Some new analytical techniques for use in conservation. *Journal of the American Institute for Conservation* 33(3):171–84.

Derrick, M. R., C. Druzik, and F. Preusser
1988. FTIR analysis of authentic and simulated black lacquer finishes on eighteenth-century furniture. In *Urushi: Proceedings of the Urushi Study Group, June 10–27, 1985, Tokyo,* ed. N. S. Brommelle and P. Smith, 227–34. Marina del Rey, Calif.: Getty Conservation Institute.

Derrick, M. R., J. M. Landry, and D. C. Stulik
1991. *Methods in Scientific Examination of Works of Art: Infrared Microspectroscopy*. Marina del Rey, Calif.: Getty Conservation Institute.

Derrick, M. R., L. Souza, T. Kieslich, D. Stulik, and H. Florsheim
1994. Embedding paint cross-section samples in polyester resin: Problems and solutions. *Journal of the American Institute for Conservation* 33(3):227–45.

Derrick, M. R., D. C. Stulik, J. M. Landry, and S. P. Bouffard
1992. Furniture finish layer identification by infrared linear mapping microspectroscopy. *Journal of the American Institute for Conservation* 31(2):225–36.

Derrick, M. R., D. C. Stulik, and E. Ordonez
1993. Deterioration of cellulose nitrate sculptures made by Gabo and Pevsner. In *Symposium '91: Saving the Twentieth Century—the Degradation and Conservation of Modern Materials, Abstracts,* 169–82. Ottawa: Canadian Conservation Institute.

Domaslowski, W., and M. Kesy-Lewandowska
1988. Identification of deposits on stone historical objects. In *2a conferenza internazionale sulle prove non distruttive metodi microanalitici e indagini ambientali per lo studio e la conservazione delle opere d'arte . . . Perugia, 17–20 aprile 1988.* Vol. 1. Rome: Istituto Centrale per il Restauro.

Dow Chemical Co.
1982. *Vikane Fumigation Manual.* Indianapolis: Dow Agro Sciences.

Duraiswamy, D., K. Ramaswamy, and R. Venkatachalapathy
1992. Infrared spectroscopic studies of archaeological materials: Determination of firing temperature. In *Role of Chemistry in Archaeology: 1st International Colloquium Held in Hyderabad (India), 15–18 November 1991,* ed. M. C. Ganorkar and N. Rama Rao, 47–54. Hyderabad, India: Birla Institute of Scientific Research.

Eastes, J. W., and J. W. Salisbury
1986. Spectral properties of sulfated limestone and marble: Implications for in situ assessment of atmospheric pollution damage to carbonate rock building materials. *Applied Spectroscopy* 40(7):954–59.

Edge, M., N. S. Allen, M. Hayes, P. N. K. Riley, C. V. Horie, and J. Luc-Gardette
1990. Mechanisms of deterioration in cellulose nitrate base archival cinematograph film. *European Polymer Journal* 26(6):623–30.

Erhardt, D., W. Hopwood, M. Baker, and D. von Endt
1988. A systematic approach to the instrumental analysis of natural finishes and binding media. In *Preprints of Papers Presented at the Sixteenth Annual Meeting of the AIC, New Orleans, Louisiana, June 1–5, 1988,* 67–84. Washington, D.C.: American Institute for Conservation of Historic and Artistic Works.

Fahrenfort, J.
1962. Attenuated total reflection—A new principle for production of useful infrared reflection spectra of organic compounds, molecular spectroscopy. In *Proceedings of the 4th International Meeting on Molecular Spectroscopy, Bologna, 1959,* vol. 2, ed. A. Mangini. London: Pergammon.

Faraone, G.
1987. La spettroscopia in reflettanza diffusa: Un efficace metodo di indagine delle superfici di materiali lapidei (Diffuse reflectance spectroscopy: An effective method for investigating stone material surfaces). *Restauro e città* 7:156–57.

Felger, R. S., and M. B. Moser
1985. *People of the Desert and Sea: Ethnobotany of the Seri Indians.* Tucson: University of Arizona Press.

Feller, R. L.
1954. Dammar and mastic infrared analysis. *Science* 120(3130):1069–70.

1959. Identification and analysis of resins and spirit varnishes. In *Application of Science in Examination of Works of Art—Proceedings of the Seminar, September 15–18, 1958.* Boston: Museum of Fine Arts.

Feller, R. L., ed.
1986. *Artists' Pigments: A Handbook of Their History and Characteristics.* Vol. 1. Cambridge: Cambridge University Press; Washington, D.C.: National Gallery of Art. (Contains articles with IR spectra on Indian yellow; cobalt yellow; barium sulfate; cadmium yellows, oranges, and reds; red lead and minium; green earth; zinc white; chrome yellow and other chromate pigments; lead antimonate yellow; carmine.)

Fellgett, P.
1970. The origins and logic of multiplex, Fourier and interferometric methods in spectrometry. In *Proceedings for Aspen International Conference on Fourier Spectroscopy*, ed. G. A. Vanasse, A. T. Stair, and D. J. Baker, 139–42. U.S. Air Force document no. AFCRL-71-0019. Bedford, Mass.: Air Force Cambridge Research Laboratories.

Ferraro, J. R., and L. J. Basile
1979. The diamond anvil cell as a sampling device in IR. *American Laboratory* 11:31–36.

Ferrus, R., R. Pages, and C. Diez
1981. Análisis cuantitativo de estucados caolin-caseina mediante espectroscopia de infrarrojos o reflexión (Quantitative analysis of kaolin-casein coatings by means of infra-red reflection spectroscopy). *Investigación y técnica del papel* 18(67):80–91.

Fischer, G., and M. Bader
1994. Melt film technique for IR spectroscopy. *American Laboratory* (November):32c.

Follette, T.
1985. Amber in goldworking. *Archaeology* 38(2):64–65.

Francis, S. A., and A. H. Ellison
1959. Infrared spectra of monolayers on metal mirrors. *Journal of the Optical Society of America* 49(1):131–38.

Frediani, P., and U. Matteoli
1978. A methodological approach to the study of the deterioration of the statues in the Boboli's garden (Pitti Palace, Florence) 3: Rapid determination of gypsum on marble by infrared spectroscopy. In *ICOM Committee for Conservation 5th Triennial Meeting, Zagreb, 1–8 October 1978, Preprints.* Paris: International Council of Museums.

Freeman, K. L.
1979. Techniques used in the analysis of polymers. *ICCM Bulletin* 5(2):12–27.

Fromageot, D.
1990. Photovieillissement de revêtements de consolidation à base de polyamides solubles (Light aging of consolidation coatings based on soluble polyamides). In *Technologie industrielle. Conservation, restauration du patrimoine culturel. Colloque AFTPV/SFIIC. Nice, 19–22 septembre 1989*, 28–33. Puteaux, France: Erec.

Fromageot, D., and J. Lemaire
1991. The prediction of the long-term photo-aging of soluble polyamides used in conservation. *Studies in Conservation* 36(1):1–8.

Fuchs, D. R., H. Romich, and H. Schmidt
1990. Glass-sensors: Assessment of complex corrosive stresses in conservation research. In *Materials Issues in Art and Archaeology 2: Symposium Held April 17–21, 1990, San Francisco, California, USA*, ed. P. B. Vandiver, J. Druzik, and G. S. Wheeler, 239–51. Pittsburgh: Materials Research Society.

Fujinaga, T., T. Takenaka, and T. Muroga
1976. Origin of foreign and Japanese ambers studied by elemental analyses and infrared spectroscopy. *Bunseki Kagaku* 25(11):795–99.

Fuller, M. P., and P. R. Griffiths
1978a. Diffuse reflectance measurements by infrared Fourier transform spectrometry. *Analytical Chemistry* 50(13):1906–10.

1978b. Infrared analysis by diffuse reflectance spectrometry. *American Laboratory* 10(10):69–80.

1980. Infrared microsampling by diffuse reflectance Fourier transform spectrometry. *Applied Spectroscopy* 34(5):533–39.

Fuoco, R., K. H. Shafer, and P. R. Griffiths
1986. Capillary gas chromatography/Fourier transform infrared microspectrometry at subambient temperature. *Analytical Chemistry* 58:3249–54.

Garrido, J. M. C.
1969. *El Azul Maya.* Madrid: Instituto de Conservación y Restauración de Obras de Arte, Arqueología y Etnología.

Gettens, R. J.
1932. A microsectioner for paint films. *Technical Studies in the Field of the Fine Arts* 1:20.

1936. The cross-sectioning of paint films. *Technical Studies in the Field of the Fine Arts* 5:18.

1952. Science in the art museum. *Scientific American* 187(1):22–27.

1959. The identification of pigments and inerts on paintings and other museum objects. In *Application of Science in Examination of Works of Art: Proceedings of the Seminar, September 15–18, 1958,* 31–49. Boston: Museum of Fine Arts.

Gettens, R. J., and G. L. Stout
1966. *Painting Materials: A Short Encyclopaedia.* New York: Dover Publications.

Gettens, R. J., and E. West-FitzHugh
1974. Malachite and green verditer. *Studies in Conservation* 19(1):2–23.

Gettens, R. J., E. West-FitzHugh, and R. L. Feller
1974. Calcium carbonate whites. *Studies in Conservation* 19(3):157–84.

Giangrande, C.
1988. Identification of bronze corrosion products by infrared absorption spectroscopy. In *Recent Advances in the Conservation and Analysis of Artifacts: Jubilee Conservation Conference, London, 6–10 July 1987,* comp. James Black, 135–48. London: University of London, Institute of Archaeology, Summer Schools Press.

Gianno, R., D. W. von Endt, D. Erhardt, K. M. Kochummen, and W. Hopwood
1987. The identification of insular southeast Asian resins and other plant exudates for archaeological and ethnological application. In *Recent Advances in the Conservation and Analysis of Artifacts: Jubilee Conservation Conference, London, 6–10 July 1987,* comp. James Black, 229–38. London: University of London, Institute of Archaeology, Summer Schools Press.

Gillard, R., S. Hardman, R. Thomas, and D. Wilkinson
1994. The detection of dyes by FTIR microscopy. *Studies in Conservation* 39:187–92.

Glauert, Audrey M.
1972. Fixation, dehydration, and embedding of biological sections. In *Practical Methods in Electron Microscopy.* Amsterdam: North-Holland Publishing Co.

Godla, J.
1990. The use of wax finishes on pre-industrial American furniture. Master's thesis, Antioch University, West Townsend, Mass.

Goos, F., and H. Hänchen
1947. Ein neuer und fundamentaler Versuch zur Totalreflexion. *Ann. Physik* 1:333.

Gorgoni, C., L. Lazzarini, and O. Salvadori
1992. Minero-geochemical transformations induced by lichens in the biocalcarenite of the Selinuntine monuments. In *Proceedings of the 7th International Congress on Deterioration and Conservation of Stone, Held in Lisbon, Portugal, 15–18 June 1992,* ed. J. D. Rodrigues, F. Henriques, and F. T. Jeremias, 531–39. Lisbon: Laboratório Nacional de Engenharia Civil.

Grant, C.
1993. *The Rock Painting of the Chumash: A Study of a California Indian Culture.* Santa Barbara, Calif.: Santa Barbara Museum of Natural History.

Grattan, D. W., and M. Bilz
1991. The thermal aging of Parylene and the effect of antioxidant. *Studies in Conservation* 36(1):44–52.

Griffiths, P. R., and J. A. de Haseth
1986. *Fourier Transform Infrared Spectrometry.* New York: Wiley-Interscience.

Griffiths, P. R., and M. P. Fuller
1982. *Mid Infrared Spectrometry of Powdered Samples.* Vol. 9 of *Advances in Infrared and Raman Spectroscopy,* ed. R. J. H. Clark and R. E. Hester. London: Heyden.

Grimaldi, D.
1993. The care and study of fossiliferous amber. *Curator* 36(1):31–49.

Grosjean, D.
1985. Wall loss of gaseous pollutants in outdoor Teflon chambers. *Environmental Science and Technology* 19(11):1059–65.

Guineau, B.
1983. Applications de la spectrometrie (rayons x, infra-rouge, raman) à la mise en évidence de lazurite et de bleu de Prusse dans des poudres minerales (Applications of spectrometry [x-ray, infrared, raman] for revealing azurite and Prussian blue in mineral powders). In *La peinture funeraire, caractérisation, datation, technique de la peinture antique* (Funerary painting: Characterization, dating, technique of ancient painting), 75–90. Valbonne, France: Apdca.

Guralnik, D. B., ed.
1984. *Webster's New World Dictionary.* 2d college ed., s.v. "infrared."

Hadzi, D., and F. Cvek
1978. Smolni Kit in Premaz za Zare (Resinous putty and paint on urns). *Arheoloski vestnik*, 128–33.

Hansen, E. F.
1993. The effects of relative humidity on some physical properties of modern vellum: Implications for the optimum relative humidity for the display and storage of parchment. *Journal of the American Institute for Conservation* 31(3):325.

Hansen, E. F., M. R. Derrick, M. R. Schilling, and R. Garcia
1991. The effects of solution application on some mechanical and physical properties of thermoplastic amorphous polymers used in conservation: Poly(vinyl acetate)s. *Journal of the American Institute for Conservation* 30(2):203–13.

Hansen, E. F, and W. S. Ginell
1989. The conservation of silk with Parylene-C. In *Historic Textile and Paper Materials: Conservation and Characterization*, ed. H. L. Needles and H. S. Zeronian, 108–33. Advances in Chemistry series 212. Washington, D.C.: American Chemical Society.

Hansen, E. F., R. Lowinger, and E. Sadoff
1994. Consolidation of porous paint in a vapor saturated atmosphere: A technique for minimizing changes in the appearance of powdering, matte paint. *Journal of the American Institute for Conservation* 32(1):1–14.

Hansen, E. F., A. Wallert, and M. Derrick
1995. An organic colorant used in painted ancient Maya architectural sculpture at Nakbe, Peten, Guatemala. Internal report, Getty Conservation Institute, Marina del Rey, Calif.

Harrick, N. J.
1959. Reflection of infrared radiation from a germanium-mercury interface. *Journal of the Optical Society of America* 49:376.

1971. Determination of refractive index and film thickness from interference fringes. *Applied Optics* 10(10):2344–49.

1979. *Internal Reflection Spectroscopy*. Ossining, N.Y.: Harrick Scientific Corp.

Harrick Scientific Corp.
1987. *Optical Spectroscopy: Sampling Techniques Manual*. Ossining, N.Y.: Harrick Scientific Corp.

Harrington, J. P.
1942. Culture element distributions 19: Central California coast. *University of California Anthropological Records* 7(1):1–42.

Harthcock, M. A., and S. C. Atkin
1988. Infrared microspectroscopy: Development and applications of imaging capabilities. In *Infrared Microspectroscopy: Theory and Applications,* ed. R. G. Messerschmidt and M. A. Harthcock. New York: Marcel Dekker.

Hedley, G., M. Odlyha, A. Burnstock, J. Tillinghast, and C. Husband
1990. A study of the mechanical and surface properties of oil paint films treated with organic solvents and water. In *Cleaning, Retouching and Coatings—Technology and Practice for Easel Paintings and Polychrome Sculpture: Preprints of the Contributions to the Brussels Congress, 3–7 September 1990,* ed. J. S. Mills and P. Smith. London: International Institute for Conservation of Historic and Artistic Works.

Hergert, H. L.
1971. Infrared spectra. In *Lignins: Occurrence, Formation, Structure and Reactions,* ed. K. V. Sarkanen and C. H. Ludwig, 267–95. New York: Wiley-Interscience.

Herschel, W.
1800. Experiments on the refrangibility of the invisible rays of the sun. *Philosophical Transactions of the Royal Society* 90:284–92.

Hexter, M. R., and W. R. Hopwood
1992. Investigation of decorative coating on Kongo ceramics using U.V. light and FTIR analysis. In *Materials Issues in Art and Archaeology 3: Symposium Held 27 April–1 May 1992, San Francisco,* ed. P. B. Vandiver, J. R. Druzik, G. S. Wheeler, and I. C. Freestone, 291–307. Pittsburgh: Materials Research Society.

Hill, H. E.
1983–84. Chemical analysis of pottery residues. *Bulletin of the Experimental Firing Group* 2:86–89.

Hill, J.
1989. Micro-scalpel. *Western Association for Art Conservators Newsletter* 11(1):7–8.

1993. Conversation with author, Getty Conservation Institute, Marina del Rey, Calif., summer.

Hodson, J., and J. A. Lander
1987. The analysis of cured paint media and a study of the weathering of alkyd paints by Fourier transform infra-red/photoacoustic spectroscopy. *Polymer* 28(2): 251–56.

Holcomb, T., and C. Dean
1993. Letter to author, including samples, November.

Hon, D. N. S.
1989. Fourier transform IR spectroscopy and electron spectroscopy for chemical analysis: Use in the study of paper documents. In *Historic Textile and Paper Materials: Conservation and Characterization*, ed. H. L. Needles and H. S. Zeronian, 349–61. Advances in Chemistry series 212. Washington, D.C.: American Chemical Society.

Humecki, H. J.
1988. Polymers and contaminants by infrared microspectroscopy. In *Infrared Microspectroscopy,* ed. R. G. Messerschmidt and M. A. Harthcock, 51–72. New York: Marcel Dekker.

1995a. Microsample preparation techniques. In *Practical Guide to Infrared Microspectroscopy*, ed. H. Humecki. New York: Marcel Dekker.

Humecki, H., ed.
1995b. *Practical Guide to Infrared Microspectroscopy*. New York: Marcel Dekker.

Humphrey, B.
1984. The application of Parylene conformal coating technology to archival and artifact conservation. *Studies in Conservation* 29(3):117.

1986. Vapor phase consolidation of books with the Parylene polymers. *Journal of the American Institute for Conservation* 25(1):15–29.

Ito, M., and M. Takahashi
1985. IR reflection-absorption spectroscopic study of benzotriazole on copper. *Surface Science* 158(1–3):609–15.

Jacquinot, P.
1954. 17e Congrès du GAMS, Paris.

Jakes, K. A., and J. H. Howard III
1968. Formation of textile fabric pseudomorphs. In *Proceedings of the 24th International Archaeometry Symposium*, ed. J. S. Olin and M. J. Blackman, 165–77. Washington, D.C.: Smithsonian Institution Press.

Jakes, K. A., J. E. Katon, and P. A. Martoglio
1990. Identification of dyes and characterization of fibers by infrared and visible microspectroscopy: Application to Paracas textiles. In *Archaeometry '90: Proceedings of the 27th Annual Archaeometry Symposium*, ed. E. Pernicka and G. Wagner, 305–15. Basel, Switzerland: Birkhauser Verlag.

Jakes, K. A., L. R. Sibley, and R. Yerkes
1994. A comparative collection for the study of fibres used in prehistoric textiles from eastern North America. *Journal of Archaeological Science* 21:641–50.

Jayme, G., and G. Traser
1971. Infrarotspektroskopische Untersuchungen an gestrichenen Papieren: Die IR-Spektren der Rohstoffe fur Streichmassen (Infrared spectroscopic investigations on coated papers: IR-spectra of raw materials used for paper coatings). *Das Papier* 25(7):356–66.

Johnson, M., and E. Packard
1971. Methods used for the identification of binding media in Italian paintings of the fifteenth and sixteenth centuries. *Studies in Conservation* 16:145–64.

Karreman, M. F. S.
1989. Toepassing van enkele analytische technieken ten behoeve van preventieve conservering en van incidentele vraagstellingen (Application of analytical techniques in solving problems in conservation science and some incidental questions). In *Mogelijkheden en grenzen van natuurwetenschappelijk onderzoek aan en ten behoeve van voorwerpen van kunst en wetenschap*, ed. W. G. T. Roelofs and J. A. Mosk. 65–72. Amsterdam: Centraal Laboratorium voor Onderzoek van Voorwerpen van Kunst en Wetenschap.

Katon, J. E., and A. J. Sommers
1992. IR microspectroscopy: Routine IR sampling methods extended to the microscopic domain. *Analytical Chemistry* 64(19):931A–940A.

Katon, J. E., A. J. Sommers, and P. L. Lang
1989–90. Infrared microspectroscopy. *Applied Spectroscopy Reviews* 25(3/4):173–211.

Kauppinen, J. K., D. J. Moffatt, H. Mantsch, and D. G. Cameron
1981. Fourier self-deconvolution: A method for resolving intrinsically overlapped bands. *Applied Spectroscopy* 35(3):271–76.

Kenjo, T.
1976. Effect of pH on the hardening of Japanese lacquer films. *Science for Conservation* 15:1–8.

1978. Studies on the analysis of lacquer, 2: Infrared spectrometry of lacquer films. *Kobunkazai no Kagaku (Scientific Papers on Japanese Antiques and Art Crafts)* 23(11):32–39.

Kirby, J., and R. White
1996. The identification of red lake pigment dyestuffs and a discussion of their use. *National Gallery Technical Bulletin* 17:56–74.

Kirillov, A. L., and E. A. Mikolajchuk
1990. Quantitative estimation of archaeological wood degradation degree by infrared Fourier transform spectroscopy. In *ICOM Committee for Conservation 9th Triennial Meeting, Dresden, German Democratic Republic, 26–31 August 1990, Preprints,* ed. K. Grimstad, 239–40. Paris: ICOM Committee for Conservation.

Kobus, H. J.
1987. The use of Fourier transform infra-red spectroscopy and energy dispersive X-ray microanalysis techniques for the analysis of small paint samples. *ICCM Bulletin* 13(3/4):9–16.

Koenig, Jack L.
1992. *Spectroscopy of Polymers.* Washington, D.C.: American Chemical Society.

Koestler, R. J.
1993. Insect eradication using controlled atmospheres and FTIR measurement for insect activity. In *ICOM Committee for Conservation 10th Triennial Meeting, Washington, D.C., USA, 22–27 August 1993, Preprints,* ed. J. Bridgland, vol. 2, 882–86. Paris: ICOM Committee for Conservation.

Koltai, O.
1984. Romai kori hagymafejes exust fibula restauralasa (Restoration of a bulbiform silver fibula of the Roman Age). *Muzeumi Mutargyvedelem* 13:275–84.

Korte, E. H., and H. Staat
1989. Regular-reflection IR spectroscopy of historical varnishes. *Proceedings of SPIE (International Society for Optical Engineering)* 1145:296–97.

Kortum, G.
1969. *Reflectance Spectroscopy.* New York: Springer-Verlag.

Kosek, J., and L. R. Green
1992. A survey of oil paintings and sketches on paper in the collection in the British Museum: An assessment of stability. In *The Institute of Paper Conservation: Conference Papers, Manchester 1992,* ed. S. Fairbrass, 96–102. Worcester, U.K.: Institute of Paper Conservation.

Kosik, M., V. Luzakova, and V. Reiser
1972. Study of the thermal degradation of cellulose and its derivatives. *Cellulose Chemistry and Technology* 6(5):589–97.

Krishnan, K., J. R. Powell, and S. L. Hill
1995. Infrared microimaging. In *Practical Guide to Infrared Microspectroscopy.* ed. H. Humecki. New York: Marcel Dekker.

Kubelka, P.
1948. New contributions to the optics of intensely light-scattering materials: Part 1. *Journal of the Optical Society of America* 38(5):448–57.

Kühn, H.
1960. Detection and identification of waxes, including punic wax by infrared spectrography. *Studies in Conservation* 5(2):1–79.

1970. Verdigris and copper resinate. *Studies in Conservation* 15(1):12–36.

Lang, P. L., J. E. Katon, and A. S. Bonanno
1988. Identification of dust particles by molecular microspectroscopy. *Applied Spectroscopy* 42:313–17.

Lang, P. L., J. E. Katon, J. F. O'Keefe, and D. W. Schiering
1986. The identification of fibers by infrared and raman microspectroscopy. *Microchemical Journal* 34:319–31.

Lang, P. L., M. V. Orna, L. J. Richwine, T. F. Mathews, and R. S. Nelson
1992. The visible and infrared microspectroscopic characterization of organic red pigments removed from three medieval Byzantine manuscripts. *Microchemical Journal* 46:234–48.

Langenheim, J. H., and C. W. Beck
1965. Infrared spectra as a means of determining botanical source of amber. *Science* 149:52–55.

1968. Catalogue of infrared spectra of fossil resins (ambers), 1: North and South America. *Botanical Museum Leaflets, Harvard University* 22(3):65–120.

Launer, P. J.
1962. *Perkin-Elmer Instrumentation News* 13(3):10.

Laurie, A. P.
1914. *The Pigments and Mediums of the Old Masters*. London: Macmillan.

Laver, M. E., and R. S. Williams
1978. The use of a diamond cell microsampling device for infrared spectrophotometric analyses of art and archaeological materials. *Journal of the IIC–Canadian Group* 3(2):34–39.

Lear, J. B.
1981. Analysis of paint films. *Journal of Coatings Technology* 53(675):63–65.

Lee, G. K.
1991. Analytical chemistry tracks poachers. *Analytical Chemistry* 63(9):513A–515A.

Leyden, D. E., and R. S. Murthy
1986. Surface selective sampling techniques in Fourier transform infrared spectroscopy. *Spectroscopy* 2(2):27.

Leyshon, W. E., and R. N. Roberts
1983. Contaminated aerosol dusters. *Abbey Newsletter* 7(2):26.

Liang, C. Y.
1972. Infrared spectroscopy and physical properties of cellulose. In *Instrumental Analysis of Cotton Cellulose and Modified Cotton Cellulose*, ed. R. T. Conner. New York: Marcel Dekker.

Lippincott, E. R., C. E. Weir, A. van Valkenburg, and E. N. Bunting
1960. Studies of infrared absorption spectra of solids at high pressures. *Spectrochimica Acta* 16:58–73.

Low, M. J. D., and N. S. Baer
1977. Application of infrared Fourier transform spectroscopy to problems in conservation. *Studies in Conservation* 22:116–28.

1978. Advances in the infrared spectroscopic examination of pigments. In *ICOM Committee for Conservation 5th Triennial Meeting, Zagreb, 1–8 October 1978, Preprints*. Paris: International Council of Museums.

Lubec, G., G. Nauer, K. Seifert, E. Strouhal, H. Porteder, J. Szilvassy, and M. Teschler
1987. Structural stability of hair over three thousand years. *Journal of Archaeological Science* 14(2):113–20.

Ludwig, F. J.
1965. Analysis of microcrystalline and paraffin waxes by means of infrared spectra in the molten state. *Analytical Chemistry* 37:1737–41.

Majors, R. E.
1992. Nomenclature for sampling in analytical chemistry. *LC•GC: Magazine of Liquid and Gas Chromatography* 10(7):500–506.

Malis, T. F., and D. Steele
1990. Ultramicrotomy for materials science. *Materials Research Society Symposium Proceedings* 199:3–42.

Maniatis, Y., A. Katsanos, and M. E. Caskey
1982. Technological examination of low-fired terracotta statues from Ayia Irini, Kea. *Archaeometry* 24(2):191–98.

Martin, G.
1988. The identification of modern polymer systems using F.T.I.R. In *Preprints of Contributions to the Modern Organic Materials Meeting, Edinburgh, 14–15 April 1988*, 47–56. Edinburgh: Scottish Society for Conservation and Restoration.

Martoglio, P. A., S. P. Bouffard, A. J. Sommer, J. E. Katon, and K. A. Jakes
1990. Unlocking the secrets of the past: The analysis of archaeological textiles and dyes. *Analytical Chemistry* 62(21):1123A–1128A.

Masschelein-Kleiner, L., and J. B. Heylen
1968. Analysis of ancient red lacquers. *Studies in Conservation* 13(2):87–97.

Masschelein-Kleiner, L., J. Heylen, and F. Tricot-Marckx
1968. Contribution à l'analyse des liants, adhésifs et vernis anciens. *Studies in Conservation* 13:105–21.

Masschelein-Kleiner, L., and L. Maes
1978. Ancient dyeing techniques in eastern Mediterranean regions. In *ICOM Committee for Conservation 5th Triennial Meeting, Zagreb, 1–8 October 1978, Preprints*. Paris: International Council of Museums.

Masschelein-Kleiner, L., and F. Tricot-Marckx
1965. La détection de polysaccharides dans les matériaux constitutifs des oeuvres d'art. *Institut royal de patrimoine artistique—Bulletin* 8:180–91.

Matsuda, Y.
1986. Dyeing behaviour and photofading of berberine as a dyestuff. *Scientific Papers on Japanese Antiques and Art Crafts* 31:24–31

Matsuda, Y., and T. Miyoshi
1989. Non-destructive analysis of Japanese ancient fabrics. *Scientific Papers on Japanese Antiques and Art Crafts* 34:1–8.

Matteini, M., A. Moles, and S. Giovannoni
1994. Calcium oxalate as a protective mineral system for wall paintings: Methodology and analyses. In *La conservazione dei monumenti nel bacino del Mediterraneo: atti del 3° simposio internazionale, Venezia, 22–25 giugno 1994* (The conservation of monuments in the Mediterranean basin: Proceedings of the 3rd international symposium), ed. V. Fassina, H. Ott, and F. Zezza, 155–62. Venice: Soprintendenza ai Beni Artistici e Storici di Venezia.

Mattiello, J.
1941. *Raw Materials for Varnishes and Vehicles*. Vol. 1 of *Protective and Decorative Coatings: Paints, Varnishes, Lacquers, and Inks*. New York: J. Wiley.

McCawley, J. C.
1975. Diamond cell infrared spectroscopy in the analysis of paints and pigments. In *ICOM Committee for Conservation 4th Triennial Meeting, Venice, 13–18 October, Preprints*, 5–11. Paris: International Council of Museums.

McClure, A., J. Thomson, and J. Tannahill
1968. Infrared spectra of ninety-six organic pigments. *Journal of the Oil and Colour Chemists' Association* 51(7):580–635.

McCormick-Goodhart, Mark
1989. Research on collodion glass plate negatives: Coating thickness and FTIR identification of varnishes. In *Topics in Photographic Preservation,* comp. R. E. Siegel. Vol. 3. Washington, D.C.: Photographic Materials Group of the American Institute for Conservation of Historic and Artistic Works.

McCrone, W. C.
1982. The microscopical identification of pigments. *Journal of the IIC–Canadian Group* 7(1/2):11–34.

McCrone, W. C., and J. G. Delly
1973. *The Particle Atlas.* 2d ed. Vol. 1. Ann Arbor, Mich.: Ann Arbor Science Publishers.

McDonald, R. S., and P. A. Wilks
1988. JCAMP.DX: A standard form for exchange of infrared spectra in computer readable form. *Applied Spectroscopy* 42(1):151–62.

McGovern, P. E., and R. H. Michel
1990. Royal purple dye: Its identification by complementary physicochemical techniques. *MASCA Research Papers in Science and Archaeology* 7:69–76.

Meikle, R. W., and D. Stewart
1962. The residue potential of sulfuryl fluoride, methyl bromide and methanesulfonyl fluoride in structural fumigations. *Journal of Agricultural and Food Chemistry* 10(5):393–97.

Meilunas, R. J.
1986. The application of diffuse reflectance infrared Fourier transform spectroscopy to the analysis of paint surfaces. Bachelor's thesis, Massachusetts Institute of Technology, Boston.

Meilunas, R. J., J. G. Bentsen, and A. Steinberg
1990. Analysis of aged paint binders by FTIR spectroscopy. *Studies in Conservation* 35:33–51.

Messerschmidt, R. G.
1988. Photometric considerations in the design and use of infrared microscope accessories. In *Infrared Microspectroscopy: Theory and Applications,* ed. R. G. Messerschmidt and M. A. Harthcock. New York: Marcel Dekker.

Messerschmidt, R. G., and M. A. Harthcock, eds.
1988. *Infrared Microspectroscopy: Theory and Applications.* New York: Marcel Dekker.

Michelson, A. A.
1891. Visibility of interference-fringes in the focus of a telescope. *Philosophical Magazine* 5(31):256–59.

1902. *Light Waves and Their Uses.* Chicago: University of Chicago Press.

Mikolajchuk, E. A., N. G. Gerassimova, A. A. Leonovich, A. V. Obolenskaya, I. Y. Levdik, and S. Y. Kazanskaya
1987. Examination of waterlogged archaeological oak wood. In *Conservation of Wet Wood and Metal: Proceedings of the ICOM Conservation Working Groups on Wet Organic Archaeological Materials and Metals,* ed. I. D. MacLeod, 95–107. Paris: International Council of Museums.

Miller, J., E. Moffatt, and J. Sirois
1990. *CCI Native Materials Project: Final Report.* Ottawa: Canadian Conservation Institute, Analytical Research Services.

Mills, J. S., and R. White
1977. Natural resins of art and archaeology: Their sources, chemistry and identification. *Studies in Conservation* 22:12–31.

1994. *The Organic Chemistry of Museum Objects.* 2d ed. London: Butterworth-Heineman.

Morris, N. M.
1991. Comparison of sampling techniques for the characterization of cotton textiles by infrared spectroscopy. *Textile Chemist and Colorist* 23(4):19–22.

Moser, F. H.
1973. Ultramarine pigments. In *Pigment Handbook,* ed. T. C. Patton. Vol. 1. New York: Wiley.

Newman, G. A., and G. W. W. Stevens
1977. Identification of an early microphotographic emulsion by means of infrared spectroscopy. *Laboratory Practice* 26(10):781–82.

Newman, M.
1994. Immunological analysis of a paint fragment from a Chumash site, Santa Barbara County, California. Analysis report to David Scott, Getty Conservation Institute, Los Angeles.

Newman, R.
1980. Some applications of infrared spectroscopy in the examination of painting materials. *Journal of the American Institute for Conservation* 19(1):42–62.

1996. Conversation with author, Museum of Fine Arts, Boston, fall.

Newman, R., C. Weston, and E. Farrell
1980. Analysis of watercolor pigments in a box owned by Winslow Homer. *Journal of the American Institute for Conservation* 19:103–4.

Nowlin, T. E., D. F. Smith, and G. S. Cieloszyk
1980. Thermal oxidative stability of poly p-xylylenes. *Journal of Polymer Science* 18: 2103–19.

Nyquist, Richard A.
1968. Factors affecting infrared group frequencies: Carbonyl stretching absorption bends. *Applied Spectroscopy* 40(3):336–39.

Olin, J. S.
1966. The use of infrared spectrophotometry in the examination of paintings and ancient artifacts. *Instrument News* 17:1.

Olin, J. S., M. E. Salmon, and C. H. Olin
1969. Investigations of historical objects utilizing spectroscopy and other optical methods. *Applied Optics* 8(1):29–39.

Omecinsky, D., and G. W. Carriveau
1982. IR investigation of resinous and synthetic varnishes. In *Preprints of Papers Presented at the 10th Annual Meeting, Milwaukee, Wisconsin, 20–30 May 1982,* 141–49. Washington, D.C.: American Institute for Conservation of Historic and Artistic Works.

Oosterbroek, M., R. J. Lammers, L. G. J. van der Ven, and D. Y. Perera
1991. Crack formation and stress development in an organic coating. *JCT: Journal of Coatings Technology* 63(797):55–60.

Orna, M. V., P. L. Lang, J. E. Katon, T. F. Mathews, and R. L. Nelson
1989. Applications of infrared microspectroscopy to art historical questions regarding medieval manuscripts. In *Archaeological Chemistry* 4, ed. R. Allen, 265–88. Washington, D.C.: American Chemical Society.

Palenik, S.
1992. Modern microscopy on the light side: The FTIR microscope. *Microscopy Today* 7:4–5.

Paralusz, C.
1974. Internal reflection spectroscopy applied to the analysis of adhesive tapes. *Journal of Colloid and Interface Science* 47(3):719–46.

Parra, E., and A. Serrano

1990. Chemical analysis of wax seals and dyed textile attachments from parchment documents: Preliminary investigations. In *ICOM Committee for Conservation 9th Triennial Meeting, Dresden, German Democratic Republic, 26–31 August 1990, Preprints,* ed. K. Grimstad, 62–67. Paris: ICOM Committee for Conservation.

Payne, K. J., and A. Veis

1988. Fourier transform IR spectroscopy of collagen and gelatin solutions—deconvolution of the amide I band for conformational studies. *Biopolymers* 27:1749–60.

Pecina, H., and R. Kommert

1985. Badania spektroskopowe w podczerwieni drewna z roznych epok historycznych (Infra-red spectral studies of wood from different historical periods). *Folia forestalia polonica, b (drzewnictwo)* 16:31–43.

Perron, J.

1989. The use of FTIR in the study of photographic materials. In *Topics in Photographic Preservation,* comp. R. E. Siegel. Vol. 3. Washington, D.C.: Photographic Materials Group of the American Institute for Conservation of Historic and Artistic Works.

Pey, E. B. F.

1987. The Hafkenscheid Collection: A collection of pigments and painting materials dating from the first half of the 19th century. *Maltechnik-Restauro* 93(2):23–33.

Plahter, L. E., E. Skug, and U. Plahter

1974. Gothic painted altar frontals from the church of Tingelstad: Materials, technique, restoration. *Mediaeval Art in Norway.* Vol. 1. Oslo: Universitetsforlaget.

Plesters, J.

1954. The preparation and study of paint cross-sections. *Museums Journal* 54(4):97.

1956. Cross-sections and chemical analysis of paint samples. *Studies in Conservation* 2(3):110–57.

Poole, J. B., and R. Reed

1962. The preparation of leather and parchment by the Dead Sea Scrolls community. *Technology and Culture* (winter):1–26.

Poslusny, M., and K. E. Daugherty

1988. Nondestructive adhesive analysis on stamps by Fourier transform infrared spectroscopy. *Applied Spectroscopy* 42(8):1466–69.

Pratt, E.

1991. An investigation into a cast resin sculpture by Jean Dubuffet. In *Selected Student Papers of the Seventeenth Annual Conservation Training Programs Conference, April 25–27, 1991.* Buffalo: Department of Conservation, State University of New York at Buffalo.

Pretzel, B.

1994. Silicon carbide sampling for FT-IR. Presentation at Infrared User's Group meeting, Philadelphia Museum of Art, Philadelphia.

Rao, G. B.

1982. Effect of atmospheric pollutants (SO_2/ NO_2/ Cl_2/ H_2S) on marble. *Birla Archaeological and Cultural Research Institute Technical Bulletin* 4:11–21.

1984. The effect of product of relative humidity and concentration of sulfur dioxide on marble. *Journal of Archaeological Chemistry* 2:39–40.

Rayleigh, Lord

1891. On the interference bands of approximately homogeneous light: In a letter to Prof. A. Michelson. *Philosophical Magazine* 5(34):407–11.

Reedy, T. J., and C. L. Reedy

1988. *Statistical Analysis in Art Conservation Research.* Los Angeles: J. Paul Getty Trust.

Reffner, J. A.
1989. Infrared spectral mapping of polymers by FT-IR microscopy. In *Microbeam Analysis—1089,* ed. P. E. Russell, 167–76. San Francisco: San Francisco Press.

1993. The commercial development of FT-IR microspectroscopy. *Microscopy Today* 93-3:6.

Reffner, J. A., and P. A. Martoglio
1995. Uniting microscopy and spectroscopy. In *Practical Guide to Infrared Microspectroscopy*. ed. H. Humecki, 41–84. New York: Marcel Dekker.

Reffner, J. A., W. T. Wihlborg, and S. W. Strand
1991. Chemical microscopy of surfaces by grazing angle and internal reflection FTIR microscopy. *American Laboratory* 23(April):46–50.

Reid, N.
1972. Ultramicrotomy (part 2). In *Practical Methods in Electron Microscopy*. Amsterdam: North-Holland Publishing Co.

Reilly, J. A.
1991. Celluloid objects: Their chemistry and preservation. *Journal of the American Institute for Conservation* 30(2):145–62.

Riederer, J.
1969. Infrarotspektrographische Untersuchung der gelben und roten Eisenoxidpigmente. *Deutsche Farben-Zeitschrift* 23(12):569–77.

1974. Recently identified Egyptian pigments. *Archaeometry* 16(1):102–9.

1977. Technik und Farbstoffe der fruhmittelalterlichen Wandmalereien Ostturkistans (Technique and materials of early medieval mural paintings from East Turkistan). In *Beitraege zur Indienforschung—Ernst Waldschmidt zum 80 Geburtstag gewidmet* (Contributions to Indian research in honour of Ernst Waldschmidt's 80th birthday). Veroeffentlichen des Museums für Indische Kunst Berlin 4. Berlin: Museum für Indische Kunst.

Robinson, N., R. P. Evershed, W. J. Higgs, K. Jerman, and G. Eglinton
1987. Proof of a pine wood origin for pitch from Tudor (Mary Rose) and Etruscan shipwrecks: Application of analytical organic chemistry in archaeology. *Analyst* 112:637–44.

Roelofs, W. G. T.
[1989]. Het onderzoek naar vernissen, bindmiddelen en organische kleurstoffen op voorwerpen van kunst en wetenschap (Identification of varnishes, media, and organic dyestuffs occurring on objects of art and science). In *Mogelijkheden en grenzen van natuurwetenschappelijk onderzoek aan en ten behoeve van voorwerpen van kunst en wetenschap*, ed. W. G. T. Roelofs and J. A. Mosk, 24–33. Amsterdam: Centraal Laboratorium voor Onderzoek van Voorwerpen van Kunst en Wetenschap.

Roelofs, W. G. T., P. B. Hallebeek, J. H. Hofenk De Graaff, and R. F. S. Karreman
1987. The analysis of natural dyestuffs and organic pigments: A comparative study into the possibilities and limits of various methods. In *ICOM Committee for Conservation 8th Triennial Meeting, Sydney, Australia, 6–11 September 1987, Preprints,* ed. K. Grimstad, vol. 2, 709–17. Marina del Rey, Calif.: Getty Conservation Institute.

Rogers, P. G., R. Cameron, N. S. Cartwright, W. H. Clark, J. S. Deak, and E. W. W. Norman
1976. The classification of automotive paint by diamond window infrared spectrophotometry, part 1: Binders and pigments. *Journal of the Canadian Society for Forensic Science* 9:1–14.

Rothe, A.
1991. Analysis request form, Getty Conservation Institute, Marina del Rey, Calif., spring.

1992. Mantegna's paintings in distemper. In *Andrea Mantegna*, ed. J. Martineau. London: Royal Academy of Arts; New York: Metropolitan Museum of Art.

Roush, P. B., ed.
1987. *The Design, Sample Handling, and Applications of Infrared Microscopes: A Symposium*. ASTM Special Technical Publication 949. Philadelphia: American Society for Testing Materials.

Roy, A., ed.
1993. *Artists' Pigments: A Handbook of Their History and Characteristics*. Vol. 2. Washington, D.C.: National Gallery of Art. (Contains articles with IR spectra on azurite and blue verditer; ultramarine blue; lead white; lead-tin yellow; smalt; verdigris and copper resinate; vermilion and cinnabar; malachite and green verditer; and calcium carbonate whites.)

Ryland, S. G.
1995. Infrared microspectroscopy of forensic paint evidence. In *Practical Guide to Infrared Microspectroscopy*, ed. H. Humecki. New York: Marcel Dekker.

Saltzman, M.
1992. Identifying dyes in textiles. *American Scientist* 80:474–81.

Saltzman, M., and A. Keay
1972. Colorant identification. In *Fundamentals and Problems of Color*. Philadelphia: Federation of Societies for Paint Technology.

Saltzman, M., A. Keay, and J. Christensen
1963. Identification of colorants in ancient textiles. *Dyestuffs* 44(8):241–51.

Sands, J.
1993. "An effect which far exceeds any conception . . ." *Colonial Williamsburg: The Journal of the Colonial Williamsburg Foundation* (spring):72–73.

Schreiner, M., and M. Grasserbauer
1985. Microanalysis of art objects: Objectives, methods and results. *Fresenius' Journal of Analytical Chemistry* 322:181–93.

Schrenk, J. L.
1991. Corrosion and past "protective" treatments of the Benin "bronzes" in the National Museum of African Art. In *Materials Issues in Art and Archaeology 2: Symposium Held April 17–21, 1990, San Francisco, California, USA*, ed. P. B. Vandiver, J. Druzik, and G. S. Wheeler, 805–12. Pittsburgh: Materials Research Society.

Schulz, H., and B. Kropp
1993. Microspectroscopy FTIR reflectance examination of paint binders on ground chalk. *Fresenius' Journal of Analytical Chemistry* 346:114–22.

Schweppe, H.
1979. Identification of dyes on old textiles. *Journal of the American Institute for Conservation* 19(1):14–23.

1989. Identification of red madder insect dyes by thin layer chromatography. In *Historic Textile and Paper Materials: Conservation and Characterization*, ed. H. L. Needles and H. S. Zeronian, 153–74. Advances in Chemistry series 212. Washington, D.C.: American Chemical Society.

Schwochau, K., T. E. Haevernick, and D. Ankner
1963. Zur infraotspecktrosckopischen Herkunftsbestimmung von Bernstein (Using infra-red spectroscopy to determine origins of amber). *Jahrbuch des Romisch-Germanischen Zentrals-Museums Mainz* 10.

Scott, D. A.
1994. Analysis request form, Getty Conservation Institute, Marina del Rey, Calif., March.

Scott, D. A., and W. D. Hyder
1993. A study of some California Indian rock art pigments. *Studies in Conservation* 38:155–73.

Scott, D. A., M. Newman, M. Schilling, M. Derrick, and H. Khanjian
1996. Blood as a binding medium in a Chumash Indian pigment cake. *Archaeometry* 38(1):103–12.

Selwitz, C.
1988. *Cellulose Nitrate in Conservation.* Research in Conservation 2. Marina del Rey, Calif.: Getty Conservation Institute.

Sharpless, T. W.
1980. Corrosion: The problem of storage. *The Numismatist* (October):2450–54.

Shearer, G.
1987. Use of diffuse reflectance Fourier transform infrared spectroscopy in art and archaeological conservation. In *Recent Advances in the Conservation and Analysis of Artifacts: Jubilee Conservation Conference, London, 6–10 July 1987,* comp. James Black, 253–56. London: University of London, Institute of Archaeology, Summer Schools Press.

Shearer, G. L., and S. Doyal
1991. Use of FT-IR in the conservation of twentieth-century objects. In *Materials Issues in Art and Archaeology 2: Symposium Held April 17–21, 1990, San Francisco, California, USA,* ed. P. B. Vandiver, J. Druzik, and G. S. Wheeler, 813–23. Pittsburgh: Materials Research Society.

Shearer, J. C., D. C. Peters, G. Hoepfner, and T. Newton
1983. FTIR in the service of art conservation. *Analytical Chemistry* 55(8):874–76, 878–80.

Shelton, C.
1994. Letter to author, including samples, November.

Shoeib, A. S. A., M. Roznerska, and K. Boryk-Jozefowicz
1990. Weathering effects on an ancient Egyptian limestone, which has been affected by salt. In *The Conservation of Monuments in the Mediterranean Basin—The Influence of Coastal Environment and Salt Spray on Limestone and Marble: Proceedings of the 1st International Symposium, Bari, 7–10 June 1989/La conservazione dei monumenti nel bacino Mediterraneo: Influenza dell ambiente costiero e dello spray marino sulla pietra calcareo e sul marmo. Atti del 1 Simposio internazionale, Bari, 7–10 giugno 1989,* ed. F. Zezza, 203–8. Brescia, Italy: Grafo Edizioni.

Shreve, O. D., M. R. Heether, H. B. Knight, and D. Sween
1950. Infrared absorption spectra: Some long chain fatty acid esters and alcohols. *Analytical Chemistry* 22:1498–1501.

Siesmayer, B., A. Giebelhausen, J. Zambelli, and J. Riederer
1975. Application of infrared spectroscopy to the examination of mineral pigments of historical works of art in comparison with recent European deposits. *Fresenius' Journal of Analytical Chemistry* 277(3):193–96.

Smith, A. L.
1979. *Applied Infrared Spectroscopy: Fundamentals, Techniques and Analytical Problem Solving.* Chemical Analysis series, vol. 54. New York: John Wiley and Sons.

Sommer, A. J., and J. E. Katon
1988. In *Proceedings of the 23rd Annual Conference of the Microbeam Analysis Society*, ed. D. E. Newbury, 207–14. San Francisco: San Francisco Press.

Souza, L., and M. Derrick
1995. The use of FT-IR spectrometry for the identification and characterization of gesso-glue grounds in wooden polychromed sculptures and panel paintings. In *Materials Issues in Art and Archaeology 4: Symposium Held May 16–21, 1994, Cancun, Mexico,* ed. P. B. Vandiver, J. R. Druzik, J. L. G. Madrid, I. C. Freestone, and G. S. Wheeler. Pittsburgh: Materials Research Society.

Srámek, J.
1978. The non-destructive investigation of chemical alterations in the surface of stones. In *Alteration et protection des monuments en pierre: Colloque international, Paris, 5–9 juin 1978/Deterioration and Protection of Stone Monuments: International Symposium, June 5–9, 1978*. Paris: n.p.

Staats, H., E. H. Korte, and D. Kolev
1989. Infrared reflection studies of historical varnishes. In *Recent Developments in Molecular Spectroscopy,* ed. R. Jordanov, 64–74. Teaneck, N.J.: World Scientific Publishing Co.

Stimson, M. M., and M. J. O'Donnell
1952. *Journal of the American Chemical Society* 74:1805.

Stodulski, L.
1994. Telephone conversation with author, summer.

Strauss, I.
1984. Uebersicht ueber synthetisch organische Kuenstlerpigmente und Moeglichkeiten ihrer Identifizierung (A survey of synthetic organic artists' pigments and their possible identification). *Maltechnik-Restauro* 90(4):29–44.

Stringari, C., and E. Pratt
1993. The identification and characterization of acrylic emulsion paint media. In *Symposium '91: Saving the Twentieth Century—the Degradation and Conservation of Modern Materials, Abstracts,* 411–40. Ottawa: Canadian Conservation Institute.

Susi, H., J. S. Ard, and R. J. Carroll
1975. Hydration and denaturation of collagen as observed by infrared spectroscopy. *Journal of the American Leather Chemists Association* 66(11):508–19.

Sutton, M. Q.
1990. Notes on creosote lac scale insect resin as a mastic and sealant in the southwestern Great Basin. *Journal of California and Great Basin Anthropology* 12(2):262–68.

Taylor, D.
1967. The sodalite group of minerals. *Contributions to Mineralogy and Petrology* 16:172.

Teetsov, A. S.
1977. Techniques for small particle manipulation. *The Microscope* 25:103–13.

1995. Techniques for nanogram samples. In *Practical Guide to Infrared Microspectroscopy*. ed. H. Humecki. New York: Marcel Dekker.

Tennent, N. H., and K. M. Antonio
1981. Bronze disease: Synthesis and characterisation of botallackite, paratacamite and atacamite by infra-red spectroscopy. In *ICOM Committee for Conservation 6th Triennial Meeting, Ottawa, 21–25 September 1981, Preprints,* 3–11. Paris: ICOM Committee for Conservation.

Tennent, N. H., and T. Baird
1985. The deterioration of Mollusca collections: Identification of shell efflorescence. *Studies in Conservation* 30(2):73–85.

Tobe, H., N. Morito, K. Monma, and W. Suetaka
1974. Spectrographic investigation of the corrosion inhibition of copper by benzimidazole. *Journal of the Japan Institute of Metals* 38(8):770–76.

Todd, J. M., M. H. Eichel, C. W. Beck, and A. Macchiarulo
1976. Bronze and iron age amber artifacts in Croatia and Bosnia-Herzegovina. *Journal of Field Archaeology* 3(3):313–27.

Tranter, G. C.
1976. Patination of lead: An infra-red spectroscopic study. *British Corrosion Journal* 11(4):222–24.

Tsang, J., and R. Cunningham
1991. Some improvements in the study of cross-sections. *Journal of the American Institute for Conservation* 30(2):163–77.

Tungol, M. W., E. G. Bartick, and A. Montaser
1991. Analysis of single polymer fibers by Fourier transform infrared microscopy: The results of case studies. *Journal of Forensic Sciences* 36(4):1027–43.

1995. Forensic examination of synthetic textile fibers by microscopic infrared spectrometry. In *Practical Guide to Infrared Microspectroscopy,* ed. H. Humecki. New York: Marcel Dekker.

Twilley, J.
1984. The analysis of exudate plant gums in their artistic applications: An interim report. In *Archaeological Chemistry 3*, ed. J. B. Lambert, 357–94. Advances in Chemistry series 205. Washington, D.C.: American Chemical Society.

van der Loeff, L. S., and K. Groen
1993. The restoration and technical examination of Gérard Dou's *The Young Mother* in the Mauritshuis. In *ICOM Committee for Conservation 10th Triennial Meeting, Washington, D.C., USA, 22–27 August 1993, Preprints,* ed. J. Bridgland, 98–103. Paris: ICOM Committee for Conservation.

van't Hul-Ehrnreich, E. H.
1970. Infrared microspectroscopy for the analysis of old painting materials. *Studies in Conservation* 15:175–82.

van Zelst, L., D. W. von Endt, and M. Baker
1988. Non-destructive and micro-sample FTIR spectrometric analysis of organic materials in art objects. In *2nd international conference on non-destructive testing, microanalytical methods and environment evaluation for study and conservation of works of art, Perugia, 17–20 April 1988,* 30.1–30.17. Rome: Istituto Centrale per il Restauro—Associazione Italiana Prove Non Distruttive.

Venkataraman, K., ed.
1977. *Analytical Chemistry of Synthetic Dyes.* New York: John Wiley and Sons.

Vincent, R. K., and G. R. Hunt
1968. Infrared reflectance from matte surfaces. *Applied Optics* 7(1):53.

Walker, P. L., and T. Hudson
1993. *Chumash Healing: Changing Medical Practices in an American Indian Society.* Banning, Calif.: Malki Museum Press.

Wannemacher, R.
1980. IR spectrophotometry: Valuable tools for textile chemists. *American Dye Stuff Reporter* 69:26–28, 35.

Ward's Natural Science
1990. *Embedding in Bio-Plastic.* 6th ed. Ward's Curriculum Aid no. 246-0005. Rochester, N.Y.: Ward's Natural Science.

Warren, J. R., W. E. Smith, and W. J. Tillman
1969. Internal reflectance spectroscopy and the determination of the degree of denaturation of insoluble collagen. *Journal of the American Leather Chemists Association* 64:4–11.

Watchman, A.
1993. Perspectives and potentials for absolute dating prehistoric rock paintings. *Antiquity* 67:58–65.

Webster, R.
1983. *Gems: Their Sources, Descriptions and Identification* 4th ed. London: Butterworths.

Weiner, S., Z. Kusanovich, E. Gil-Av, and W. Traub
1980. Dead Sea Scrolls parchments: Unfolding of the collagen molecules and racemization of aspartic acid. *Nature* 28:820–23.

Weir, C. E., E. R. Lippincott, A. van Valkenburg, and E. N. Bunting
1960. Infrared studies in the 1- to 15-micron region to 30,000 atmospheres. *Journal of Research of the National Bureau of Standards—A, Physics and Chemistry* 63(1):55–62.

West-FitzHugh, E., ed.
1997. *Artists' Pigments*. Vol. 3. Washington, D.C.: National Gallery of Art. (Contains articles with IR spectra on Egyptian blue, orpiment and realgar, indigo and woad, madder and alizarin, gamboge, Vandyke brown, Prussian blue, emerald green and Scheele's green, chromium oxide greens, and titanium dioxide whites.)

White, J.
1964. New method for measuring diffuse reflectance in the infrared. *Journal of the Optical Society of America* 54(11):1332.

White, R. L., and J. Ai
1992. Variable-temperature diffuse reflectance infrared studies of calcium oxalate monohydrate. *Applied Spectroscopy* 46(1):93–99.

Wilks, P. A., Jr.
1967. Applications of internal reflection spectroscopy: Identification of gas chromatographic fractions, pyrolysis, and skin analysis. In *Modern Aspects of Reflectance Spectroscopy, Proceedings Symposium*, 192–206. New York: Plenum Press.

Wilks, P. A., Jr., and M. R. Iszard
1965. The identification of fibers and fabrics by internal reflection spectroscopy. In *Developments in Applied Spectroscopy,* vol. 14, ed. E. N. Davis, 141–49. New York: Plenum Press.

Williams, R. S., J. B. Waddington, and J. Fenn
1990. Infrared spectroscopic analysis of Central and South American amber exposed to air pollutants, biocides, light and moisture. *Collection Forum* 6(2):65–75.

Wolbers, R.
1984. Developing Fourier transform infrared spectroscopy (FTIR) as an aid in the detection of asphalt-containing paint films. In *Art Conservation Training Programs Conference, May 1–3, 1983, Postprints,* 103–22. Cooperstown, New York: Department of Conservation of Historic and Artistic Works, State University of New York at Buffalo.

Wright, N., and L. W. Hersher
1947. *Journal of the Optical Society of America* 37:211.

Zappia, G., C. Sabbioni, M. G. Pauri, and G. Gobbi
1992. Effects of SO_2-enriched atmosphere on ancient and modern building materials. *Materials Engineering* 3(3):445–58.

Zolotovich, G., and A. Popov
1969. Cosmetics from a Thracian tomb in Bulgaria (1st century A.D.). *Soap, Perfum. Cosmet.* 42(6):425–32.

Zuber, G. E., R. J. Warren, P. P. Begosh, and E. L. O'Donnell
1984. Direct analysis of thin-layer chromatography spots by diffuse reflectance Fourier transform infrared spectrometry. *Analytical Chemistry* 56(14):2935.

Index

Note: Italicized letters *f* and *t* following page numbers indicate figures and tables, respectively.

D

E

F

About the Authors

Michele R. Derrick is a chemist and conservation scientist with more than twenty years' experience analyzing and characterizing materials. She graduated from Oklahoma State University in 1979 with an M.S. degree in analytical chemistry, then worked at the University of Arizona Analytical Center. She next worked for twelve years as a conservation scientist at the Getty Conservation Institute. Derrick currently has a dual role as a consultant for the Museum of Fine Arts, Boston, and as a chemist at a commercial analytical laboratory. Derrick has published more than thirty articles in journals and proceedings. In 1995 she received the Kress Foundation conservation publication fellowship that funded this volume. She currently holds a grant from the National Center for Preservation Technology and Training to compile a database on materials used in the production and conservation of historic and artistic objects and sites.

Dusan Stulik graduated from Charles University in Prague with B.S. and M.S. degrees in chemistry. He subsequently obtained a Ph.D. degree in physics from the Czechoslovakia Academy of Sciences. He is currently a senior scientist in the Scientific Program at the Getty Conservation Institute, where his research focuses on the application of modern scientific methods to conservation science.

James M. Landry is a professor of chemistry and director of the Natural Science Program at Loyola Marymount University in Los Angeles. He received his B.S. and M.S. degrees in chemistry from Xavier University in Cincinnati. He then obtained his Ph.D. degree in inorganic chemistry from Miami University of Ohio, specializing in molecular spectroscopy. His research interests include projects in IR microspectroscopy and gas chromatography–mass spectrometry. He has recently been involved in developing a formulation and application process for a novel coating that blocks the transmission of heat through a variety of materials. He has been involved in the Los Angeles Collaborative for Teacher Excellence (LACTE), an association of ten colleges funded by the National Science Foundation that works to improve the preparation of future K–12 science and math teachers.